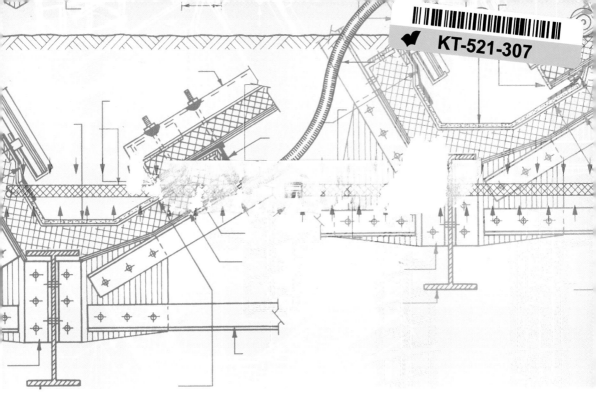

5th edition

Construction Technology

ROY CHUDLEY • ROGER GREENO • MIKE HURST • SIMON TOPLISS

ALWAYS LEARNING

PEARSON

Published by Pearson Education Limited, Edinburgh Gate, Harlow, Essex, CM20 2JE.

www.pearsonschoolsandfecolleges.co.uk

Text © Pearson Education Limited 2011
Typeset by Phoenix Photosetting, Chatham
Original illustrations by the authors
Illustrations new to this edition by KJA Artists and Phoenix Photosetting
All other illustrations courtesy of Roy Chudley and Roger Greeno
Cover design by Brian Melville

The rights of Roy Chudley, Roger Greeno, Mike Hurst and Simon Topliss to be identified as authors of this work have been asserted by them in accordance with the Copyright, Designs and Patents Act 1988.

First published (as *Construction Technology*) 1973 (Volume 1), 1974 (Volume 2)
Second edition 1987

Third edition (published as a single volume, with revisions by Roger Greeno) 1999

Reprinted 2002 (revised edition), 2003 (twice) (third edition update)

Fourth edition 2005

Reprinted 2006 (twice), 2007, 2009

This edition first published 2011

15 14
10 9 8 7 6 5 4 3

British Library Cataloguing in Publication Data
A catalogue record for this book is available from the British Library

ISBN 978 0 435 04682 8

Printed in Malaysia (CTP-VVP)

Every effort has been made to contact copyright holders of material reproduced in this book. Any omissions will be rectified in subsequent printings if notice is given to the publishers.

Websites
Pearson Education Limited is not responsible for the content of any external internet sites. It is essential for tutors to preview each website before using it in class so as to ensure that the URL is still accurate, relevant and appropriate. We suggest that tutors bookmark useful websites and consider enabling students to access them through the school/college intranet.

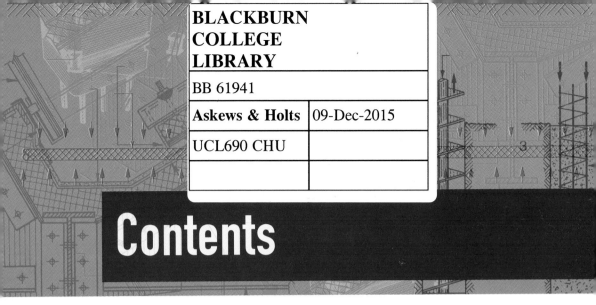

Contents

Preface

This book is updated and revised from the 4th edition published in 2004. *Construction Technology* was first published in 1973 and conceptualised initially in four volumes by Roy Chudley. It is an essential and invaluable working tool and reference for those working in the Building sector and has run into several editions. This new edition incorporates a substantial amount of material from the first four editions where much of the updating was done by Roger Greeno keeping the spirit and endeavour of the first edition alive.

Since the 4th edition of *Construction Technology* in 2005, there have been many advances in the field with the drive towards a carbon neutral constructed building that is sustainable and does not add in its construction to climate change. This edition retains and enhances many of the features of the previous editions with the unique and detailed drawings and sketches that enhanced the core text.

The introduction of site investigation, an essential sustainable construction technique, begins the construction process. Initial site set up and temporary works has been revised to include the requirements of the CDM regulations and the substructure section has been restricted to low rise developments. Superstructure now incorporates the sustainable techniques of thin joint masonry, timber-framed construction and volumetric prefabrication of buildings; less used materials and techniques have been revised and removed. A new chapter on energy and efficiency in construction now incorporates the use of sustainability within energy conservation. Roofing now includes the use of green roof technologies and the roof finishes of turf and sedum as a living sustainable roof. Single sheet skin roof finishes have been added with the recent developments in seeming sheet materials to use as a sustainable roof finish. Access for the disabled has been revised in accordance with the changes to Part M of the Building Regulations which altered many of the dimensional requirements for access to buildings and dwellings. Framed buildings now incorporates the use of steel portal framed technology and construction, steel roof trusses has been revised with the introduction of prefabricated roof truss technology.

All of the British Standards and other codes have been revised in accordance with the new BS EN Codes and standards throughout the text.

In conjunction with this edition's companion volume, the revised *Advanced Construction Technology*, the reader will gain valuable understanding of the construction of low rise, commercial and industrial construction techniques.

Mike Hurst and Simon A Topliss

Acknowledgements

The book originated in 1973, conceived and authored by Roy Chudley with invaluable assistance from Colin Basset as General Editor. This book and its companion *Advanced Construction Technology* have been continually updated and reprinted and have gained a world-wide readership. Roger Greeno worked on updating the content from the second to the fourth editions maintaining the agreeable style and quality of illustration and clear and accessible text.

Roy Chudley, MCIOB.
Roy is a chartered builder and was formerly senior lecturer in construction at Guildford College. Roy's prominently illustrated building technology textbooks have become the established standard reference resource for students. His numerous books represent building practice and procedures in a unique style of comprehensive text supplemented with simple illustrations. This interpretation of the subject combined with an understanding of student needs evolved from many years on site and in the design office.

Roger Greeno, BA(Hons.), FCIOB, FCIPHE, FRSA.
Roger is a chartered builder and registered plumber. He has lectured in several further and higher education colleges and Portsmouth University. Roger is a published author, editor, writer and illustrator of numerous construction papers, including well-known and accredited course readers and textbooks as study guides and industry references. He has been an examiner and moderator for many examination boards, including City & Guilds, Edexcel and the Chartered Institute of Building with ongoing commitments to the College of Estate Management, University of Reading and the Open University.

Mike Hurst MSc. BSc.(Hons) ICIOB CertEd MIfL
Mike is currently a senior lecturer in Civil Engineering at the University of East London and has over 25 years teaching experience of a variety of construction and civil engineering qualifications at both FE and HE level. He is also an FE

External Verifier for Edexcel and an HE External Examiner for a number of Universities. His working career spans both local authority and private practice firstly in building control then in various engineering consultancies. With his interest in historic buildings, he is currently involved in research into environmental and energy issues of upgrading current building stock, building pathology and the use of traditional construction materials and skills. He has contributed to a number of construction related publications, textbooks and study guides including the BTEC Level 3 Construction & the Built Environment standard course text.

Simon Topliss BSc (Hons) PGCE MIFL CIEA
Simon is the Team Leader for Construction at the Grimsby Institute for Further and Higher Education in North East Lincolnshire, where he has taught on the Edexcel BTECs at Levels 2, 3 and 4 for over 15 years. He has written extensively for Pearson Education, contributing to a suite of construction textbooks from the Foundation Learning Tier through to Level 3. He currently holds the posts of the Senior Standards Verifier for Construction, Chief Examiner and Moderator on the Level 2 Construction and Built Environment Diploma and External Examiner for the Higher Education Construction and Civil Engineering programmes. He has written a substantial number of specification units for Edexcel assisting with the rewriting from the NQF to QCF framework

Pearson Education would like to thank the following:

Roy Chudley as founding author and Roger Greeno for his work in updating and revising over the years.

Mike Hurst and Simon Topliss for their dedication and enthusiasm in working tirelessly on this new edition.

Damian McGeary from The College of Haringey, Enfield and North London and Peter Lakin from the University of East London for their perceptive and careful technical reviews.

Sarah Christopher for the initial work with the authors in preparing the manuscript for this edition.

Mike Hurst would like to thank his wife and daughters for their patience, understanding and encouragement.

Simon Topliss would like to thank his wife Linda for her unwavering support and Paul Monroe who introduced him to Edexcel. He would especially like to thank Paul Brown and Steve Shelley for their support and interest.

Pearson Education would like to thank the following for their kind permission to use their material in illustrations:

Balfour Beatty Plant & Fleet Services (p.43); Wernick Group (p.44); Mabey Hire Services (p.55 and p.75); Ground Force (p.56); Youngman Group (p.65); Taylor & Francis (Hughes, P. and Ferrett, E., *Introduction to Health & Safety in Construction 3rd ed.*, Butterworth-Heinemann, Figure 9.6, page 139) (p.72);

Roger Bullivant (p.102, p.103); Environmesh (p.115); Ready Made Basements (p.125); BRE Group (p.128, p.134 (Copyright BRE, reproduced with permission from *BRE Good Building Guide 72*) p.530); Manthorpe Group (p.190); BM Trada (p.199); BSI (p.201); Kingspan Insulation Ltd. (p.210); Delston Roofing (www.delston.co.uk) (p.291); Safeguard Europe (p.293); Umicore (p.297); Zero Waste Scotland (p.303); NHBC Foundation (pp.310-311); Ecohouse Solar (p.337); BSRIA (p.343); Eurocell (p.381); British Gypsum (p.409, p.410, p.430); Centre for Accessible Environments (p.454, p.458); Corrugated Beam Form Ltd (p.494); PERI Ltd. (p.500, p.503); Taylor & Francis Elliott, K.S. (2002) *Precast Concrete Structures*. Butterworth-Heinemann (p.506); Genie Ltd. (p.527); Steel Construction Institute (p.545); Ideal Heating/Keston UK & Eire (p.560).

Introduction

There are two general approaches to the construction of buildings:

- conventional or traditional methods;
- modern or industrialised methods.

Conventional or traditional methods are studied in the first two years of most construction courses, with the intention of forming a sound knowledge base before proceeding to studies of advanced techniques in the final years. There is, nevertheless, an element of continuity and overlap between traditional and contemporary, and both are frequently deployed on the same building: for example, traditional brick facing to a prefabricated steel-framed commercial building or to a factory-made timber-framed house.

Initial studies of building construction concentrate on the smaller type of structure, such as a domestic dwelling of one or two storeys built by labour-intensive traditional methods. Generally it is more economic to construct this type of building by these methods, unless large numbers of similar units are required on the same site. In these circumstances, economies of scale may justify factory-manufactured, prefabricated elements of structure; however, in recent years it is increasingly common to find the use of industrialised prefabricated elements in many domestic buildings, as they have become cheaper and easier to install. These industrialised methods are usually a rationalised manufacturing process used to produce complete elements (floors, walls, roof frames, etc.) in modules or standardised dimensional increments of 300 mm.

Very few building contractors in the UK and other developed countries employ many staff directly. They are therefore relatively small companies when compared with the capital value of the work they undertake. This is partly due to the variable economic fortunes of the construction industry and the need for flexibility. Hence most practical aspects of building are contracted out to specialist subcontractors, (such as bricklayers, electricians, carpenters) in response to the main contractor's workload. The main contractor is effectively a building management company, which could be engaged on a variety of work, including major serial developments for the same client, maintenance work or aftercare programmes, extensions to existing structures, or possibly just small one-off projects.

It is essential that all students of building are aware of the variable methods of construction and application of materials in traditional, modern and industrial practice, in order to adapt their career patterns to the diverse expectations of the industry.

THE BUILDING TEAM

Building is essentially a team process in which each member has an important role to play. The function of each member is outlined below.

- **Building owner or client** The person or organisation that finances and commissions the work. They directly, or indirectly, employ all other personnel, with particular responsibility for appointing the planning supervisor co-ordinator (sometimes the architect) and nominating the principal contractor – see Construction (Design and Management) Regulations 1994.

- **Architect** Engaged by the building owner as agent to design, advise and ensure that the project is kept within cost and complies with the design.

- **Clerk of works** Employed on large, traditional contracts as the architect's on-site representative. Their main function is to liaise between the architect and main contractor and to ensure that construction proceeds in accordance with the design. They can offer advice, but directives must be through the architect.

- **Private quantity surveyor** In traditional contracts they work closely with the architect. They are engaged to prepare cost evaluations and bills of quantities, check tenders, prepare interim valuations, effect cost controls and advise the architect on the cost of variations.

- **Consulting engineers** Engaged to advise on and design a variety of specialist installations, such as structural, services and security aspects. They are employed to develop that particular aspect of the design within the cost and physical parameters of the architect's brief.

- **Construction design and management co-ordinator** Engaged to ensure that the building project complies with health and safety at every stage, from the design through to the construction and ultimately the safe use of the facility by the end user.

- **Principal or main contractor** Employed by the client on the advice of the architect, by nomination or competitive tendering. They are required to manage and administer the construction programme within the architect's direction.

- **Contracts manager** or **divisional manager** In large companies this person would oversee a number of construction contracts and be the formal link between the head office and the construction site.

- **Site manager, project manager** or **site agent** On large projects, the main contractor's representative on site, with overall responsibility for ensuring that work proceeds effectively and efficiently, in accordance with the design specification and to time. On small or domestic projects this person is sometimes known as the **general foreman**.

- **Contractor's quantity surveyor** or **cost control surveyor** Employed by the main contractor to check progress and assist the private quantity surveyor in the preparation of interim valuations for stage payments and final accounts. They prepare and submit accounts to clients and make payments to suppliers and subcontractors. They may also be required to measure work done for bonus and subcontractor payments.

- **Site engineer** Employed in large contracts to accurately set out the buildings using specialist surveying equipment and high-powered software. Also where new work, such as basement construction, may affect the stability of existing neighbouring structures, the site engineer may be called in to monitor its movement.

- **Estimator** Prepares unit rates for the pricing of tenders, and carries out pre-tender investigations into the cost aspects of the proposed contract.

- **Buyer** Orders materials and obtains quotations for the supply of materials and services.

- **Administrator** or **document controller** Organises the general clerical duties of the contractor's office for the preparation of contract documents and payment of salaries, subcontractors' and suppliers' invoices, insurances and all necessary correspondence.

- **Assistant contract manager** Assists with the general responsibility for administering site proceedings. Often a trainee, in the process of completing professional examinations.

- **Nominated subcontractor** Engaged by the client or architect for specialist construction or installation work, such as lifts or air conditioning.

- **Domestic subcontractor** Employed by the principal contractor to assist with the general construction: for example, groundworkers or bricklayers.

- **Operatives** The main subcontractors or direct workforce onsite; includes craftsmen, apprentices and labourers.

How these roles interlink and the composition of the building team depend on a number of factors, such as the contractual framework for a given project, the scale of the project and how the risk is apportioned between the client, the building designer and the contractor.

The size of the building firm or the contract will determine the composition of the construction team. For small or medium-sized contracts, some of the above functions may be combined, such as those of quantity surveyor and estimator. It used to be common for the architect to oversee the project using traditional contractual arrangements, with most of the risk held by the client. However, with the increasing need for efficiencies, better communication and the sharing of contractual risk, the contractor often takes a leading role much earlier in the programme. This is why the 'design-and-build' contract has become increasing popular. Furthermore, many design-and-build practices have been created by combining the professional expertise of architect, builder and consultants. The objective is to improve communications and create better working relationships, to provide the client with a more efficient and cost-effective service.

THE CONSTRUCTION LIFECYCLE

The construction processes to create and maintain a building comprise a series of linked, sequential stages, often referred as the lifecycle of the project. With the increasing emphasis on the reduction of waste and sustainable living, it is important to realise that construction activities do not cease when the building has been built. It is conceived as a need by the client, developed into a design by the architect and specialist consultants, built by the contractor, used by the client and hopefully maintained in good order, refurbished and altered to suit changing needs, and when uneconomical to do so is demolished, with the materials recycled and land redeveloped.

Figure A on the following page shows the lifecycle of a construction project.

A number of protocols map this lifecycle process, including the Chartered Institute of Building (CIOB) *Code of Practice for Project Management for Construction and Development* and the British Standard BS6079-1:2000. The most commonly used protocol is the Royal Institute of British Architects (RIBA) *Plan of Work*, which sets out the planning and control of the design and production process in several stages. The following is an outline description of the various stages of the RIBA Plan of Work:

Stage A: Appraisal

The client's requirements and any possible constraints on development need to be determined at this initial stage. The client usually needs assistance in this task by the architect or a suitably qualified design consultant. Studies are prepared to enable the client to decide whether to proceed and to select the probable procurement method. The latter is a particularly important decision, as it will determine how project resources, responsibilities and risks are apportioned between the client and their consultants, and contractors.

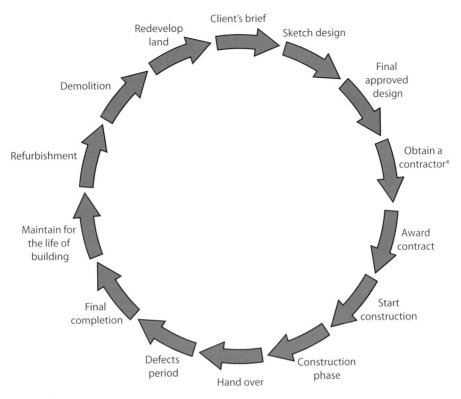

Figure A Lifecycle of a project.

*Note that this shows a traditional contractual arrangement; in a design-and-build or two-stage contract, the contractor may be brought in earlier to advise at the scheme or design stage.

Stage B: Design brief

Here the general outline of the client's requirements and the planning of future action for the client are clarified and confirmed. The architect will identify the procedures, organisational structure and range of other consultants needed to assist in the design of the project. This strategic brief is a key output from this stage, and becomes the guiding document for the design of the project.

Stage C: Concept

The design team, led by the architect in a traditional contract, will provide the client with the appraisal and recommendation so that they may determine the form in which the project is to proceed. They will ensure that it is feasible functionally, technically and financially. At this point the development of the strategic brief into the full project brief begins, and outline design proposals and cost estimates are prepared.

Stage D: Design development

With the client's approval, the design team determines the general approach to the layout, design and construction in order to obtain authoritative approval of the client on the outline proposals. The project brief will be fully developed and detailed proposals will be made and compiled, generally in a 'Stage D' report. The application for full development control approval will be made at this point with the submission of the planning drawings.

Stage E: Technical design

The various design consultants will work up their technical designs and specifications, including sufficient information to co-ordinate elements of the project to satisfy statutory approvals, such as Building Regulation and Full Planning approvals.

Stage F: Production information

In a traditional contact, all the detailed designs and specifications are finalised to enable works to be tendered. By this stage all statutory approvals should be obtained. In a design-and-build contract, where the client appoints a principal contractor to procure the design based on a performance specification, the design information for the whole project is not required at this stage. This does allow an earlier start to the construction work, but places considerable risk on the contractor.

Stage G: Tender documentation

This stage involves the preparation and collation of the tender documentation in sufficient detail to enable tenders to be obtained for the construction of the project. It should be noted that this stage (and stage H) is much more relevant to traditional forms of procurement than, for example, design-and-build or management contracts; here the main contractor or project manager is brought into the process by the client much earlier, at the preparation or design stage.

Stage H: Tender action

Under a traditional contract, this is where the potential contractor is identified and evaluated. It is important that the contractors' understanding of, and commitment to, the project vision and its sustainability is tested at this stage. Other important factors to be considered include their environmental and health and safety record.

Stage J: Mobilisation

Under a traditional contract, the principal contractor is awarded the contract and takes possession and responsibility for the site and is issued with all the production information for the project, such as the drawings and specifications. The contactor has to finalise the planning of the necessary resources, set up the site offices and temporary services, and place orders for work with subcontractors and material or component suppliers. They also have to undertake notification of the project to all the statutory authorities, such as the local authority and the Health and Safety Executive (for projects that fall under the CDM Regulations, see the CDM section under 'Legislation' below).

Stage K: Construction to practical completion

The contractor manages the project and secures all the necessary labour, plant and materials, including liaison with all the subcontractors. The client or their representative – normally the architect in traditional procurement – administers the building contract up to and including practical completion. Further information is supplied to the contractor as and when reasonably required.

Stage L: Post practical completion

This stage is clearly separated from the construction phase. Final inspections are made to ensure specifications have been met, and the building is handed over to the client for occupation. Any defects will have to be remedied before the principal contractor's final account is settled. Finally, all parties to the contract will undertake a review of the project to see what can be learned in order to improve future performance.

LEGISLATION

Many statutes made by Act of Parliament affect the construction of buildings and associated work. The most significant are:

- the Health and Safety at Work, etc. Act 1974;
- the Building Act 1984;
- the Disability Discrimination Act 1995–2004;
- the Town and Country Planning Act 1990;
- the Environmental Protection Act 1990;
- the Party Wall Act 1996.

The Health and Safety at Work, etc. Act 1974

This requires employers and their agents to implement an overall duty of care through a safe system of working. This applies to people in the workplace and others who could be affected by the work activity. The Act provides an all-embracing standard supporting a framework of statutory instruments and regulations administered by the Health and Safety Executive (HSE) through specialist inspectors operating from local offices. The inspectorate has powers to access premises, review a company's safety records, issue improvement and prohibition notices and, if necessary, effect prosecutions for non-compliance.

There are many regulations that relate to health and safety in construction. The following are a selection of the most common principal statutory regulations made under the Health and Safety at Work, etc. Act. They are legally binding on the client, architect and builder contractor.

The Construction (Design and Management) Regulations 2007

The Construction (Design and Management) Regulations (CDM) set out clear responsibilities for health and safety for all parties involved in the design and construction process. CDM also ensures that all work is planned and carried out by competent professionals, who work together through all the stages of the project to reduce health and safety risks to workers and those people who will eventually use and maintain the building until the end of its life.

The Regulations require that the client takes reasonable steps to ensure that arrangements are made to appoint a competent person, known as the CDM co-ordinator, who will ensure all the construction work can be carried out so far as is reasonably practicable without risk to the health and safety of any person. The CDM co-ordinator will work with the designer to ensure that risks are minimised, then will work with the contractor to ensure that the risks in the proposed methods of construction are reduced to a satisfactory and manageable level. The document that forms the basis of this collaborative health and safety risk assessment is known as the Health and Safety Construction Plan.

The main responsibilities for these key people are as follows.

The client

- to ensure that arrangements are in place for the proper management of health and safety of the project by competent persons;

- to make available all necessary pre-construction information about the project to the relevant people.

The CDM co-ordinator

- to notify the HSE of the project details;

- to ensure co-operation between designers;

- to co-ordinate the design with regard to avoidance of undue risks;

- to prepare a pre-tender construction phase plan and confirm the production of a health and safety manual (file) for the client occupier;

- to ensure that suitable arrangements are implemented for the co-ordination of health and safety measures during planning and preparation for the construction phase;
- to provide updates on health and safety issues.

The project designers

- to ensure that design work does not commence until the client is aware of the client's responsibilities under the CDM Regulations (by appointing a suitable CDM co-ordinator);
- to foresee and minimise any health and safety risks when preparing designs relating to the construction and maintenance of the project.

The principal contractor

- to manage, plan and monitor the construction work, so far as is reasonably practicable, without risks to health and safety;
- to ensure site personnel comply with the pre-tender construction phase/ plan;
- to develop and update the health and safety plan throughout the duration of the work, including production of risk assessed method statements;
- to provide workers with any relevant information and training they need for particular work to be safely carried out;
- to exclude from site unauthorised and uninsured persons.

Any non-domestic project that is longer than 30 days or involves more than 500 person-days of construction work falls under the CDM Regulations, and has to be notified to the Health and Safety Executive (HSE), which is the government body that enforces all the UK health and safety legislation.

The CDM Regulations (2007) also now incorporate legislation to establish objectives for the well-being of personnel on a building site throughout the duration of work. The main areas for these objectives include: good order and site security; safe access to the place of work; temporary timbering and other non-permanent support facilities; fire detection and fire fighting; provision for fresh air and lighting; prevention of drowning; defined traffic and pedestrian routes; emergency lighting and power; temperature and weather protection for workers; demolition or dismantling and the use of explosives. Also, provision must be made for welfare facilities, to include sanitation, hot and cold water supply, first aid equipment/personnel, protective clothing, and facilities to dry clothes, and appropriate accommodation for meals. Implicit is good site management, with regard for organisation and planning.

Control of Substances Hazardous to Health Regulations 2002 (COSHH)

These redress the balance in favour of health, as the introduction of the Health and Safety at Work, etc. Act has greater emphasis on safety. Manufacturing companies are obliged to monitor and declare health risks of their products, and building contractors must provide operatives with protective clothing and/or a well-ventilated environment if required to use them. COSHH has promoted the removal of harmful substances from materials, e.g. toxins, irritants and solvents, but finding substitutes is not always possible. Timber preservatives, welding fumes, dust from cement and plaster and insulating fibres are, as yet, a few of the unavoidable harmful constituents in building materials. Where these are applied, employers are obliged to monitor exposure levels, retain records, identify personnel who could be at risk and document the facilities provided for their protection.

The Manual Handling Operations Regulations 1992

These determine the employer's responsibility to ensure that employees under their control are not expected to undertake manual tasks that impose an undue risk of injury. The client, or their agent the main contractor, must assess all manual handling operations and reduce risk of injury to the lowest practicable level. This will be manifest in the provision of work systems, such as cranes and hoists, that the employee is obliged to make full use of. Due regard must be applied to:

- **tasks** space availability, manipulating distance, body movement, excess pulling and pushing and prolonged physical effort;
- **load** unwieldy or bulky, eccentric, excessive, liability to shift, temperature and sharpness of finish;
- **work environment** surface finish (slippery, uneven or variable), lighting, ventilation and temperature variation;
- **individual capacity** unusual weight or dimension, health problems/limitations, special training and provision of personal protective clothing.

Work at Height Regulations 2005

Falls from heights in the construction industry remain the single biggest cause of workplace deaths and one of the main causes of major injury. These regulations apply to all work at height where there is a risk of a fall liable to cause personal injury. The Regulations cover the range of systems used to work safely at height and their inspection regimes, which include:

- collective fall prevention, such as guard rails and toe boards;
- working platforms, such as stability of scaffolding;
- collective fall arrest, such as restraining nets and airbags;
- personal fall protection, such as work restraint harnesses, fall arrest and rope access systems.

In general, collective fall prevention systems (such as scaffold platforms) are preferred to personal fall protection systems (such as harnesses); items such as ladders and stepladders are not acceptable as working platforms unless suitable risk assessments have been carried out.

The Building Act 1984

This Act is a consolidating Act of primary legislation relating to building work. It contains enabling powers and a means for the Department of Communities and Local Government to produce building regulations for the outline purpose of:

- maintaining the health, well-being and convenience of people using and relating to buildings;

- promoting the comfort of building occupants, with due regard for the conservation of fuel and efficient use of energy within the structure;

- preventing the misuse of water, excessive consumption and contamination of supply.

The Building Regulations or statutory instruments currently in place under the Building Act are:

- the Building Regulations applicable to England and Wales;

- the Building (Approved Inspectors, etc.) Regulations;

- the Building (Local Authority Charges) Regulations;

- the Building (Inner London) Regulations;

- the Building (Disabled People) Regulations;

- the Energy Performance of Buildings (Certificates and Inspections) Regulations.

The Building Regulations contain minimum performance standards expected of contemporary buildings. They are supported by a series of Approved Documents that are not mandatory, but give practical guidance on compliance with the requirements of the Regulations. This guidance often incorporates British Standards, Building Research Establishment, British Board of Agrément and other authoritative references.

Control of the Building Regulations is vested in the local council authority. However, a developer/builder can opt for private certification, whereby the developer and an approved inspector jointly serve an initial notice on the local authority. This describes the proposed works, which the local authority can reject (if they have justification) within 10 days of application. With this option, the responsibility for inspecting plans, quality of work, site supervision and certification of satisfactory completion rests with the approved inspector, as defined in the Building (Approved Inspectors, etc.) Regulations.

Building legislation is regionally divided throughout the UK, each area having its own regulations:

- the Building Regulations applicable to England and Wales;
- the Building Standards (Scotland) Regulations;
- the Building Regulations (Northern Ireland);
- the Building (Inner London) Regulations.

There are also statutory provisions for the Channel Islands, the Isle of Man and the Republic of Ireland.

Approved Documents

The current Building Regulations came into effect in 2010. They are constantly under review in response to new and changing technologies, public demands and environmental directives. The specific guidance that the Regulations set out is contained with a supporting series of Approved Documents. Those made under the Regulations applicable to England and Wales are currently listed from A to P (I and O omitted):

- A – Structural safety;
- B – Fire safety;
- C – Site preparation and Resistance to contaminants and moisture;
- D – Toxic substances;
- E – Resistance to the passage of sound;
- F – Ventilation;
- G – Sanitation, hot water safety and water efficiency;
- H – Drainage and waste disposal;
- J – Combustion appliances and fuel storage systems;
- K – Protection from falling, collision and impact;
- L1 – Conservation of fuel and power in dwellings;
- L2 – Conservation of fuel and power in buildings other than dwellings;
- M – Access to and use of buildings;
- N – Glazing safety;
- P – Electrical safety.

There is another Approved Document made under Regulation 7, Workmanship and materials. This requires any building subject to the Building Regulations be carried out with proper materials and in a workmanlike manner.

The Approved Documents provide practical and technical guidance for satisfying the Building Regulations. There is no obligation to adopt any of these, provided the Building Regulations' performance requirements are shown to be satisfied in some other way. This may include European Technical Approvals, British Board of Agrément certification, CE marking of products, and calculations in accordance with acceptable structural standards for selection of components.

The Disability Discrimination Act 1995–2004

This Act was introduced in three stages between 1995 and 2004. It is designed to assist and benefit an estimated eight million UK residents who suffer some sort of disability. In principle, the Act requires that all new, adapted and refurbished buildings be constructed with unobstructed access and facilities that can be used by wheelchair occupants. This requirement extends to service providers and owners of public buildings. Details of this regulation are explored more fully in Part 9 on pages 451–468.

The Town and Country Planning Act 1990

This Act establishes the procedures for development of land and construction of buildings. It is administered through the hierarchy of government, regional and local offices:

- central government;
- county planning departments;
- local authority planning departments.

Central government issues departmental circulars and policy planning statements (PPSs) to local planning departments. There are currently 25 policy planning statements in existence, for example:

- PPS1 – 'Delivering Sustainable Development';
- PPS2 – 'Green Belts';
- PPS3 – 'Housing';
- PPS5 – 'Planning for the Historic Environment';
- PPS22 – 'Renewable Energy'.

These are prepared by the Government after public consultation to explain statutory provisions and provide strategic objectives and guidance to local authorities and others on planning policy and the operation of the planning system. They also explain the relationship between planning policies and other policies that have an important bearing on issues of development and land use, such as development on a large scale, as with the Thames Gateway development. Full details of all the PPSs can be found by contacting the Department of Communities and Local Government (www.communities.gov.uk).

On the basis of the PPSs, the county planning departments formulate development policy within the framework determined by central government. This is known as the **structure plan**. This takes into account the social and economic demands for housing, the need for commercial/industrial development and recreational facilities, and viable communications and infrastructure projects such as transport and services. Structure plans must be prepared with regard to the use of brownfield sites (reuse of redundant industrial buildings or land) and green belt (rural land that should not be developed). Structure plans must take into account the views of the local community, and are subject to public consultation and once established they normally remain in place for 15 years.

Local planning departments or local authorities then establish a **local plan** for their district. This is produced within the framework of the county's structure plan with regard for an economic, social and practical balance of facilities for the various communities under their administration. To maintain a fair and equitable interest, local plans are subject to public and central government consultation. In certain parts of the country, such as London and the metropolitan cities, National Parks, and in a few non-metropolitan unitary areas, authorities produce unitary development plans (UDPs), which combine the functions of structure and local plans and include minerals and waste policies. Local authorities are also responsible for processing applications for development in their area. Applications range from a nominal addition to an existing building to substantial estate development. Procedures for seeking planning consent vary depending on the scale of construction. All require the deposit of area and site plans, building elevations, forms declaring ownership or nature of interest in the proposal, and a fee for administration. If an application is refused, the applicant has a right of appeal to central government.

Owners can make certain types of minor changes to their house without needing to apply for planning permission. These are known as 'permitted development rights' and derive from a general planning permission granted by Parliament under specific legislation: the Town and Country Planning (General Permitted Development) Order 1995 (last amended in 2008). Generally, permitted development rights do not apply to flats, maisonettes or other buildings. This Order grants permitted development to existing houses provided that the following conditions are met.

1. The total area of ground covered by buildings within the curtilage of the house would exceed 50 per cent of the total area of the curtilage (excluding the ground area of the original house).

2. The height of the part of the house enlarged, improved or altered would exceed the height of the highest part of the roof of the existing house.

3. The height of the eaves of the part of the house enlarged, improved or altered would exceed the height of the eaves of the existing house.

4. The enlarged part of the house would extend beyond a wall which fronts a highway and forms either the principal elevation or a side elevation of the original house.

5. The enlarged part of the house would have a single storey and extend beyond the rear wall of the original house by more than 4 metres in the case of a detached house, or 3 metres in the case of any other house, or exceed 4 metres in height.

6. The enlarged part of the house would have more than one storey and extend beyond the rear wall of the original house by more than 3 metres, or be within 7 metres of any boundary of the curtilage of the house opposite the rear wall of the house.

In any case it is best to consult with the local planner to ensure that there are no misunderstandings with regard to permitted development, as where these conditions are not met then a planning application to the local authority will be required. In certain situations, permitted development rights may be withdrawn by issuing an Article 4 direction. This often happens in new high-density housing development where the size of the plots are small, historic conservation areas or National Parks. There are also different requirements if the property is a listed building, where an additional consent has to be granted under the Planning (Listed Building and Conservation Areas) Act 1990.

The Environmental Protection Act 1990

The construction industry impacts on the natural and built environment in many ways. There are five main categories where care is required to reduce the impact of construction processes:

- air emissions – such as dust from earthworks, demolition activities or emissions from plant and equipment;
- land contamination – such as from working on previously contaminated industrial sites;
- noise pollution – from construction operating plant and equipment;
- waste disposal – from spoil, off-cuts and other building materials;
- water discharges – such as dewatering excavations and pipe testing.

A series of UK legislation has set out controls and checks on these activities. The major Act was the Environmental Protection Act 1990, which set out the principal legislation followed by the Environment Act 1995 which, among other issues, established the Environment Agency to enforce and refine the protection legislation. It also places a broad duty of care on importers, producers, carriers, keepers, treaters or disposers of waste to prevent unauthorised or harmful activities to the environment. Other Acts have followed, such as the Clean Neighbourhoods and Environment Act 2005, which set out to regulate and license the acceptable disposal of household, industrial and commercial waste, including construction waste and contaminated spoil material.

The Environment Agency has also produced a range of UK-wide Pollution Prevention Guidelines (PPGs); each PPG is targeted at a particular activity to provide practical advice on legal responsibilities and good environmental practice. PPGs can be found at www.netregs.gov.uk/netregs/businesses/construction/62405.aspx

The Site Waste Management Plans Regulations 2008

For construction managers, this is the main regulation concerning waste management. These Regulations require a site waste management plan to be prepared and implemented by clients and principal contractors for all construction projects with an estimated cost greater than £300,000. The plans must record details of the construction project, estimates of the types and quantities of waste that will be produced, and confirmation of the actual waste types generated and how they have been managed. Failure to comply can lead to a fine of up to £50,000 or a conviction.

The Party Wall Act 1996

A party wall or structure, such as a floor, is a shared element of a building which separates the ownership either side of that element. The Party Wall Act came into force on 1 July 1997 in England and Wales, and it provides a legal framework for preventing and resolving disputes in relation to development work to party walls, boundary walls and excavations near neighbouring buildings. Anyone intending to carry out work involving an existing party wall as listed below must give the adjoining owners notice of their intentions. If the adjoining owners object, then a dispute under the Act is recognised and professional help is required by the appointment of a specialist party wall building surveyor. In most cases the party wall surveyor is appointed to act on behalf of the wall to ensure that an equitable and fair agreement between the adjacent owners is reached.

Examples of such works covered in the Act include to:

- cut into a wall to take the bearing of a beam (for example, for a loft conversion);
- insert a damp proof course all the way through the wall;
- raise the height of the wall and/or increase the thickness of the party wall;
- demolish and rebuild the party wall;
- underpin the whole thickness of a party wall.

BRITISH STANDARDS

BSI British Standards is the UK's National Standards Body and represents UK social and economic interests across all of the European and international standards organisations. It is a not-for-profit organisation that works with manufacturing and service industries, businesses, governments and consumers to facilitate the production of British, European and international standards. Within construction, it offers testing, assessment and certification for products and services.

Standards function as support documents to building design and practice. Many are endorsed by the Building Regulations and provide recommendations for minimum material and practice specifications. They are published in a number of possible formats.

- British Standards (BS)

These set out each standard as a specification, method or guide. Where good practice is described, it is referred to as a Code of Practice (CP), although within construction these have become very limited.

- Drafts for Public Comment (DPC)

These are draft standards published so that people can submit comments to the relevant committee for them to consider, before finalising the standard. A DPC is usually current for only six months, after which it is withdrawn.

- Drafts for Development (DD)

These are issued instead of codes of practice or standard specifications where there is insufficient data or information to make a firm or positive recommendation. They have a maximum life of two years, during which time sufficient data may be accumulated to upgrade the draft to a standard specification.

- Published Documents (PD)

This is a catch-all category for standard-type documents that do not have the same status as a BS. For example, they can be documents derived from British Standards that conflict with ENs (see on the next page), but are still needed by industry, such as PD 5500: *Pressure vessels*.

- Publicly Available Specifications (PAS)

These are documents developed by BSI British Standards but commissioned by an external organisation, such as a trade association or private company.

- Amendments (AMD)

Standards are reviewed every five years. Between reviews, it is sometimes necessary to issue amendments – approved supplementary material that alters or adds to previously agreed technical provisions, without changing the date of publication.

European standards are administered by the Comité Européen de Normalisation (CEN), which incorporates bodies such as the British Standards Institution (BSI) from member states. The BSI is obliged to adopt all European Standards and to withdraw any national standards that might conflict with them. They are published in the UK as BS ENs. Many British Standards have been harmonised in this way. For example, BS EN 196-2: 1995: *Chemical analysis of cement* has replaced BS 4550: Part 2: 1970: *Methods of testing cement; chemical tests*. However, the most recent major harmonisation has been the development of the structural design Eurocodes. The British Standards Institute withdrew 54 British Standards for the design of buildings and civil engineering structures in May 2010 as part of the implementation of the new Eurocodes. This implementation of the structural Eurocodes is the biggest change to codified structural design ever experienced in the UK.

THE EN EUROCODES

The EN Eurocodes and associated National Annexes are a pan-European set of reference structural design codes for construction products and engineering services that are mandatory for European public works and likely to become the de-facto standard for the private sector. They satisfy two major EU directives: the Public Procurement Directive (2009/81/EC) and the Construction Products Directive (93/68/ECC).

Structural Eurocodes comprise 10 standards in 58 separate documents for the design of all types of structures including buildings, bridges, chimneys, masts and towers, dams, etc. To enable local variation to be incorporated, there are National Annexes (nationally determined parameters) which set out details such as, for example, loading conditions for differing snow loads and wind loads. They also include Published Documents (Non-Contradictory Complementary Information), Product Standards (specification of construction products) and Execution Standards (details of construction requirements and workmanship).

The ten structural Eurocodes
Structural safety, serviceability and durability

- Eurocode: Basis of structural design   BS EN 1990

Loadings on structures

- Eurocode 1: Actions on structures   BS EN 1991

Design and detailing of structures

- Eurocode 2: Design of concrete structures   BS EN 1992
- Eurocode 3: Design of steel structures   BS EN 1993

- Eurocode 4: Design of composite steel and concrete structures   BS EN 1994
- Eurocode 5: Design of timber structures   BS EN 1995
- Eurocode 6: Design of masonry structures   BS EN 1996
- Eurocode 9: Design of aluminium structures   BS EN 1999

Geotechnical and seismic design

- Eurocode 7: Geotechnical design   BS EN 1997
- Eurocode 8: Design of structures for earthquake resistance   BS EN 1998

The new Eurocodes provide a single suite of standards with a common understanding and vocabulary. They use common design criteria based on performance evidence and statistical methods as encompassed by limit state design codes (as opposed to the traditional permissible stress design codes, which are gradually being withdrawn). It is envisaged that Eurocodes will encourage shared investment in software and design development using the latest materials technologies, and will promote Europe-wide research and design investment. In terms of best practice, the Eurocodes are based on the most up-to-date research and provide a rational and consistent framework.

INTERNATIONAL STANDARDS

The International Standards Organisation (ISO) also produces documents compatible with British Standards. The BSI has the option of adopting any international standard as a British Standard, but does not have to do so. These are then republished as BS ISO, BS IEC or BS ISO/IEC.

AGRÉMENT CERTIFICATES

The purpose of these is to establish the quality and suitability of new products and innovations not covered by established performance documents, such as British (BSI) or European Standards (CEN), and are administered by the British Board of Agrément. The Board assesses, examines and tests materials and products with the aid of the Building Research Establishment and other research centres, to produce reports. These reports are known by the acronym MOATS (Methods Of Assessment and Test). Products satisfying critical examination relative to their declared performance are issued with an Agrément Certificate.

MODULAR CO-ORDINATION

Since the building industry adopted the metric system of measurement during the early 1970s, there has been a great deal of modification and rationalisation as designers and manufacturers have adapted themselves to the new units.

Most notable has been the co-ordination of product developments about the preferred dimension of 300 mm. The second preference is 100 mm, with subdivisions of 50 mm and 25 mm as third and fourth preferences. Many manufacturers now base their products on the 100 mm module, and co-ordinate their components within a three-dimensional framework of 100 mm cubes. This provides product compatibility and harmonisation with international markets that are also metric and dimensionally co-ordinated. Kitchen fitments provide a good example of successful development, with unit depths of 600 mm and widths in 300, 600, 1200 mm and so on. Less successful has been the move to metric bricks in lengths of 200 or 300 mm long x 100 mm wide x 75 mm high. They are just not compatible with existing buildings, and bricklayers have found them less suited to the hand. Another example is the use of board materials used in dry lining and internal partitions that span from floor to ceiling without the need for being cut.

Although the metric system is now well established, a large proportion of the nation's buildings were designed and constructed using imperial units of measurement, which are incongruous with their metric replacements. Many building materials and components will continue to be produced in imperial dimensions for convenience, maintenance, replacement and/or refurbishment; where appropriate, some manufacturers provide imperial–metric adaptors (for example, with plumbing fittings).

1

Site investigations

Importance of
site investigations 1.1

Before construction work can go ahead, it is often necessary to undertake a thorough site investigation. This is where an adequate budget should be devoted to ground investigation, in order to remove many of the uncertainties when constructing on unknown ground. Gaining as much information as possible about a site before construction begins will help to avoid any unforeseen works; this in turn will avoid contract variations appearing after the award of the contract that could be costly for a client, in terms of increased expenditure and delays.

A thorough site investigation is therefore an essential requirement of any proposed construction project for a variety of reasons, including:

■ the sustainable use of brownfield sites that have had an unknown previous use which may contain contamination;

■ to reduce the risks associated with the financing of a project.

Both of these are sustainable decisions that will have to be made if a project is to be taken through to the outline design and detailed design planning stages. Any uncertainty needs to be fully investigated and ascertained.

Clients need to be briefed thoroughly about any hidden risks or uncertainties that may become apparent. A well-planned and prepared site investigation can put clients' minds at rest and lead to a successful project for all parties concerned.

Buildability 1.2

Checking that a construction project can be built – its 'buildability' – involves reviewing the various processes involved during the pre-construction phase. It is important to identify any obstacles or potential problems before a project is actually built to reduce or prevent errors and delays, and to make sure that the project stays within budget.

A well-planned and properly financed site investigation will provide a number of benefits for a client and contractor by:

- removing any uncertainties within a substructure design;
- confirming budgets for design and construction;
- enabling fixed decisions to be made at the feasibility stages of design;
- allowing the contractor to adjust programme constraints;
- enabling resources to be allocated.

The site investigation is a vital component of the design and construction of a project, as the superstructure depends on the substructure. Thorough site investigation must be acknowledged as a key element, and should be supported by adequate resources and finance.

The site investigation findings will need careful discussion with the client and their architect or designer, as the findings may impact on the feasibility of the project and the design or aesthetics that the client wishes for their project, due to the nature of the ground it has to be built upon.

Buildability is a sustainable process that, when employed on a project, produces many benefits for both the client and the contractor, whose expertise is valuable.

Elements of a
site investigation 1.3

A site investigation is generally carried out using the two following methods:

- a desktop survey, involving a document search of pre-existing information;
- a site survey, involving firstly a walkover survey where reconnaissance of the site is undertaken, followed by the use of detailed subsoil investigation by means of trial pits and borehole records.

The investigation work within the subsoil covers factors such as:

- **contaminated ground** – knowing the type of contamination will help ascertain the remedial works that may be required to make the site safe;
- **groundwater level** – this will have an effect on the design of any substructure works required to resist water pressure and water contamination;
- **obstructions** – brownfield sites often contain obstructions such as petrol tanks, concrete foundations and services;
- **bearing capacity of the soil** – this is the amount of safe loads that a soil can resist without settlement;
- **sloping building sites** – landslip planes within soil strata can cause gravitational rotation of a site down a slope;
- **chemical composition** – soils that contain sulphates can interfere with the bonding of cement in concrete.

DESKTOP SURVEY

ONLINE MAPS

Access to satellite mapping online means that anyone can obtain digital imagery of a potential site. The UK has been successfully mapped by the

Ordnance Survey (OS), Great Britain's national mapping agency. OS maps are available in digital format in a range of scales, including a 1:500 scale series; maps at the largest scales give details of any buildings, structures and boundaries, as well as surrounding roads, footpaths and properties.

This will often reveal:

- topographical details – relative heights can often be estimated by the use of shadows within the photograph;
- existing features – boundary fence lines, rights of way and existing buildings;
- watercourses – strict planning regulations are now enforced on any development within the reach of a floodplain;
- trees – which trees have preservation orders can be established by talking to the local conservation officer, using the photographic map for reference;
- archaeological features – finding any items or areas of archaeological interest will delay a project and increase costs, as the developer often has to pay for work to be undertaken.

HISTORICAL MAPS AND LOCAL ARCHIVES

Many local authority reference libraries hold local maps, as copies or originals. These can reveal the history of a potential site and what industrial and commercial applications it has been subjected to. This can provide valuable information on:

- the type of potential contamination, by identifying any industrial processes that used to be carried out on the site;
- any underground obstructions, by identifying the size of any earlier buildings;
- the locations of boundaries, footpaths and roadways.

Local Records Offices are a great source of local historical information. Many historical documents are contained within these archives including photographs, old maps, street index records and other information that can be used to research a particular property or site. Local Records Offices also may hold a range of oblique and orthographic aerial photography undertaken from the 1940s. These can clearly show the previous history of the site and hence disclose evidence of previous use and possible contamination. The Land Registry also holds information on the ownership and deeds associated with a plot of land and can prove a valuable source of information.

BUILDING AND PLANNING APPLICATIONS

These are a valuable desktop resource. Many local authority websites have a link on the website which covers planning conservation and the environment. This link will guide you to the planning division, which gives you access to an online planning application database. Here planning applications can be searched for an address. This enables any historical planning applications to be viewed and valuable information gained about the site with regard to:

- what type of application has been passed for the site previously;
- existing drawings for current use;
- any planning conditions or restrictions associated with the potential site;
- the level of neighbourhood support for the potential site.

ADDRESS WEB SEARCH

By entering the address of the site into any internet browser, you can undertake a web search to establish if there are any documents connected with that address. This would cover aspects such as:

- County Court judgements;
- historical values of the property and average values for the area;
- previous use of a site or building;
- previous site investigation reports.

THE SITE SURVEY

WALKOVER SURVEY

As its name suggests, a walkover survey involves a physical walk over the site of a proposed project. By walking over the site you will be able to pick up visual information that will contribute to the desktop survey, on things such as:

- existing structures;
- existing services;
- trees, hedges and boundaries;
- indication of topography and levels;
- evidence of groundwater levels and flooding;
- overhead obstructions, such as pylons and telephone lines.

The walkover survey can be recorded on a sketch or detailed plan, as in Fig. 1.3.1.

TRIAL PITS AND TRENCHES

Trial pits and slit trenches are used to examine the structure of the soils that the proposed development will rest upon. A trial pit is a square pit,

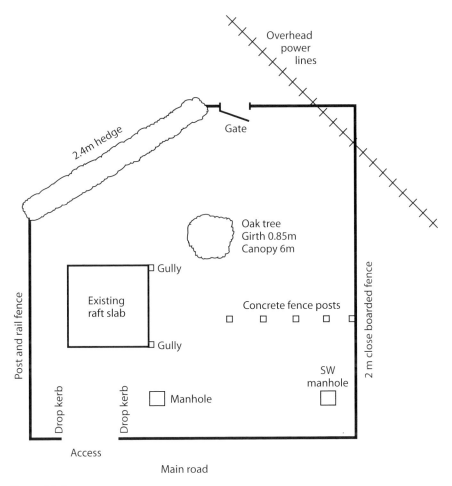

Figure 1.3.1 Example of a walkover survey.

normally 2 m x 2 m x 2 m, where an engineer can visually inspect the strata below, establish the groundwater level and take any samples from the wall of the pit for further testing. Trial pits are normally excavated using a back-acter excavator.

Slit trenches cover a longer area than trial pits and provide a soil engineer with a greater range of subsoils for visual inspection.

The location of any trial pit or slit trench must be recorded accurately on a scale plan or map and numbered, so that soil test results can be linked back to the relevant trial pit or trench.

A detailed record should be made of the trial pit in a log (see Fig. 1.3.3), including information such as:

■ the depth and type of the different strata;

■ the level of the groundwater;

■ any testing undertaken and at what depth it is carried out.

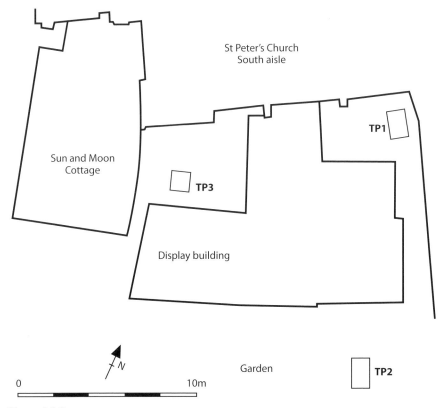

Figure 1.3.2   Scale plan showing location of trial pits.

Care must be taken with the health and safety associated with entering the confined space of the slit trench or trial pit. If required, additional earthwork support should be provided in unstable ground.

With any excavation work, the requirements of the Confined Spaces Regulations 1997 and the Health and Safety at Work, etc. Act 1974 must be adhered to.

BOREHOLE INVESTIGATION

A borehole is a method of investigating the subsoil strata using a drilling rig (see Fig. 1.3.4). Using a borehole can enable the collection of information from greater depths than when using trial pits. The drilling rig is normally towed behind a four-wheel-drive vehicle and erected on site in the form of a tripod.

The rig operates by using a winch and cable, which is attached to a motor and a winch clutch. When the clutch is released the cable drops with the heavy cutting head which gradually works its way down into the ground. This method is known as percussion drilling using a shell and auger. Various samples can be taken at specified depths and loose samples can be taken from the cutting head as it is emptied.

URS

TRIAL PIT LOG

Project Name and Site Location		Client		TRIAL PIT No
Orchard Retail Park		Defran Investments Ltd		**TP1**

Job No	Date		Ground Level (m)	Co-Ordinates ()	
49327965	Start Date 18-02-09 End Date 18-02-09				

Contractor	Method / Plant Used	Sheet
Whippet Plant	JCB 3CX	1 of 1

SAMPLES & TESTS				STRATA			
Depth	Type & Ref No	Test Result	Water	Reduced Level	Legend	Depth (Thickness)	DESCRIPTION
							Brown clayey SAND. Frequent rootlets. (TOPSOIL)
						(0.50)	
						0.50	
0.5	B 1						Light brown slightly gravelly medium to coarse SAND. Gravel is subrounded and coarse quartzite. (GLACIAL SAND AND GRAVEL)
1.0	B 2					(1.30)	
1.5						1.80	
2.0	B 3						Brown and light grey sandy subrounded and rounded fine to coarse GRAVEL. Frequent subrounded and rounded cobbles. (GLACIAL SAND AND GRAVEL)
2.5						(1.00)	
						2.80	
3.0	B 4					3.00	Firm to stiff grey mottled orange brown silty CLAY. Occasional black woody fragments. (GLACIAL SAND AND GRAVEL)
							Firm red brown slightly sandy CLAY. (MERCIA MUDSTONE)
3.5						(0.80)	
						3.80	
							End of trial pit

TRIAL PIT INFORMATION				GENERAL REMARKS
Dimensions (m)	Stability	Groundwater Observations	Remarks	
x	Unstable below 1.30m Support: None	Seepage at 0.70m into north end of pit		

All dimensions in metres Scale 1:31.25		Logged By LJ	Approved By AH

URS GEOTECH TRIAL PIT ORCHARD RETAIL PARK TRIAL PITS.GPJ AGS3 ALL.GDT 27/03/09

Figure 1.3.3 A trial pit log.

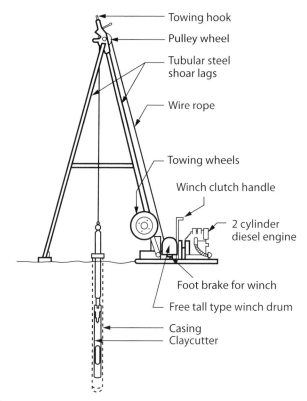

Figure 1.3.4 **A borehole drilling rig.**

A borehole log of the subsoil strata can quickly be established. Several boreholes would be required across a potential site to check for the consistency of the strata and to provide information for a final foundation design.

Boreholes can be undertaken at shallow depths using a hand auger. This is turned by hand and drills a suitable hole from which disturbed and in-situ soil samples can be taken.

SOIL TESTING

A number of different tests can be performed on both loose and in-situ samples. These cover different properties of the soils such as:

- density and compaction;
- chemical composition;
- grading profile (fine through to coarse) using a sieve analysis;
- moisture content and permeability;
- shear strength;
- Californian bearing ratio – a test where a known weight is applied to a point penetrating the soil to a recorded depth.

Contaminated ground and waste disposal 1.4

The Contaminated Land Regulations, Planning Regulations and Waste Management Regulations are just three of the vast array of regulations and EU Directives that now control development on contaminated land.

The UK Planning Regulations require that any development that may involve contaminated ground be tested and a remediation report prepared, identifying the actions needed to bring the land back to a habitable state. This may often involve the removal in bulk of the contamination by licensed carriers to a tip licensed to receive it. The waste's origin, who transported it and which tip received it must all be clearly identified. This is done so the waste is traceable back from the waste disposal site to the point of origin, should any issues occur with incorrect paperwork or the level of contamination.

There are other methods for dealing with waste and site contamination which involve site treatment and reuse, or the installation of a horizontal capping layer to act as a barrier. The choice of which method to select depends on the type and toxicity of the waste present – details of these methods will be explored in *Advanced Construction Technology*, under 'Contaminated land remediation'.

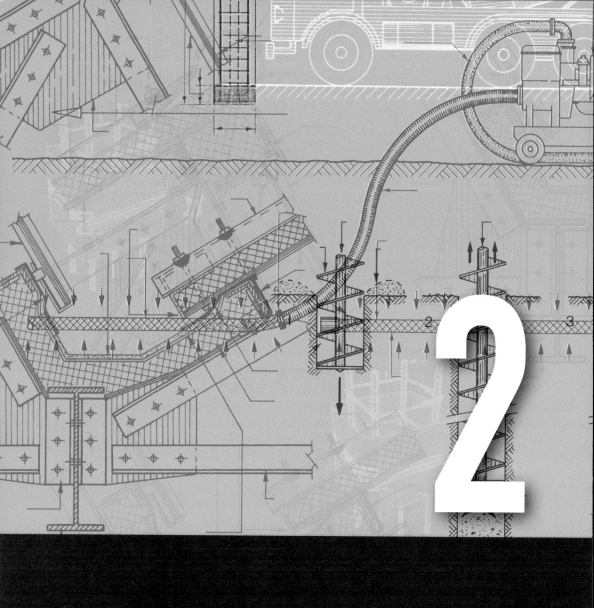

Site and temporary works

2

Site establishment 2.1

The client will procure a contactor to undertake the construction work on their behalf through a tendering process. When tenders have been checked, the contract is awarded to the winning contractor judged against the procurement criteria. Once the contract has been awarded and accepted, the contractor normally has a period of time for mobilisation of the resources required to fulfill it. Within this period they must:

- obtain all the contract documents, including drawings, specifications;
- organise site accommodation;
- schedule material deliveries;
- secure the site boundaries;
- plan the sequence of work;
- organise labour and plant resources;
- organise the administration and management of the contract.

SITE SETUP

A construction site will need to be set up in accordance with the Construction (Design and Management) Regulations 2007, which detail under Schedule 2 the welfare facilities that are to be provided for construction workers and staff. These include:

- toilets;
- washing facilities;
- drinking water;
- changing rooms and lockers;
- rest rooms.

The setting up of the site will involve the temporary use of:

- accommodation;
- fencing, lighting and security;
- waste disposal;
- services;
- car parking;
- storage.

SITE CLEARANCE

The site may require clearance, enabling works to be undertaken before the main construction work can commence. This may involve:

- demolition and removal of existing buildings and structures;
- grubbing out of bushes and trees;
- removal of soil to reduce levels.

Demolition is a skilled occupation and should be tackled only by an experienced demolition contractor in accordance with the CDM Regulations. The removal of trees can be carried out by hand using chain saws by a specialist arborist, or by the use of an excavator of sufficient size.

Site clearance of topsoil has to be undertaken to remove the 150 mm of soil that contains seeds, plants and weed, which would not be sound material to construct on.

Building Regulation Approved Document C states: 'Vegetable matter such as turf and roots should be removed from the ground to be covered by the building at least to a depth to prevent later growth.' This is in effect to sterilise the ground, because the top 300 mm or so will contain plant life and decaying vegetation. This means that the topsoil is easily compressed and would be unsuitable to support foundations. Topsoil is valuable as a dressing for gardens, and will be retained for reinstatement when the site is landscaped. The method chosen for conducting the site clearance work will be determined by the scale of development, and by consideration for any adjacent buildings.

SETTING OUT THE SITE

The first task is to establish a baseline from which the whole of the building can be set out. This will often be provided on a setting-out drawing for the foundations. The position of this line must be clearly marked on site so that it can be re-established at any time. For onsite measuring a steel tape should be used (30 m would be a suitable length); plastic-coated synthetic tapes are also available. After the baseline has been set out, marked and checked, the main

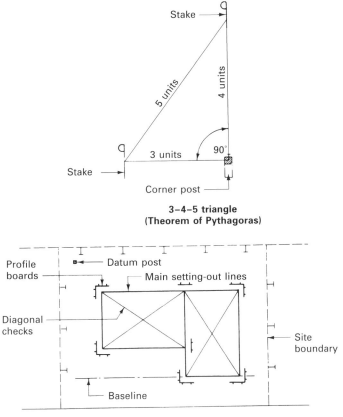

3–4–5 triangle
(Theorem of Pythagoras)

Figure 2.1.1   **Setting out and checking methods.**

lines of the building can be set out, each corner being marked with a corner peg. A check should now be made of the setting-out lines for right angles and correct lengths. The best method to establish this is with a calculated diagonal length, thus ensuring that the set-out building or structure is square. The setting-out procedure and the methods of checking the right angles are illustrated in Fig. 2.1.1.

Modern methods now employ the use of a total station. This is an electronic piece of equipment that can store the digital plan of a building and replicate it using a prism target on site. Each corner of a building can be established electronically on site from a known point.

After the setting out of the main building lines has been completed and checked, profile boards are set up as shown in Fig. 2.1.2. These are set up clear of the foundation trench positions to locate the trench, foundations and walls. Profile boards are required at all trench and wall intersections. The nails inserted onto the boards can then have string lines secured to them, to establish the faces of the building and trench.

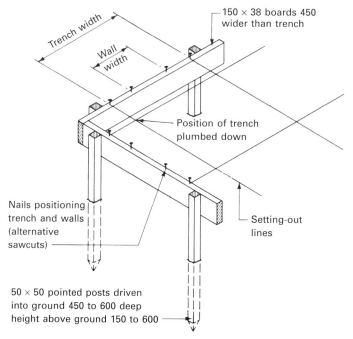

Trench width

Wall width

150 × 38 boards 450 wider than trench

Position of trench plumbed down

Nails positioning trench and walls (alternative sawcuts)

Setting-out lines

50 × 50 pointed posts driven into ground 450 to 600 deep height above ground 150 to 600

Figure 2.1.2   **Typical profile board.**

ESTABLISHING A TEMPORARY BENCHMARK

It is important that all levels in a building are taken from a fixed point called a **datum**. This is a known point that is often related to an Ordnance Survey benchmark (OSBM). When transferred from the OSBM it is commonly known as a temporary benchmark (TBM) on site. The design drawings will have floor levels set to the OS datum by the designer and the TBM must relate to this.

An OS benchmark is illustrated in Fig. 2.1.3 and is a horizontal mark above three pointing arrows. The centreline of the horizontal is the actual level indicated on an Ordnance Survey map. Benchmarks are found cut or let into the sides of walls and notable buildings such as churches. Where there are no benchmarks on or near the site, a suitable datum must be established. A site datum or temporary benchmark could be a post set in concrete or a concrete plinth set up on site. The OS datum level is established by referring to the OS 1:500 series maps which are now obtained online; this is the height above mean sea level at Newlyn in Cornwall marked in metres.

TAKING LEVELS

The equipment used is an engineer's level and a levelling staff. The level is a telescope fitted with cross-hairs to determine alignment. The telescope rotates

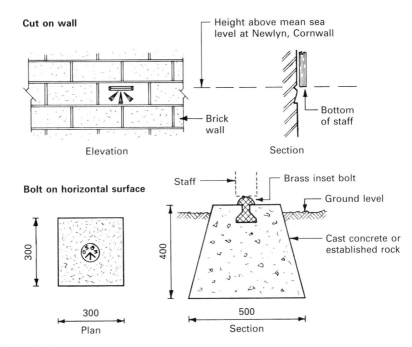

Cut on wall

Elevation

Section

Height above mean sea
level at Newlyn, Cornwall

Brick
wall

Bottom
of staff

Bolt on horizontal surface

Staff

Brass inset bolt

Ground level

Cast concrete or
established rock

300

300

Plan

400

500

Section

Figure 2.1.3 **Common types of permanent benchmark.**

on a horizontal axis plate, mounted on a tripod. The staff is usually 4 m long in folding or extendable sections. The 'E' pattern shown in Fig. 2.1.4 is generally used, with graduations at 10 mm intervals. Readings are estimated to the nearest millimetre, although digital levels with unique bar-coded staff are now more popular.

Normally a grid of levels is established on the site before excavation work commences. This and the formation level of the excavation are then used to calculate how much material will have to be cut from the site and taken to a tip. Due to the financial implications of landfill tax and regulations, site levels are often co-ordinated to reduce the amount of excavation, and the material is often landscaped into earth bunds (see Fig. 2.1.6).

Levelling commences with a backsight to a benchmark from the instrument stationed on firm ground. Staff locations are located at measured intervals such as a 10 m grid. From these, instrument readings are taken, as shown in Fig. 2.1.5. The level differentials can then be combined with plan area calculations to determine the volume of site excavation or cut and fill required to level the site.

MEASURING ANGLES

Where setting out involves an angle that is not 90°, an instrument known as a theodolite can be used to set out any angle accurately. This instrument is

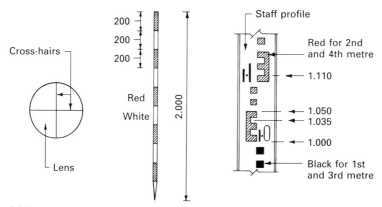

Figure 2.1.4 Levelling 'scope, ranging rod and 'E' pattern staff.

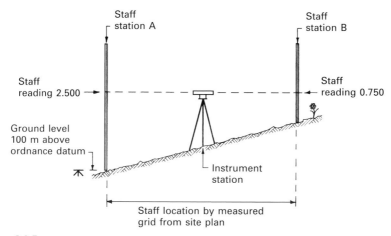

Figure 2.1.5 Principles of levelling (difference in height = difference in readings).

basically a focusing telescope with cross-hairs, mounted on horizontal index plates over a tripod. A vertical measurement circle with index is built into one side of the telescope. With the instrument firmly stationed, the telescope and horizontal (vertical if appropriate) plate are rotated from an initial sighting through the required angle, which can be read directly from the LCD display. A check can be made by rotating the telescope through 180° vertically and horizontally for a second reading; this uses the face left and face right principle, and averages out the errors in the measurement of the read angle and provides a check on the collimation. Angles are recorded in degrees (°), minutes (') and seconds ("), the extent of accuracy determined by the quality of instrument and the number of repeated observations and skill of the user.

SLOPING SITES

Very few sites are level, and therefore before any building work can be commenced the area covered by the building must be levelled. In building terms this operation is called **grading surfaces**. Three methods can be used, and it is the most economical that is usually employed.

- **Cut and fill** The usual method because, if properly carried out, the amount of cut will equal the amount of fill, therefore saving on transport and landfill.

- **Cut** This method has the advantage of giving undisturbed soil over the whole of the site, but has the disadvantage of the cost of removing the spoil from the site.

- **Fill** A method not to be recommended because, if the building is sited on the filled area, either deep foundations would be needed or the risk of settlement at a later stage would have to be accepted. The amount of fill should never exceed a depth of 600 mm.

The principles of the above methods are shown in Fig. 2.1.6.

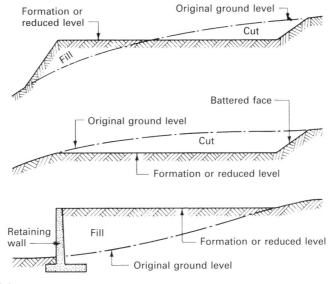

Figure 2.1.6 **Sloping sites.**

Accommodation, storage and security 2.2

The contractor's site accommodation is a temporary structure that remains in place during the length of the construction phase on site. The length of time on site will therefore dictate the type of site accommodation that is provided. It is, however, within the contractor's interest to provide the welfare facilities that are economically possible for any particular contract.

Site accommodation for longer contracts often takes the form of portable buildings that connect together to form an office complex. The provision of site accommodation must also include the administration of the contract through office staff to run the accommodation, and cleaners to keep it up to an acceptable standard. The better the facilities and amenities provided on a construction site, the greater will be the contentment of the site staff, which will ultimately lead to higher productivity.

ACCOMMODATION

The Construction (Health, Safety and Welfare) Regulations 2007 is a statutory instrument that provides details of the minimum requirements through Schedule 2 to be provided on a construction site. An extract from the current regulations is shown in Table 2.2.1, which indicates absolute minimum standards for guidance only. You will note that the number required is now 'suitable and sufficient', with no guidance given as to how many you have to provide.

Units of staff accommodation usually come in one of three forms:

- semi-portable jack leg units;
- mobile cabins on wheels;
- modular office accommodation.

Table 2.2.1 Site-based health and welfare requirements according to the CDM Regulations 2007.

Item	Content	Method of providing
Toilets	No reference to quantity, only suitable and sufficient toilets should be provided All should be kept clean and orderly Separate rooms for male and female	Portable single occupancy unit Mobile toilet block with direct drainage connection
Washing facilities	Clean hot and cold water Soap Towels or other means of drying Must be lit and ventilated	Using a mobile toilet block with plumbed-in sinks and electric water heater
Drinking water	Adequate supply of drinking water shall be provided Needs to be marked and identified Readily accessible Cups must be provided	Stand-alone drinking fountains either bottled or mains fed
Changing rooms	Must be provided for workers who have to wear special clothing and can not change elsewhere Separate rooms for male and female Seating to be provided Drying facilities required Lockers must be provided for work and own clothing and personal effects	Drying room accommodation with background heating and locker provision
Rest rooms	Protect non-smokers Provide tables and seating Provide meal preparation facilities Provide boiling water Provide heating	Provide a smoking shelter Purpose-built meal room with cooker, sink, bin, tables and benching Electric wall-mounted heating Instantaneous hot water heater

Taken and adapted from the CDM Regulations 2007: www.legislation.gov.uk/uksi/2007/320/schedule/2/made

The site manager will need to undertake an initial sketch to organise their site layout, including the position of temporary site accommodation, storage areas, secure areas, temporary roads and car parking. Offices need to be weatherproof, heated, insulated, lit and furnished with desks, work surfaces, plan, racks and chairs to suit the office activity. A typical semi-portable site office is shown in Fig. 2.2.1. The same basic units can be used for all accommodation, including meal rooms and toilets equipped as indicated in Table 2.2.1. Where site space is limited, most semi-portable units are designed to be stacked, with a stairway module bolted to them for access to the units above.

Semi-mobile cabins are available in a wide variety of sizes, styles and configurations. The outer construction is generally of galvanised sheet steel over a structural steel frame, suitably insulated and finished internally with plasterboard walls and ceiling.

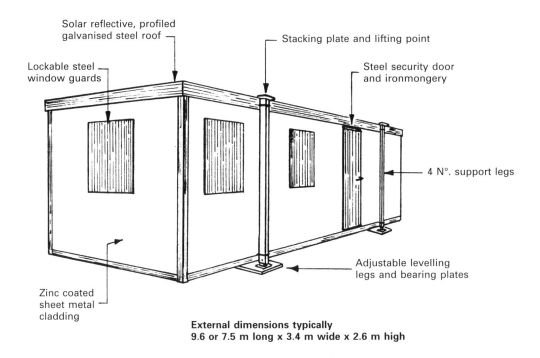

Solar reflective, profiled galvanised steel roof

Stacking plate and lifting point

Lockable steel window guards

Steel security door and ironmongery

4 N°. support legs

Zinc coated sheet metal cladding

Adjustable levelling legs and bearing plates

**External dimensions typically
9.6 or 7.5 m long x 3.4 m wide x 2.6 m high**

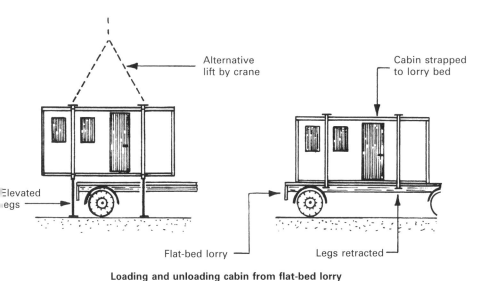

Alternative lift by crane

Cabin strapped to lorry bed

Elevated legs

Flat-bed lorry

Legs retracted

Loading and unloading cabin from flat-bed lorry

Figure 2.2.1   Semi-mobile portable cabin site accommodation.

Floor finishes vary from vinyl to carpeted boards. To prevent over-heating, the roof is covered with a solar reflective material on profiled galvanised steel. Units may be hired or purchased, usually pre-wired and plumbed as appropriate for connection to mains supplies and drains. Inclusion of furniture is also an option. Cabins are transported to site by flat-bed lorry as shown in Fig. 2.2.1 and craned into position. The industry is working towards safer delivery of these units, avoiding the need for any work at height. A series of cabins can be linked together to form an office complex, with units delivered fully serviced and ready to fit together.

Fig. 2.2.2 illustrates a self-contained and fully serviced modern portable unit that can be placed directly onto site where no services exist. Fig. 2.2.3 shows a modern modular office layout that uses modular units that bolt together to form the site accommodation; the illustration shows ground and first floor layouts.

The Joint Fire Code is designed into site accommodation for safety with many manufacturers now incorporating this code into fire-rated temporary buildings and temporary accommodation.

Ref. *Fire Prevention On Construction Sites – The Joint Code of Practice on the Protection From Fire of Construction Sites and Buildings Undergoing Renovation*; published by the Construction Confederation RISC (Risk Insight, Strategy and Control Authority) Authority and the Fire Protection Association.

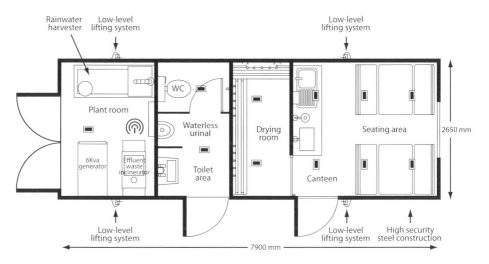

Figure 2.2.2 **Self-contained site accommodation.**

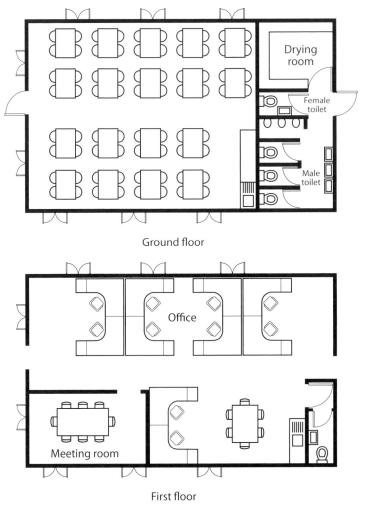

Ground floor

First floor

Figure 2.2.3 Multi-storey site accommodation (stair unit omitted).

STORAGE

The type of storage facilities required of any particular material will depend upon the following factors:

▪ durability (will it need protection from the elements?);

▪ vulnerability to damage from weather;

▪ financial value.

Cement and plaster supplied in bag form require a dry store free from draughts, which can bring in moist air and may cause an air set of material;

they also need to be raised off the ground. These materials should not be stored for long periods on-site: therefore provision should be made for rotational use so that the material being used comes from the older stock.

Loose materials such as gravel for drainage fill and sand should be stored on a clean and firm base to prevent contamination when they are loaded by shovel from the base of the stockpile.

The use of pre-mixed materials has now reduced the need for delivery and stockpiling of loose aggregates such as sand. At the press of a button, a dry mortar silo produces consistently mixed mortar ready to lay bricks. Bricks and blocks are now delivered shrink-wrapped in polythene and banded with steel bands, which both protect and keep the product in a secure pack. The bricks should be stacked in stable piles on a level and well-drained surface in a position where double handling is reduced to a minimum. Both are available with a crane offload delivery facility that allows placement onto the construction site at the point of use. The shrink-wrapping also serves to reduce the amount of moisture through rain that can be absorbed by the product. Facing bricks and light-coloured bricks can become discoloured by site mud and/or adverse weather conditions; in these situations the brick stacks should be covered with tarpaulin or polythene sheeting, adequately secured to prevent dislodgement.

Roof tiles have a greater resistance to load when they are laid on their edge: for this reason tiles should be stacked on edge and in pairs, head to tail, to give protection to the nibs that they hang on. An ideal tile stack would be five to seven rows high, with end tiles laid flat to provide an abutment. Again the use of shrink-wrapping and banding ensures that tiles are secure until loaded onto a roof for fixing.

Drainage is manufactured in either clayware or plastic. The latter is obviously less prone to physical damage. Where clay drainage is specified, suitable precautions should be taken. Plain-ended clay pipes are delivered to site on pallets with pipes separated by timber battens recessed to suit the pipe profile. Pipes should be stored in this way until they are required. Fittings should be kept separate and those, such as gullies, which can hold water, should be placed upside down.

Timber is a hygroscopic material, and therefore to prevent undue moisture movement it should be stored in such a manner that its moisture content remains fairly constant by allowing airflow around it. A rack of scaffold tubulars with a sheet roof covering makes an ideal timber store: the various section sizes allow good airflow around the timber, and the roof provides protection from the rain and snow. MDF timber products which are wrapped in polythene are less prone to hygroscopic movement by moisture, but must be kept under cover.

Ironmongery, power and hand tools are some of the most vulnerable items on a building site. Small items such as locks, power drills and cans of paint should be kept in a locked hut and issued only against an authorised stores requisition. Large items such as baths can be kept in the compound and suitably protected; it is also good practice only to issue materials from the compound against a requisition order.

With all construction materials the use of 'just in time' (JIT) delivery processes reduces the amount of site storage required and, therefore, the potential losses that could occur.

SECURITY AND PROTECTION

TEMPORARY FENCING

Health and safety legislation now requires that the site be fenced off. This is to prevent any unauthorised personnel entering the site, and to protect the public from the works, which is owed under the 'general duty of care'. The fence fulfils several functions:

- it defines the limit of the site or compound;
- it acts as a physical barrier preventing access and theft.

Two common forms of fencing are used to surround a construction site:

- solid sheet materials;
- open welded-mesh fabric panels.

A fence can be constructed to provide a physical barrier of solid construction or a visual barrier of open-work construction. If the site is to be fenced as part of the contract, it may be advantageous to carry out this work at the beginning of the site operations using the permanent structure and repairing any damage before handover. The type of fencing chosen will depend upon the degree of security required, cost implications, type of neighbourhood and duration of contract.

The use of chestnut pales and orange PVC mesh is now considered not a suitable method of fencing a site. The use of welded galvanised-mesh fencing which clips together and fixes to concrete or PVC feet is a quick, temporary solution to fencing off construction work; along with pressed steel hoarding, these are now the temporary fence modern approach to site safety. The provision of lockable gates and pedestrian access must be considered along with, on larger sites, the use of security accommodation and personnel.

Standard fences are made in accordance with the recommendations of BS 1722, which covers 13 different forms of fencing, giving suitable methods for both visual and physical barriers: typical temporary modern fencing examples are shown in Fig. 2.2.4.

3450 panel

Fully welded tube intersection for added strength

43 mm x 261 mm mesh aperture

Coupler

2050

Rubber foot

Figure 2.2.4   **Modern site fencing.**

HOARDINGS

These are close-boarded fences or barriers erected adjacent to a highway or public footpath to prevent unauthorised persons obtaining access to the site, and to provide a degree of protection for the public from the dust and noise associated with construction operations. Under Sections 172 and 173 of the Highways Act 1980 it is necessary to obtain written permission from the local authority to erect a hoarding. The permission, which is in the form of a licence, sets out the conditions and gives details of duration, provision of footway for the public, and the need for lighting during the hours of darkness.

Two forms of hoarding are in common use:

■   vertical hoardings;

■   fan hoardings.

The vertical hoardings consist of a series of sheet panels securely fixed to resist wind loads and accidental impact loads; these panels are often painted in the construction companies' corporate colours and have safety signage attached to them. They can be free-standing, or fixed by stays to the external walls of an existing building (see Fig. 2.2.5). Scaffolding tubes and fittings are often used instead of timber posts and stakes, as these tend to rot and become unstable.

The Construction (Design and Management) Regulations 2007 require that a construction site must be safe. In order to protect the public on footpaths and roadways a fan hoarding is often used. This fan hoarding is usually constructed in scaffolding and directs any falling debris back towards the building or scaffold (see Fig. 2.2.5). It is fully boarded and often lined.

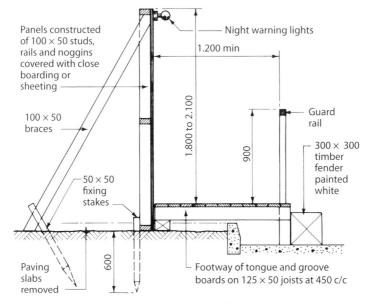

Panels constructed of 100 × 50 studs, rails and noggins covered with close boarding or sheeting

Night warning lights

1.200 min

100 × 50 braces

1.800 to 2.100

Guard rail

900

300 × 300 timber fender painted white

50 × 50 fixing stakes

Paving slabs removed

600

Footway of tongue and groove boards on 125 × 50 joists at 450 c/c

Typical free-standing vertical hoarding

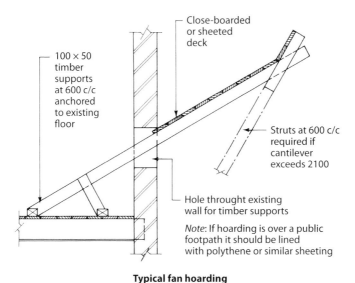

Close-boarded or sheeted deck

100 × 50 timber supports at 600 c/c anchored to existing floor

Struts at 600 c/c required if cantilever exceeds 2100

Hole throught existing wall for timber supports

Note: If hoarding is over a public footpath it should be lined with polythene or similar sheeting

Typical fan hoarding

Figure 2.2.5 **Timber hoardings.**

Subsoil drainage 2.3

The presence of groundwater on a potential construction site and its level in relation to the foundations is often a critical factor in the design and construction of the substructure, in terms of both cost and how this will be achieved technologically.

This section of the Building Regulations covers the resistance of the construction to moisture. It requires that subsoil drainage shall be provided if it is needed to avoid:

- the passage of ground moisture to the interior of the building;
- damage to the fabric of the building.

Fig. 2.3.1 illustrates a site that will not require any treatment, as the water table is well below the formation level of the foundations,

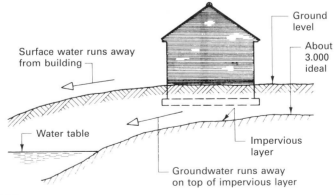

Surface water runs away from building

Ground level

About 3.000 ideal

Water table

Impervious layer

Groundwater runs away on top of impervious layer

Figure 2.3.1   The ideal site.

which can be constructed safely and economically. The water table is the level at which water occurs naturally below the ground, and this level will vary with the seasonal changes from summer to winter. Sites with a high-water table will require some form of subsoil drainage to reduce groundwater pressure on any construction.

The object of subsoil drainage is to lower the water table to a level such that it will comply with the above Building Regulations, so it will not rise to within 0.25 m of the lowest floor of a building. Subsoil drainage also has the advantages of improving the stability of the ground, lowering the humidity of the site and improving its horticultural properties.

MATERIALS

Traditional land drainpipes used in subsoil drainage were usually laid dry-jointed in clayware and were either porous or perforated pipes. With the development of plastics, a continuous roll of perforated pipe can be introduced into a trench, and can adequately drain away any groundwater. The plastic pipes are available with a series of fittings to join them together. Careful consideration must be made as to where the outfall of the land drains will be discharged, into either a drainage ditch, surface water drain or soakaway.

Suitable pipes

- Perforated clayware (BS EN 295-5).
- Profiled and slotted polypropylene or uPVC (BS 4962).
- Perforated uPVC (BS 4660).

DRAINAGE LAYOUTS

Pipework is arranged in a pattern to cover as much of the site as is necessary. Typical arrangements are shown on the plans in Fig. 2.3.2. Trenches will need filling with a drainage medium that captures any silt, which would reduce efficiency. The system is terminated at a suitable outfall such as a river, stream or surface-water sewer (see Figs. 2.3.3 and 2.3.4). In all cases permission must be obtained from the Environment Agency before discharging a subsoil system. The outfall may need protection to prevent erosion from the flow of groundwater into the ditch. If discharge is into a tidal river or stream, precautions should be taken to ensure that the system will not work in reverse, by providing a non-return valve to prevent the rising tide from entering the system. This is normally a hinged flap on the end of the drain that closes when the tide pushes against it.

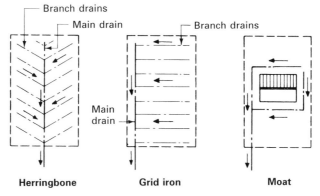

Herringbone   **Grid iron**   **Moat**

(Branch drains spacing 6.000 to 10.000 – max length 30.000)

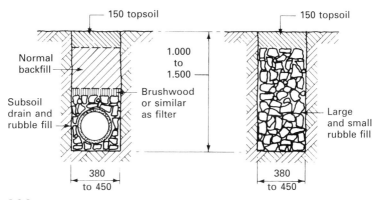

Figure 2.3.2   **Subsoil drainage systems and drains.**

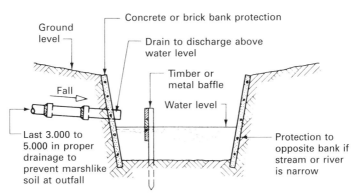

Figure 2.3.3   **Outfall to stream or river.**

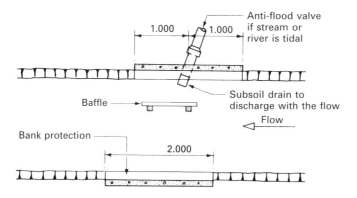

Anti-flood valve
if stream or
river is tidal

1.000   1.000

Baffle

Subsoil drain to
discharge with the flow

Flow

Bank protection

2.000

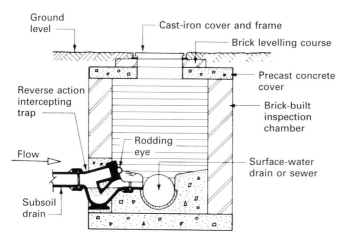

Ground
level

Cast-iron cover and frame

Brick levelling course

Precast concrete
cover

Reverse action
intercepting
trap

Brick-built
inspection
chamber

Rodding
eye

Flow

Surface-water
drain or sewer

Subsoil
drain

Figure 2.3.4   Outfall to surface-water sewer or drain (plan view).

Excavations and earthwork support 2.4

Foundations require excavated trenches to be made using a mechanical excavator or, if small enough, by hand excavation. This process of excavation must comply with the Construction (Design and Management) Regulations 2007 (CDM Regs). These Regulations cover the support of excavations and the prevention of materials and personnel from falling into an excavation, and include the inspection regime for the excavation.

The general procedure for the excavation of foundation trenches is illustrated in Fig. 2.4.1. The central rod is more commonly known as a traveller, as it moves up and down the excavation checking depth; the two site rails are placed in the ground adjacent to the excavation. All three are sighted to check depth as work proceeds.

EARTHWORK SUPPORT

Earthwork support (EWS) is a term used to cover temporary supports to the sides of excavations. The sides and top of some excavations will need support to:

- protect the operatives while working in the excavation;
- keep the excavation open by acting as a retaining wall to the sides of the trench;
- prevent anything falling into the excavation.

The type and amount of earthwork support required will depend upon the depth and nature of the subsoil. Over a short period many soils may not require any support, but weather conditions, depth, groundwater, type of soil and duration of the operations must all be taken into account, and each excavation must be risk-assessed separately.

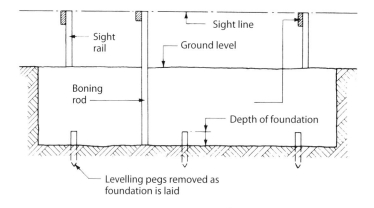

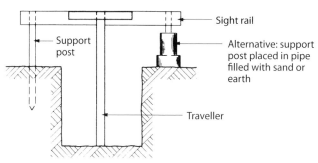

Figure 2.4.1 Trench excavations.

LIGHT SUPPORT

The use of vertical shores in stronger ground provides an additional support to the sides of any excavation trench. Lightweight aluminium vertical shores are braced apart using manual struts, which are attached to the verticals using pins so that they can rotate. Fig. 2.4.3 illustrates the use of the vertical shore.

MEDIUM SUPPORT

Fig. 2.4.2 shows the plan of a steel trench-support sheet. Rolling the sheet like this stiffens it, giving it depth and allowing it to be driven into the ground. Looking at the sides of the sheet, you will notice that each has a roll on it that can be hooked into the next one, forming a continuous wall of earthwork support.

With this system, walings are often required to support the trench sheets (see Fig. 2.4.3). The lightweight walings are suspended from the tops of the sheets using chains and hooks. Between each waling is a hydraulic strut that can be

Interlocking mechanism

Figure 2.4.2 **A steel trench-support profile.**

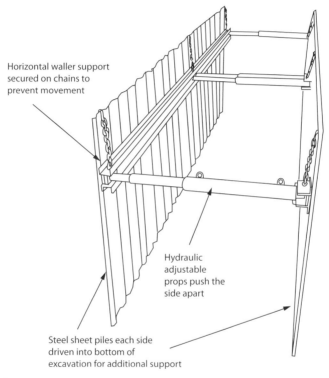

Horizontal waller support
secured on chains to
prevent movement

Hydraulic
adjustable
props push the
side apart

Steel sheet piles each side
driven into bottom of
excavation for additional support

Figure 2.4.3 **Walings.**

hand-pumped outwards to force the walings back onto the sheets, providing a
safe and secure environment to work in. Alternatively, the adjustable metal
prop illustrated in Fig. 2.4.5 can be used.

HEAVY SUPPORT

The other current alternative to steel sheet piling is the use of trench boxes.
Fig. 2.4.4 illustrates a typical trench box. This is constructed of steel sheets that
are joined together using hydraulic struts. Extra depth can be accommodated
by adding boxes on top of the initial box. A final handrail can be used to prevent

personnel and materials falling into an excavation. These boxes can be installed the full length of an excavation.

Typical details of support to trenches are shown in Fig. 2.4.5.

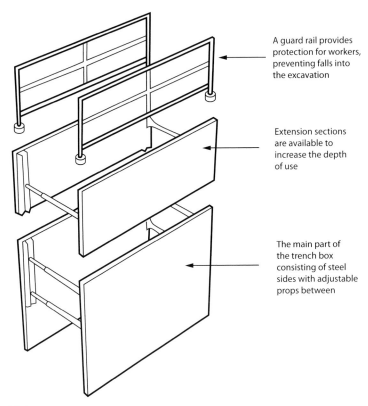

A guard rail provides protection for workers, preventing falls into the excavation

Extension sections are available to increase the depth of use

The main part of the trench box consisting of steel sides with adjustable props between

Figure 2.4.4   **A trench box.**

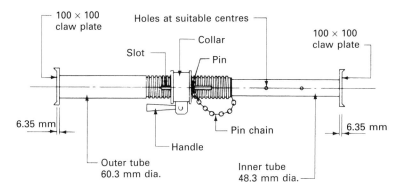

100 × 100 claw plate

Holes at suitable centres

Collar

Slot

Pin

100 × 100 claw plate

6.35 mm

Pin chain

6.35 mm

Handle

Outer tube 60.3 mm dia.

Inner tube 48.3 mm dia.

Figure 2.4.5   **Adjustable metal struts – BS 4074.**

THE CONSTRUCTION (DESIGN AND MANAGEMENT) REGULATIONS 2007

This document establishes objectives for employers, the self-employed and employees, to ensure safe working and support in excavations.

Working within excavations can be very dangerous due to several hazards that might be present:

- the atmosphere and composition of the air;
- collapse of the sides of the excavation;
- flooding and the risk of drowning.

The CDM Regulations were brought in to legislate for such hazards. The following have been taken specifically from the Regulations and cover what has to be provided for this type of substructure working.

Regulation 31 – Excavations

This Regulation within the CDM Regulations covers the health and safety aspects of undertaking excavation works that can be dangerous as they are temporary structures. A number of steps have to be undertaken in order for operatives and machinery to work in and around excavations. The following is an abridged version of the particular aspects that concern excavation work.

1. All practicable steps shall be taken, where necessary to prevent danger to any person, including, where necessary, the provision of supports or battering, to ensure that:

    a. any excavation does not collapse;

    b. no material from the sides, or above the excavation, is dislodged and falls into the excavation;

    c. no person is buried or trapped in an excavation by material which is dislodged or falls.

2. Work equipment or any accumulation of material shall be stopped from falling into any excavation.

3. Any part of an excavation or ground adjacent to it shall not be overloaded by work equipment or material.

4. Construction work shall not be carried out in an excavation where any supports or battering have been provided unless:

    a. the excavation has been inspected by a competent person:

        i. at the start of the shift in which the work is to be carried out;

ii. after any event likely to have affected the strength or stability of the excavation;

iii. after any material unintentionally falls or is dislodged;

b. the person who carried out the inspection is satisfied that the work can be carried out there safely.

5. Where the person undertaking the inspection has noted that safety work is required and this has not been undertaken, no work will be carried out in the excavation.

The CDM Regulations also make reference to providing:

■ a safe place of work (Regulation 26);

■ Inspection Reporting (Regulation 33).

A safe place of work is also a requirement under the Health and Safety at Work, etc. Act 1974.

Excavations, scaffolding and other work places may require regular inspections. These must be formally recorded and evidenced within the health and safety file.

BUILDING REGULATIONS 2010

Regulation 16

This requires that the Building Control Office of the local authority is notified by a person carrying out building work, before starting and at specific stages during construction work. The notice should be given in writing or by such means as may be agreed with the local authority.

Notices of the stages when statutory inspections are required under this Regulation occur as follows:

■ Commencement	2 days
■ Foundation excavation	1 day
■ Foundation concrete	1 day
■ Oversite preparation, before concreting	1 day
■ Damp-proof course	1 day
■ Drains before backfilling (foul and rainwater)	5 days
■ Drain test after covering	5 days after
■ Occupation before completion	5 days before
■ Completion	within 5 days

A day here means any period of 24 hours commencing at midnight and excludes any Saturday, Sunday, Bank Holiday or public holiday.

Temporary access 2.5

The Work at Height Regulations 2005 have brought legislation to bear on any work that is undertaken at height, as this is the major cause of fatalities within the construction industry. As a result of the legislation, many aspects of scaffolding and temporary access platforms have had to be reviewed. Access during construction can be provided in several different ways, by:

- scaffolding;
- mobile elevated platforms and scissor lifts;
- mobile access towers;
- low-level access platforms.

SCAFFOLDING

A scaffold is a temporary structure from which people can gain access to a place of work in order to carry out construction operations. It includes any working platforms, ladders and guard rails. Basically there are two forms of scaffolding:

- putlog scaffolds;
- independent scaffolds.

PUTLOG SCAFFOLDS

This form of scaffolding consists of a single row of uprights or standards set away from the wall at a distance that will accommodate the required width of the working platform. The uprights (**standards**) are joined together with horizontal members (**ledgers**) that run along the building and are tied to the building with cross-members (**putlogs**), which sit into the joint of the brickwork. The scaffold is erected as the building rises in lifts, and is used mostly for buildings of traditional cavity-brick construction (see Fig. 2.5.1).

INDEPENDENT SCAFFOLDS

An independent scaffold has two rows of standards, which are tied by cross-members called **transoms**. This form of scaffold does not rely on the building for support and is, therefore, independent of any structure, and can be used with different types of framed and facade structures (see Fig. 2.5.2). The development of technology in independent scaffolds has led to 'quick rig' type systems, which use a point and wedge system instead of traditional couplings to assemble and secure the scaffold.

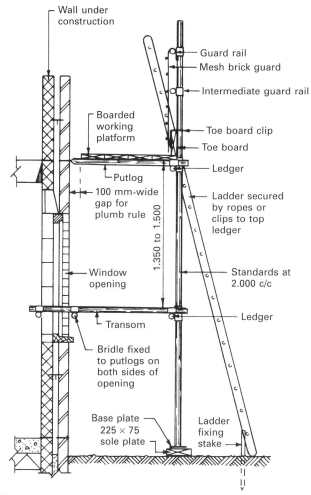

Figure 2.5.1 Typical tubular steel putlog scaffold.

Every scaffold should be securely tied to the building for stability at intervals of approximately 3.600 m vertically and 6.000 m horizontally. This can be achieved using one of the following:

- a horizontal tube called a **bridle** bearing on the inside of the wall and across a window opening, with cross-members connected to it (see Fig. 2.5.1);
- a tube with a reveal pin in the opening to provide a connection point for the cross-members (see Fig. 2.5.2).

If suitable openings are not available, the scaffold should be strutted from the ground using raking tubes inclined towards the building.

Scaffolding also requires diagonal bracing to provide support from loading while it is used, both along the face and internally as **cross-bracing** (see Fig. 2.5.2).

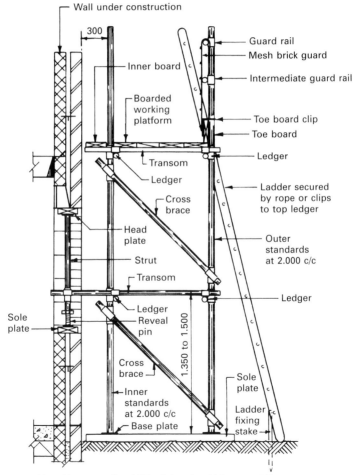

Note: Not more than 50% of ties should be reveal ties

Figure 2.5.2   Typical tubular steel independent scaffold.

INSPECTION

Under the requirements of the Working at Height Regulations 2005, all scaffolding and means of access must be inspected before use and a report must be prepared.

SCAFFOLDING MATERIALS

Scaffolding can be made from:

- tubular steel;
- tubular aluminium alloy;
- glass fibre.

Tubular steel

British Standard 1139 gives recommendations for both welded and seamless steel tubes of 48 mm outside diameter with a nominal 38 mm bore diameter. Steel tubes can be obtained galvanised (to guard against corrosion). Steel tubes are nearly three times heavier than comparable aluminium alloy tubes, but are far stronger and, as their deflection is approximately one-third that of aluminium alloy tubes, longer spans can be used.

Aluminium alloy

Seamless tubes of aluminium alloy with a 48 mm outside diameter are specified in BS 1139 for metal scaffolding. No protective treatment is required unless they are to be used in contact with materials such as damp lime, wet cement or seawater, which can cause corrosion of the aluminium alloy. Aluminium alloy tends to be used in the manufacture of lightweight mobile access platforms, where weight is a factor affecting assembly on a construction site.

Glass fibre

Glass fibre is now used to manufacture access towers and ladders. It is a product that does not conduct electricity and therefore can be used effectively within chemical environments where an ignition spark would prove fatal. They are non-corrosive and non-oxidising, and therefore have an extended lifespan and use. They are normally manufactured in a yellow resin to distinguish them from alloy materials.

Scaffold boards

These are usually boards of softwood timber, complying with the recommendations of BS 2482: 2009, used to form the working platform at the required level. They should be formed out of specified softwoods of 225 mm x 38 mm sections. To prevent the ends from splitting they should be end-bound with not less than 20 mm x 0.6 mm galvanised hoop iron, extending at least 150 mm along each edge and fixed with a minimum of seven fixings to each end.

Scaffold fittings

Here are the major fittings used in metal scaffolding.

- **Double coupler** The only real loadbearing fitting used in scaffolding; used to join ledgers to standards.

- **Swivel coupler** Composed of two single couplers riveted together so that it is possible to rotate them and use them for connecting two scaffold tubes at any angle.

- **Putlog coupler** Used solely for fixing putlogs or transoms to the horizontal ledgers.

- **Base plate** A square plate with a central locating spigot, used to distribute the load from the foot of a standard onto a sole plate or firm ground. Base plates can also be obtained with a threaded spigot and nut for use on sloping sites to make up variations in levels. Base plates should rest on either plastic or timber sole plates to spread the load.

- **Split joint pin** A connection fitting used to join scaffold tubes end to end. A centre bolt expands the two segments, which grip on the bore of the tubes.

- **Reveal pin** Fits into the end of a tube to form an adjustable strut, which expands between the reveals of an opening.

- **Putlog end** A flat plate that fits on the end of a scaffold tube to convert it into a putlog.

Typical examples of the above fittings are shown in Fig. 2.5.3.

MOBILE ACCESS TOWERS

Mobile access towers are temporary structures used for gaining access to buildings for maintenance and repair. They are preferred to conventional scaffolding for work of a relatively short duration. Access towers are also easily moved where work is continuous, such as painting building exteriors or gutter renewal.

Mobile access and working towers must have certain features, as shown in the following list.

- They are assembled from prefabricated components, interconnecting 'H'-shaped steel or aluminium tubular frames, as shown in Fig. 2.5.4.

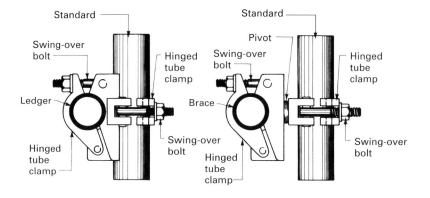

Double coupler

Swivel coupler

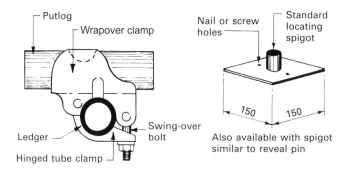

Putlog coupler

Base plate

Also available with spigot
similar to reveal pin

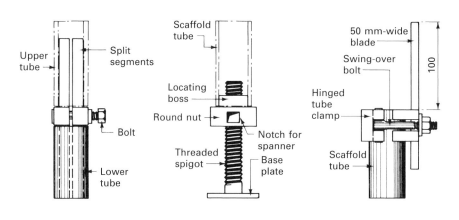

Split joint pin

Reveal pin

Putlog end

Figure 2.5.3 Typical steel scaffold fittings.

- They have the facility to be moved manually on firm, level ground.
- They have dimensions to a predetermined design.
- They are free-standing, with supplementary support optional.
- They have at least four legs; normally each leg is fitted with castors. A base plate can be used at the bottom of each leg where mobility is not required.
- The platform is accessed by a ladder or steps contained within the base dimensions of the tower. A ladder must be firmly attached to the tower, and should not touch the ground.
- The working platform should have a hinged opening for ladder access, adequate guard rails, and toe boards.

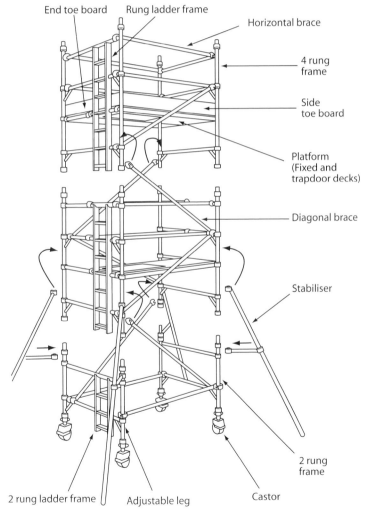

Figure 2.5.4 **Tower scaffold.**

STABILISERS, OUTRIGGERS OR DIAGONAL BRACING

Stabilisers, outriggers and diagonal bracing are optional attachments that can be adjusted to ensure ground contact where the surface is uneven, providing stability to the tower structure. One is placed at each corner extending out diagonally, ensuring that all feet are in contact with the ground, and they are fixed at two points to the tower scaffolding. They should be attached securely to enable direct transfer of loads without slipping or rotating.

ADDITIONAL SAFETY GUIDANCE

- Assembled by competent persons.
- Never move with people, equipment or materials on the platform or frame.
- Access ladders fitted within the frame.
- Castors fitted with a locking device and secured before access is permitted.
- Stable ground essential.
- Components visually inspected for damage before assembly.
- Inspected by a competent person before use and every seven days if it remains in the same place. Inspected after any substantial alteration or period of exposure to bad weather.
- Local authority Highways Department approval required prior to use on a public footpath or road. Licence or permit to be obtained.
- Barriers or warning tape used to prevent people walking into the tower. Where appropriate, illuminated.
- No work to be undertaken below a platform in use or within the tower.
- Maximum of two persons working from the platform at any one time.
- Material storage on the platform to be minimal.
- Ladders or other means of additional access must not be used from the platform or any other part of the structure.

Further details and additional reading on this topic can be obtained from BS EN 1298: *Mobile access and working towers* and BS EN 12811-1: *Temporary works equipment. Scaffolds.*

THE WORK AT HEIGHT (WAH) REGULATIONS 2005 (AMENDED 2007)

This statutory instrument is designed to ensure that suitable and sufficient safe access to and egress from every place at which any person at any time works are provided and properly maintained. Scaffolds and ladders are covered by this document, which sets objective requirements for materials, maintenance, inspection and construction of these working places. The main constructional requirements of these regulations are illustrated in Figs. 2.5.5 and 2.5.6.

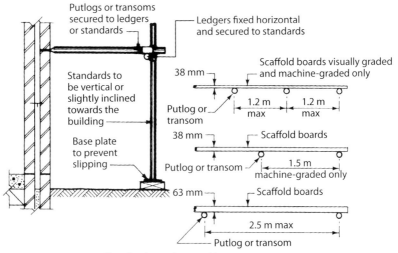

Putlogs or transoms secured to ledgers or standards

Ledgers fixed horizontal and secured to standards

Standards to be vertical or slightly inclined towards the building

Base plate to prevent slipping

Scaffold boards visually graded and machine-graded only

38 mm

Putlog or transom

1.2 m max   1.2 m max

38 mm   Scaffold boards

Putlog or transom

1.5 m machine-graded only

63 mm   Scaffold boards

2.5 m max

Putlog or transom

Standards, putlogs and transoms

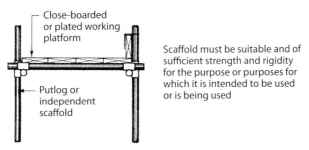

Close-boarded or plated working platform

Putlog or independent scaffold

Scaffold must be suitable and of sufficient strength and rigidity for the purpose or purposes for which it is intended to be used or is being used

Platforms, gangways and runs

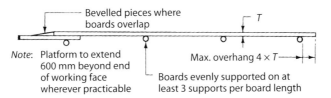

Bevelled pieces where boards overlap

T

Note: Platform to extend 600 mm beyond end of working face wherever practicable

Max. overhang 4 × T

Boards evenly supported on at least 3 supports per board length

Boards in working platforms

Figure 2.5.5   Scaffold regulations – 1.

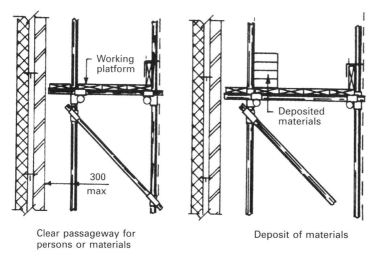

Working platform

300 max

Clear passageway for persons or materials

Deposited materials

Deposit of materials

Widths of working platforms for putlog and independent scaffolds

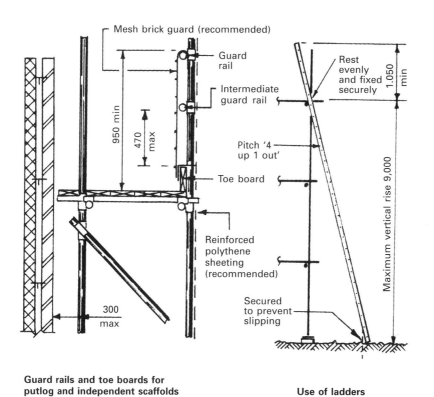

Mesh brick guard (recommended)

Guard rail

Intermediate guard rail

950 min

470 max

Pitch '4 up 1 out'

Toe board

Reinforced polythene sheeting (recommended)

300 max

Rest evenly and fixed securely

1.050 min

Maximum vertical rise 9,000

Secured to prevent slipping

Guard rails and toe boards for putlog and independent scaffolds

Use of ladders

Figure 2.5.6 Scaffold regulations – 2.

The WAH Regulations state that 'every existing place of work shall be of sufficient dimensions to permit the safe passage of persons and the safe use of any plant or materials required to be used and to provide a safe working area having regard to the work being carried out there'.

The Regulations also say that every employer must ensure that a working platform used for construction work from which a person could fall two metres or more 'is not used in any position unless it has been inspected in that position or, in the case of a mobile working platform, inspected on the site, within the previous seven days'. Records of all such inspections must be kept until the next recorded inspection in accordance with the Regulations. Each inspection report must provide the:

- name and address of the person for whom the inspection was carried out;
- location of the work equipment inspected;
- description of the work equipment inspected;
- date and time of the inspection
- details of any matter identified that could give rise to a risk to the health or safety of any person;
- details of any action taken as a result of any matter identified above;
- details of any further action considered necessary;
- name and position of the person making the report.

3

Substructure

Excavation works 3.1

Excavations are generally the first 'real' work to be undertaken as part of the new construction project once the site facilities temporary works and services have been established. Great care needs to be taken in planning and digging out the excavations, as no matter how detailed your site and soil investigation has been, there may be unseen hazards present or the subsoil may vary beyond what was expected. Traditionally, it was only when you had the excavations dug and the foundations constructed that you were assured that your construction programme could be completed on time and to budget.

SAFETY CONSIDERATIONS FOR GENERAL EXCAVATION WORK

All excavations are potential hazards and need to be considered carefully when planning excavation works. Both the Work at Height Regulations and the Construction (Design and Management) Regulations require that all excavations, no matter how shallow, need to be individually risk assessed to ensure that there are sufficient control measures in places to prevent accidents.

Here are some practical considerations to take into account when planning excavations.

- A full survey to locate existing services in the area of the excavation above and below the line of the excavation is needed, such as overhead power lines and foul sewers.

- The use of trenchless technology, such as directional drilling or pipe jacking for service pipe installations, will avoid the need to excavate a trench in the first place.

- In open trenches, to prevent the sides and the ends of trenches from collapsing, they should be battered back to a safe angle or supported with proprietary support systems, trench sheets or timbering (see Fig. 3.1.1).

- There should always be a suitable means of access into and egress from excavations in the form of tied pole ladders or proprietary steps.

- Workers should not enter any unsupported excavations and should never work ahead of the support system.

- Edge-protection fences need to be provided to prevent persons or materials falling into the excavation (see Fig. 3.1.2).

- The ground adjacent to the excavation should not be surcharged with plant, stored materials, spoil or foundation loads from existing structures.

The method of excavation and support to be used in any particular case will depend upon a number of factors:

- the nature of the subsoil, which can determine the type of plant or hand tools required and the amount of timbering necessary;

- the type and degree of contaminated soil that may be present in land that has been previously built on;

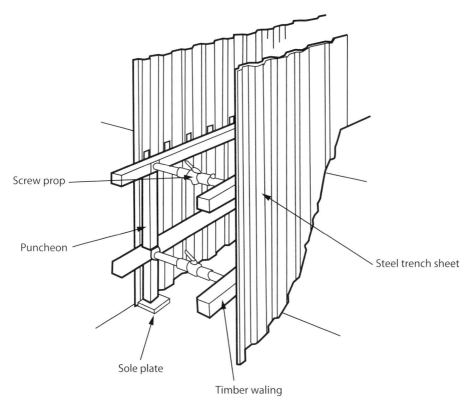

Screw prop

Puncheon

Steel trench sheet

Sole plate

Timber waling

Figure 3.1.1   Typical trench sheeting arrangement.

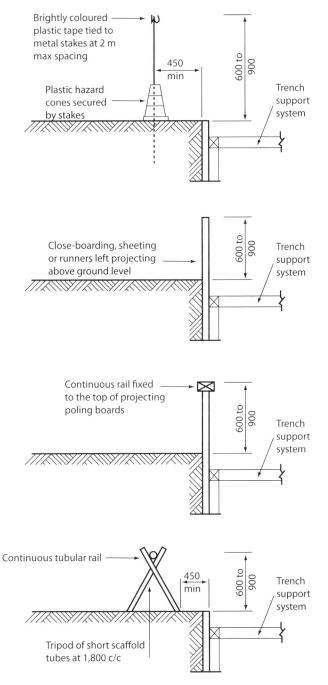

Brightly coloured plastic tape tied to metal stakes at 2 m max spacing

450 min

600 to 900

Plastic hazard cones secured by stakes

Trench support system

Close-boarding, sheeting or runners left projecting above ground level

600 to 900

Trench support system

Continuous rail fixed to the top of projecting poling boards

600 to 900

Trench support system

Continuous tubular rail

450 min

600 to 900

Trench support system

Tripod of short scaffold tubes at 1,800 c/c

Figure 3.1.2   Barriers to excavations.

- the purpose of the excavation, which can determine minimum widths, minimum depths and the placing of support members to give a reasonable working space within the excavation;

- the presence of groundwater, which may necessitate interlocking timbering, sump pits and pumps; large quantities of groundwater may prompt the use of de-watering techniques;

- any restrictions imposed by the position of the excavation, such as the need for a licence or wayleave, highway authority or police requirements when excavating in a public road;

- non-availability of the right type of plant for bulk excavation, which may mean that a different method must be used;

- the presence of a large number of services, which may restrict the use of machinery to such an extent that it becomes unsafe or uneconomic;

- the disposal of the excavated spoil, which may restrict the choice of plant because the load and unload cycle does not keep pace with the machine output.

See Part 2.4 on pages 53–54 for more details of trench excavations.

BASEMENT EXCAVATIONS

There are two methods that can be used for excavating a large pit or basement:

- complete excavation with sloping sides;

- complete excavation with temporary support to vertical sides.

Excavation for a basement on an open site can be carried out by cutting the perimeter back to the natural angle of repose of the soil. This method requires sufficient site space around the intended structure for the over-excavation. No temporary support is required, but the savings must pay for the over-excavation and consequent increase in volume of backfilling to be an economic method.

In most modern methods of construction of basements, particularly on cramped urban sites, steel sheet piles are used to form a temporary support to the soil in the form of a coffer dam. The steel sheet piles, which can be galvanised or plain, are driven vertically to overlap and to a depth below the bottom level of the excavation to provide a fixing for the base of the sheet pile.

There are a number of proprietary systems used that extend the technology used to support narrow trenches to small and medium-sized basements. The Mabey Multibrace System uses standard leg assemblies to cover excavations from 2.5 m to 10.3 m deep. The basic box configuration can be lengthened by the use of additional bracing struts and walings. In this system pinned joints at the corners allow for bracing of non-square excavations. Once the sheet piles have been driven or vibrated into position around the basement, excavation is carried out down to the level where the waling brace will be located. This multibracing system is lowered into position and adjusted so that it fits tightly against the sheet piles. Chains are then hooked over the top of the sheet piles to prevent the brace from slipping down.

Steel-braced adjustable walings held tight against the sheet piles and supported by chains hung from the top of the sheet piles

Interlocking steel sheet piles driven below formation level

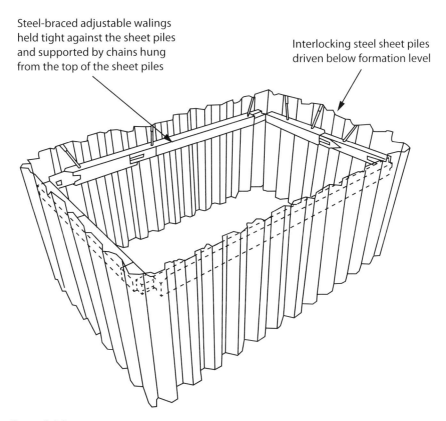

Figure 3.1.3 **Temporary support systems used in basement construction.**

Concrete materials 3.2

Concrete is a mixture of cement, fine aggregate, coarse aggregate and water. The proportions of each material control the strength and quality of the resultant concrete.

PORTLAND CEMENT

Cement is the setting agent of concrete, and the bulk of cement used in this country is Portland cement. This is made from chalk or limestone and clay, and is generally produced by the **wet process**.

In this process the two raw materials are washed, broken up and mixed with water to form a slurry. This slurry is then pumped into a steel rotary kiln, which is 3 to 4 m in diameter and up to 150 m long, and lined with refractory bricks. While the slurry is fed into the top end of the kiln, pulverised coal is blown in at the bottom end and fired. This raises the temperature at the lower end of the kiln to about 1400 °C. First the slurry passing down the kiln gives up its moisture; then the chalk or limestone is broken down into carbon dioxide and lime; finally it forms a white-hot clinker, which is transferred to a cooler before being ground. The grinding is carried out in a ball mill, which is a cylinder some 15 m long and up to 4.5 m in diameter, containing a large number of steel balls of various sizes, which grind the clinker into a fine powder. As the clinker is being fed into the ball mill, gypsum (about 5 per cent) is added to prevent a flash setting of the cement.

The alternative method for the preparation of Portland cement is the **dry process**. The main difference between this and the wet process is the reduction in the amount of water that has to be driven off in the kiln. A mixture of limestone and shale is used, which is proportioned, ground and blended to form a raw meal of low moisture content. The meal is granulated in rotating pans with a small amount of water before being passed to a grate for preheating prior to entering the kiln. The kiln is smaller than that used in the wet process but its function is the same – that is, to form a clinker, which is then cooled, ground and mixed with a little gypsum as described for the previous process.

Rapid-hardening Portland cement is more finely ground than ordinary Portland cement. Its main advantage is that it gains its working strength earlier than ordinary cement. The requirements for both ordinary Portland and rapid-hardening Portland cement are given in BS EN 197-1.

OTHER TYPES OF CEMENT

SULPHATE-RESISTING PORTLAND CEMENT – BS 4027

This is a Portland cement that is low in tricalcium aluminate (less than 3.5 per cent). As a result it has a reduced susceptibility to attack by dissolved sulphates in water or soils. It is also a low alkali, which renders it useful in making concrete with aggregates that are susceptible to alkali-silica reaction (ASR). ASR is a reaction that occurs over time between the highly alkaline cement paste and reactive non-crystalline silica, found in many common aggregates. The expansion of the affected aggregate by the formation of a swelling gel of calcium silicate hydrate exerts pressure inside the material, causing spalling of the concrete, loss of strength and finally failure.

PORTLAND BLAST FURNACE CEMENT – BS EN 197

This is made from ground granulated blast furnace slag (GGBS), which is a by-product of the steel-making process and so is highly sustainable. Concrete made with GGBS cement sets more slowly than concrete made with ordinary Portland cement, but also continues to gain strength over a longer period in production conditions. It is more durable and makes avoiding cold joints easier. It is also likely to suffer from damage caused by ASR.

HIGH-ALUMINA CEMENT

Many reinforced concrete buildings built in the 1960s and 1970s were made using high-alumina cement (HAC). This was made by firing limestone and bauxite (aluminium ore) to a molten state, casting it into pigs, and finally grinding it into a fine powder. Its rate of hardening was very rapid, and produced a concrete that was resistant to the natural sulphates found in some subsoils and had improved fire-resistance. However, high-alumina cement had its limitations: it was largely prohibited for use in new structural concrete in the UK following a few well-publicised building collapses in the 1970s. Subsequent research has shown that the primary causes of these collapses were generally poor construction details or aggressive chemical attack, rather than problems with the concrete itself.

Most HAC concrete in the UK went into precast beams; up to 50,000 buildings with similar beams remain successfully in service today in the UK. The beams can be found in public and industrial buildings such as schools, flats and business units.

Other examples of cement available include:

- low-heat Portland – BS EN 197-1;
- masonry – BS EN 413-1;
- Portland pulverised fuel ash – BS 6588;
- Portland limestone – BS 7583.

Cement should be stored on a damp-proof floor in the dry and kept for short periods only, because eventually it will harden as a result of the action of moisture in the air. This is known as **air hardening**, and any hardened cement should be discarded.

AGGREGATES

These are the materials that are mixed with the cement to form concrete. They are classed as fine or coarse aggregate: fine aggregates are those that will pass a standard 5 mm sieve, while coarse aggregates are retained on a standard 5 mm sieve. All-in aggregate is a material composed of both fine and coarse aggregates.

A wide variety of materials is available as aggregates for the making of concrete (for example, gravel, crushed stone, brick, furnace slag and lightweight substances, such as foamed slag, expanded clay and vermiculite).

In making concrete, aggregates must be graded so that the smaller particles of the fine aggregate fill the voids created by the coarse aggregate. The cement paste fills the voids in the fine aggregate, thus forming a dense mix.

Aggregates from natural sources and synthetic aggregates are defined in BS EN 12620: *Aggregates for concrete.*

WATER

The water used in the making of concrete must be clean and free from impurities that could affect the concrete. It is usually specified according to BS EN 1008: *Mixing water for concrete* as being of a quality fit for drinking. A proportion of the water will set up a chemical reaction, which will harden the cement. The remainder is required to give the mix workability, and will evaporate from the mix while it is curing, leaving minute voids. An excess of water will give a porous concrete of reduced durability and strength.

The quantity of water to be used in the mix is usually expressed in terms of the **water/cement ratio**, which is:

$$\frac{\text{total weight of water in the concrete}}{\text{weight of cement}}$$

For most mixes the ratio is between 0.4 and 0.7.

Traditionally concrete mixes were expressed as prescribed ratios of the constituent parts, for example:

- 1:2:4 = 1 part cement, 2 parts fine aggregate and 4 parts coarse aggregate.

Typical mixes were as follows:

- 1:10 – not a strong mix, but suitable for filling weak pockets in excavations and for blinding layers;
- 1:8 – slightly better than the last, suitable for paths and pavings;
- 1:3:6 – a basic mix for general-purpose mass concrete works;
- 1:2:4 – a strong mix that is practically impervious to water, used especially for reinforced concrete.

This method of specifying concrete is no longer currently specified in the Building Regulations Approved Document A but may be still heard in common usage.

MIXING AND BATCHING CONCRETE

In simple DIY projects and small domestic projects involving small quantities like shed bases and post holes, batching is done 'by volume'. However, if the fine aggregate is damp or wet, its volume will increase by up to 25 per cent. This increase in volume, called **bulking**, makes the batching and hence the water/cement ratio unreliable. In most professional projects, batching is done by 'by mass' as this is the far more accurate method of achieving the correct strength. This method involves the use of a balance to give the exact mass of the materials as they are placed in the scales. It offers greater accuracy, and the balance can be attached to the mixing machine.

Hand mixing

This should be carried out on a clean, hard surface. The materials should be thoroughly mixed in the dry state before the water is added. The water should be added slowly, preferably using a rose head, until a uniform colour is obtained.

Machine mixing

The mix should be turned over in the mixer for at least two minutes after adding the water. The first batch from the mixer tends to be harsh, because some of the mix will adhere to the sides of the drum. This batch should be used for some less important work, such as filling in weak pockets in the bottom of the excavation.

Ready-mixed

This is used for large batches with lorry transporters of up to 6 m³ capacity. It has the advantage of eliminating site storage of materials and mixing plant, with the guarantee of concrete manufactured to quality-controlled standards. Placement is usually direct from the lorry via a fold-out chute that can extend up to 6 m from the back of the truck. When using ready-mixed, site-handling facilities must be co-ordinated with deliveries, which may include the use of concrete skips manoeuvred by crane or concrete pumps facilitated by stout hydraulic hoses.

Handling

If concrete is to be transported for some distance over rough ground, the runs should be kept as short as possible, because vibrations of this nature can cause segregation of the materials in the mix. For the same reason, concrete should not be dropped from a height of more than 1 m. If this is unavoidable, a chute should be used.

Placing

If the concrete is to be placed in a foundation trench, it will be levelled from peg to peg (see Fig. 2.4.1, page 54, or if it is to be used as an oversite bed the external walls could act as a levelling guide. The levelling is carried out by tamping with a straight-edge board; this tamping serves the dual purpose of compacting the concrete and bringing the excess water to the surface so that it can evaporate. Concrete must not be over-tamped, as this will bring not only the water to the surface but also the cement paste that is required to act as the matrix. Concrete should be placed as soon as possible after mixing to ensure that the setting action has not commenced. Concrete that dries out too quickly will not develop its full strength: therefore new concrete should be protected from the drying winds and sun by being covered with damp canvas or polythene sheeting, or being regularly sprayed with water. This protection should be continued for at least three days, because concrete takes about 28 days to obtain its working strength.

Specifying concrete

Any of five methods may be used to specify concrete and are set out in BS 8500-1: *Method of specifying and guidance for the specifier* (complementary to BS EN 206-1: *Specification, performance, production and conformity of concrete*):

■ designated mix;

■ designed mix;

■ prescribed mix;

■ prescribed standard mix;

■ proprietary mix.

Designated mix

This method of specification was designed to make the specification of concrete mixes simpler and more reliable, by using given designations for the most common applications and site conditions. Grading and strength characteristics are extensive, and vary with application. Strength is specified by cube-crushing strength and workability by its slump – refer to page 84 for details.

■ **General (GEN)** Graded 0,1, 2 and 3, ranging from 8 to 20 N/mm^2 characteristic cube strength for foundations, floors and external works with slump values from 90–170 mm (see Fig. 3.2.1).

■ **Foundations (FND)** Graded 2, 2z, 3, 3z, 4, 4z with characteristic cube strength of 30 N/mm^2. These are particularly appropriate for resisting the effects of sulphates in the ground.

■ **Paving (PAV)** Graded 1 or 2 in 30 or 35 N/mm^2 strengths respectively with a lower slump of 40–110 mm. A strong concrete for use in driveways and heavy-duty pavings.

■ **Reinforced and pre-stressed concrete (RC)** In eight designations ranging from 25 to 50 N/mm^2 characteristic strength, and exposures ranging from mild to most severe.

In addition to application, the purchaser needs to specify:

■ reinforced or unreinforced;

■ pre-stressed;

■ heated or ambient temperature;

■ maximum aggregate size (if not 20 mm);

■ exposure to chemicals (chlorides and sulphates);

■ exposure to subzero temperatures (placing and curing).

Because of the precise nature of these designated mix concretes, quality control is of paramount importance: therefore producers are required to have quality assurance product conformity accreditation to BS EN ISO 9001.

Designed mix

This method offers more flexibility to the specifier than designated concretes, which do not cover every application, and is suitable for almost all applications. They may be used where the requirements are outside of those covered by designated concretes: for example, where the concrete is to be exposed to one of the chloride or sea water exposure classes.

The mix is specified by a grade corresponding to the required characteristic compressive strength at 28 days. There are 16 grades, from C10 to C115: the 'C' indicates characteristic compressive strength, and the number indicates the strength in N/mm^2 or MPa at 28 days of a standard 150 mm test cube.

Other factors that may be specified include:

- exposure classes – which takes into account the environmental or chemical attack that may be present at the location (see Table 3.2.1);
- maximum nominal upper aggregate size – which would hinder the flow of concrete around densely packed reinforcement;
- consistency of the mix – which specifies how workable the mix is when placed and general ranges from a slump of less than 40 mm for the driest mixes to greater than 100 mm for the wettest mix;
- density class – which looks at the air entrainment and weight of concrete elements.

Table 3.2.1   Exposure classes (adapted from EN 206-1 – Table 1).

Title	Class	Designation	Example
1 **No risk of corrosion or attack**	X0	For concrete without reinforcement or embedded metal: all exposures except where there is freeze/thaw, abrasion or chemical attack	For concrete with reinforcement or embedded metal: very dry concrete inside buildings with very low air humidity
2 **Corrosion induced by chlorides other than from sea water**	XD1	• Moderate humidity	• Concrete surfaces exposed to airborne chlorides
	XD2	• Wet, rarely dry	• Swimming pools • Concrete exposed to industrial waters containing chlorides
	XD3	• Cyclic wet and dry	• Parts of bridges exposed to spray containing chlorides • Pavements, car park slabs
3 **Corrosion induced by chlorides from sea water**	XS1	• Exposed to airborne salt but not in direct contact with sea water	• Structures near to or on the coast
	XS2 XS3	• Permanently submerged • Tidal, splash and spray zones	• Parts of marine structures • Parts of marine structures

Cont.

Title	Class	Designation	Example
4 Freeze/thaw attack with or without de-icing agents	XF1	• Moderate water saturation, without de-icing agent	• Vertical concrete surfaces exposed to rain and freezing
	XF2	• Moderate water saturation, with de-icing agent	• Vertical concrete surfaces of road structures exposed to freezing and airborne de-icing agents
	XF3	• High water saturation, without de-icing agent	• Horizontal concrete surfaces exposed to rain and freezing
	XF4	• High water saturation, without de-icing agent	• Road and bridge decks exposed to de-icing agents • Concrete surfaces exposed to direct spray containing de-icing agents and freezing • Splash zones of marine structures exposed to freezing

Prescribed mix

This is a recipe of constituents with their properties and quantities used to manufacture the concrete. The specifier must vouch for the creation of the mix and state the:

▪ type of cement;

▪ type of aggregates and their maximum size;

▪ mix proportions by weight;

▪ degree of consistence or workability (slump and/or water/cement ratio) and the application.

Prescribed mixes are based on established data indicating conformity to strength, durability and other characteristics. However, under BS EN 206 it is not permitted to include requirements on concrete strength, and so this option has only limited applicability. A typical example of a prescribed mix for general concreting work is 1:3:6 (1 part cement, 3 parts fine aggregate, 6 parts coarse aggregate), 20 mm max size and 75 mm slump.

Prescribed standard mix

Standardised prescribed concretes are applicable for housing and similar construction where concrete is site-batched on a small site or obtained from a ready-mixed concrete producer who does not have accredited third-party insurance/certification. Mixes are produced from one of five grades, ranging from ST1 to ST5, with corresponding 28-day strength characteristics of 8 to a limit of only 25 N/mm^2. Mix composition and details are specified by:

- cement to aggregate by weight;
- type of cement;
- aggregate type and maximum size;
- workability;
- use or omission of reinforcement.

These mixes are most suited to site production, where the scale of operations is relatively small. Alternatively, they may be used where mix design procedures would be too time-consuming, inappropriate or uneconomic.

Proprietary mix

This approach is appropriate where it is required that the concrete achieves a specific performance, using defined test methods but also complying with BS 8500-1. This is particularly useful in the development of new concrete materials and mix techniques. An example of the recent use of this is the development of self-compacting concrete, a very high-flowing concrete that requires no mechanical compaction to achieve the required density and strength, which is achieved by undisclosed use of patented additives to the basic constituents of the concrete.

TESTING OF CONCRETE

There are many different test procedures for determining the properties of concrete. These are detailed in BS 1881: *Testing concrete*, BS EN 12350: *Testing fresh concrete* and BS EN 12390: *Testing hardened concrete*. The two most common tests are the slump test, applied to wet or fresh concrete, and the compression test, applied to hardened concrete.

SLUMP TEST – BS EN 12350-2

The slump test is suitable for establishing uniformity of mixes in subsequent batches or deliveries. It is not strictly a test for workability, but it can be used as a guide when constituents are known to be constant. Mixes of the same slump can vary with regard to their cement content and grade of aggregates.

Equipment

The equipment, as shown in Fig. 3.2.1, comprises an open-ended steel frustum of a cone, a tamping rod and a rule.

Procedure

The cone is one-quarter filled with concrete and tamped 25 times. A further three layers are applied, each layer tamped as described. Surplus concrete is struck from the surface, and the cone is raised immediately.

Typical slump

Mass concrete/thick sections of reinforced concrete	50 mm
Roads, hand tamped and general use	100 mm
Thin sections of reinforced concrete	150 mm
Maximum aggregate size	40 mm

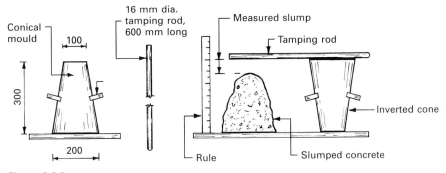

Figure 3.2.1 Slump test.

COMPRESSION TEST – BS EN 12390-1

Concrete test cubes are made from samples taken from site. Specimens are taken before and during the placing of concrete.

Equipment

As shown in Fig. 3.2.2, this involves a standard machined steel mould, 150 mm × 150 mm × 150 mm for aggregate size up to 40 mm. Cube moulds of 100 mm may be used for aggregates up to 20 mm. Internal faces of the mould are lightly oiled prior to receiving concrete. Tamping is done with a 25 mm square steel tamping rod, 380 mm long.

Procedure

Concrete is placed in the mould in 50 mm layers; each layer is tamped 35 times for 150 mm cubes or 25 times for 100 mm cubes. Alternatively, the concrete may be compacted by vibration. Surplus concrete is struck off.

Samples remain in the mould for 24 hours ± 30 minutes, covered with a damp sack or similar. After this time specimens are marked, removed from the mould and submersed in water at a temperature between 10 and 21 °C until required for testing.

The cube strength is the stress failure after seven days. If the strength specification is not achieved at seven days, a further test is undertaken at 28 days. If the specification is not achieved at 28 days, specimen cores may be taken from the placed concrete for laboratory analysis.

Typical 28-day characteristic crushing strengths (that below which not more than five per cent of the test results are allowed to fall) are graded 7.5, 10, 15, 20, 25, 30, 35 and 40 N/mm^2.

As a guide, shear stress of the concrete is taken at approximately one-tenth of the stress under compression.

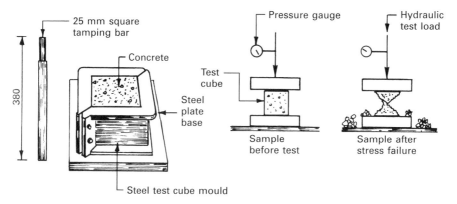

Figure 3.2.2 **Compression test.**

REINFORCING STEEL

Mass concrete can be very strong in compression when it is used in foundations and ground slabs. However, it is weak in tension, in some cases only a tenth as strong in tension as it is in compression. This can be particularly problematic in members that span openings, such as in the lower regions of beams and slabs. Therefore concrete is reinforced with steel bars to add tensile strength. Reinforcement is also added in areas of compression to reduce surface cracking, to improve the durability of the concrete.

Steel reinforcement comes in two basic forms: ribbed high-yield steel bars (in nominal diameters of 8, 10, 12, 16, 20, 25, 32 and 40 mm, all to BS 444), or as welded fabric mesh to BS 44483. The types of fabric mesh are given in Table 3.2.2.

Table 3.2.2 Types of fabric mesh.

Prefix letter	Type of fabric	Size of mesh (mm × mm)	Usual applications
A	Square mesh	200 × 200	Slabs, both suspended and on the ground
B	Structural mesh	100 × 200	Suspended slabs
C	Long mesh	100 × 400	Roads, paved areas, ground floor slabs
D	Wrapping	100 × 100	Sprayed concrete work and wrapping structural steelwork that is to be encased in concrete

Ref: Concrete on Site 2: The Concrete Society: www.concrete.org.uk

All steel reinforcement must be covered by a sufficient thickness of concrete to prevent corrosion. This distance is referred to as the 'cover' to the concrete and for foundations is usually 75 mm. The steelwork arrives on site in bundles marked with a specific steel-fixing code, which clearly shows the number of bars, type of steel (now always ribbed high-yield steel, as smooth mild steel is no longer covered by the design codes), the identifying barcode that relates to the steel fabricator's drawing, and other information about the location or spacing of the bars. A 50–75 mm layer of weak concrete, known as blinding, is laid in the bottom of the foundation excavation to form a firm, level surface from which to set out the steel reinforcing bars and to prevent them from getting muddy and greasy. The blinding layer also fills any weak pockets encountered during excavations.

Formwork may be required for the foundation. This is the temporary mould that forms the shape of the concrete foundation. Some contractors prefer to lay the blinding before assembling the formwork; the alternative is to place the blinding within the formwork and allow this to set before positioning the reinforcement and placing the concrete.

Further information about reinforced concrete technology is provided in Part 10 on page 474, which deals with framed buildings.

Foundations   3.3

A foundation is the base on which a building rests, and its purpose is to transfer the load of a building safely to a suitable subsoil.

Terminology

- **Backfill** Materials excavated from site and if suitable used to fill in around the walls and foundations.
- **Bearing capacity** Safe load per unit area that the ground can carry.
- **Bearing pressure** The pressure produced on the ground by the loads.
- **Made ground** Refuse, excavated rock or soil deposited for the purpose of filling in a depression or for raising the site above its natural level.
- **Settlement** Ground movement, which may be caused by:
  - deformation of the soil due to imposed loads;
  - volume changes of the soil as a result of seasonal conditions;
  - mass movement of the ground in unstable areas.

CHOICE OF FOUNDATION TYPE

The Building Regulations require all foundations of buildings to:

- safely sustain and transmit to the ground the combined dead and imposed loads so as not to cause any settlement or other movement in any part of the building or of any adjoining building or works;
- be of such a depth, or be so constructed, as to avoid damage by swelling, shrinkage or freezing of the subsoil;
- be capable of resisting attack by deleterious material, such as sulphates, in the subsoil.

The choice and design of foundations for domestic and small types of buildings depend mainly on three factors:

- the total loads of the building;
- the nature and bearing capacity of the subsoil;
- the ease and safety of constructing the foundation.

TOTAL LOADS OF THE BUILDING

These are determined from knowledge of the self-weight of the building materials (the dead load) and the proposed occupancy of the building (the imposed load). BS EN 1991-1-1:2002 Eurocode 1: Actions on structures. Part 1-1: *General actions, densities, self-weight, imposed loads for buildings* gives guidance on these loads. Usually a structural engineer will be employed to undertake calculations if required to meet Part A of the Building Regulations. Imposed loads vary from 1.5 kN/m^2 for domestic dwellings to 7.5 kN/m^2 for a building needing industrial storage. Dead loads are based on the characteristic density of materials, with typical values expressed as unit weights of materials: for example, concrete (24 kN/m^3), steel (75 kN/m^3) and timber (4 kN/m^3). For a domestic dwelling, the average load per metre length required to be supported by the soil can range from 30 to 50 kN/m.

NATURE AND BEARING CAPACITY OF THE SUBSOIL

Engineering subsoils are classified as the natural materials found below the surface layer of organic topsoil, which itself can vary in thickness, usually between 100 and 300 mm. These engineering subsoils support the applied loads caused by any new building or structure via a main substructural element called the foundation. They do this by spreading the loads over the plan area of the foundation causing a bearing pressure on the subsoil. Bearing capacities vary depending on the type and strength of the subsoil. Table 3.3.1 shows typical bearing capacities of subsoils.

Table 3.3.1   **Typical subsoil bearing capacities.**

Type	Bearing capacity (kN/m^2)
Rocks, granites and chalks	600–10,000
Non-cohesive soils; compact sands; loose uniform sands	100–600
Cohesive soils; hard clays Soft clays and silts	150–600 < 150
Peats Made ground/fill material	< 75 To be determined by investigation

In certain parts of the UK, where rock is encountered no foundations are necessary and the building can be constructed directly on the subsoil.

However, in most cases a foundation is required to provide a base on which the building rests. Its purpose is to distribute safely the pressure caused by the building loads (the 'bearing pressure') to a suitable subsoil of known strength that can provide a safe bearing capacity equal to the bearing pressure.

Clay is the most difficult of all subsoils to deal with. Down to a depth of about 1 m clays are subject to seasonal movement, as the clay dries and shrinks in the summer but swells in the winter with heavier rainfall. This movement occurs whenever a clay soil is exposed to the atmosphere, and special foundations may be necessary.

In cold weather, subsoils that readily absorb and hold water are subject to frost heave. This is a swelling of the subsoil due to the expansion of freezing water held in the soil; like the movement of clay soils, it is unlikely to be even, and special foundations may be needed to overcome the problem.

The specific properties, nature and bearing capacity of the subsoil can be determined by detailed site evaluations, including:

- desk study – examining previous soil reports, geological maps, aerial photographs, trial holes;

- reconnaissance survey – undertaking a physical survey of the site noting vegetation, site slope, water courses, local knowledge, condition of existing buildings, etc.;

- soil investigation – using both trial pits and boreholes for taking soil samples for in-situ and laboratory testing, such as shear strength, plasticity and settlement potential.

Details of site evaluation methods have been covered in Part 2 (see page 35).

EASE AND SAFETY OF CONSTRUCTING THE FOUNDATION

The choice of foundation must take into account the resources and labour available locally. The Construction (Design and Management) Regulations 2007 require that the designer must consider how any work proposed promotes the use of a safe system of work for the excavations to be dug and the new foundations to be constructed: for example, wherever possible ground workers should only enter open excavations that are appropriately supported or battered back to a safe angle to avoid being crushed in the event of a collapse.

It is also good practice to select a foundation that requires minimal disposal of waste spoil materials. The cost and availability of suitable landfill sites is making disposal increasingly difficult to sustain, particularly if the spoil material contains contaminated or hazardous materials, such as old industrial tars and oils. If these materials can be excavated and cleaned up, they can be reused as part of the landscaping, which will save costs. If they cannot be readily cleaned, other methods need to be employed to encapsulate or trap

the hazardous materials so that they will not affect any future inhabitants
or users.

TYPES OF FOUNDATION

Foundations are usually made of either mass or reinforced concrete. The
lowest level of the constructed foundation is called the formation level.

- **Shallow foundations** are those that transfer the loads to subsoil at a point
  near to the ground floor of the building, such as narrow strips and flat rafts.
  These foundations are often unreinforced in lightly loaded conditions,
  although reinforcement may be specified where the ground is weak.

- **Deep foundations** are those that transfer the loads to a subsoil some
  distance below the ground floor of the building, such as pad or pile
  foundation systems incorporating reinforced concrete ground beams that
  span between the pads or piles.

The principal types of foundations for buildings are:

- strip foundations;
- raft foundations;
- isolated or pad foundations;
- short (replacement or displacement) piled foundations;
- specialist foundations, such as underpinning systems for existing buildings
  that have suffered from previous ground settlement or deep-piled rafts
  where heavy loads or uplifts have to be resisted, as in major basement or
  underground civil engineering projects. These are covered in more depth in
  Advanced Construction Technology.

The detailed design of reinforced concrete foundations is covered in BS EN
1992:2004 Design of concrete structures.

STRIP FOUNDATIONS

Concrete strip foundations are used to support and transmit the loads from
walls and so are most commonly specified for dwellings and similar low-rise
structures up to three storeys high. The typical arrangements required in the
Building Regulations for strip foundations are shown in Fig. 3.3.1.

The width of the strip foundations can be calculated from first principles
as follows:

$$\textit{Minimum width of strip} = \frac{\text{total load of building per metre kN/m}}{\text{safe bearing capacity of subsoil kN/m}^2}$$

where the safe bearing capacity is determined by obtaining the actual bearing
capacity by laboratory analysis and applying a factor of safety.

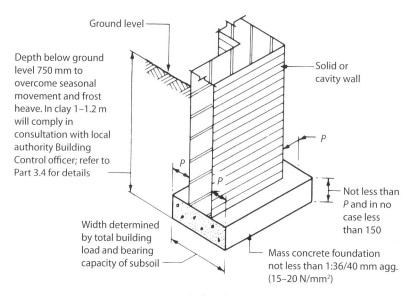

Ground level —

Depth below ground level 750 mm to overcome seasonal movement and frost heave. In clay 1–1.2 m will comply in consultation with local authority Building Control officer; refer to Part 3.4 for details

Solid or cavity wall

P

P

P

Not less than *P* and in no case less than 150

Width determined by total building load and bearing capacity of subsoil

Mass concrete foundation not less than 1:36/40 mm agg. (15–20 N/mm²)

Strip foundations

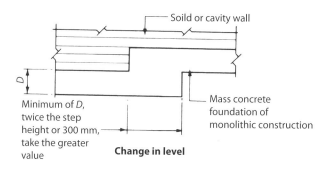

Soild or cavity wall

D

Minimum of *D*, twice the step height or 300 mm, take the greater value

Mass concrete foundation of monolithic construction

Change in level

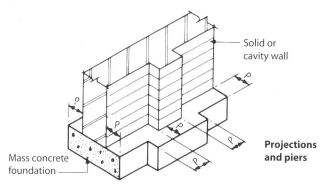

Solid or cavity wall

P

P

P

P

P

Projections and piers

Mass concrete foundation

Figure 3.3.1   Strip foundations and Building Regulations.

Having ascertained the nature and bearing capacity of the subsoil, the width of the foundation can be determined by one of the following methods:

1. Calculating the total (dead + imposed) load per metre run of foundation and relating this to the analysed safe bearing capacity of the subsoil. For example, if the total load is 40 kN/m and the subsoil safe bearing capacity is 80 kN/m^2, then the foundation width is:

$$\frac{40\ \dfrac{kN}{m}}{80\ \dfrac{kN}{m^2}} = 0.5 \text{ m or } 500 \text{ mm}$$

2. Alternatively the Building Regulations provides minimum widths in design tables, such as Table 3.3.2, where size width of foundations needed are related to subsoil type, wall loading and field tests on the soil.

Strip foundations are normally trench-filled – that is, dug to a minimum depth of 1 m and filled with mass concrete to within 150 mm of the ground level as shown in Fig. 3.3.2. Although the cost of concrete increases because of the trench fill, the labour and materials costs of constructing a wall below ground level decrease. Also the work is safer as no bricklayers need to work in the excavation trench.

In some cases these foundations may be of mass concrete fill; however, in weak soils like clay, it is common to specify a nominal reinforcement cage to stiffen up the strip. The reinforcement will also assist the strip in spanning any weak pockets of soil encountered in the excavations.

RAFT FOUNDATIONS

With pad foundations as the loads become larger, particularly in medium- and high-rise buildings, the size of individual pads needs to spread the load safely over the ground so must be made larger. Thus the principle of any major raft foundation is to spread the load over the entire area of the building. This method is particularly useful where the bearing capacity is low.

Raft foundations can be considered under three headings:

1. nominally reinforced rafts (lightly loaded low-rise situations);
2. designed RC slab rafts (heavily loaded or point-loaded situations);
3. RC beam and slab rafts.

Nominally reinforced raft foundations are often used on poor soils for lightly loaded buildings, and are considered capable of accommodating small settlements of the soil if strengthened with standard structural fabric mesh, one layer in the top and one layer in the bottom. Fig. 3.3.3 on page 96 shows a typical raft foundation for a low-rise dwelling.

Table 3.3.2 Guide to strip foundation width relative to subsoil type.

Subsoil type	Subsoil condition	Field test	Bearing capacity (kN/m²)	Minimum width of strip (mm) for a total load (kN/m) of:					
				20	30	40	50	60	70
Rock	Stronger than sandstone, limestone or chalk	Requires mechanical device to break up	> 600	Wall thickness at least					
Chalk	Solid	Requires a pick to remove	600	Wall thickness at least					
Gravel	Medium density	Requires a pick to excavate	> 600	250	300	400	500	600	650
Sand	Compact	Breaks down when dry	> 300	250	300	400	500	600	650
Clay / Sandy clay	Stiff	Pick or mechanical device required to remove	150 to 300	250	300	400	500	600	650
Sand or gravel / Silty sand / Clayey sand	Loose	Manually excavated with a spade, 50 mm square peg easily driven in	> 200	400	600	Tradional strip foundation not suitable for loading > 30 kN/m, consider reinforced strip, deep strip or short bored piling			
Silt / Clay / Sandy or silty clay	Soft	Easily excavated manually, can be moulded by hand	Lab. test needed	450	650	As above			
Silt, clay, sandy or silty clay	Very soft	Samples exude water when squeezed	Lab. test needed	Subsoil of this condition is not suitable to receive unreinforced strip foundations. Consider reinforced strip, deep strip and end bearing piled foundations to solid strata					

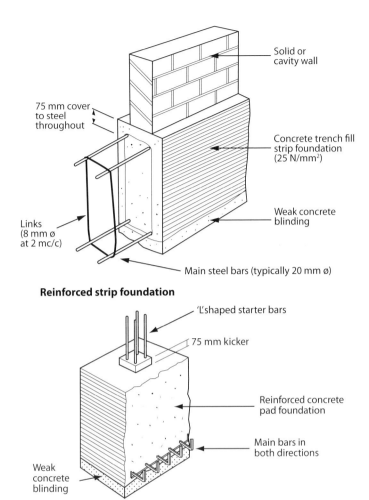

Reinforced strip foundation

Solid or cavity wall

75 mm cover to steel throughout

Concrete trench fill strip foundation (25 N/mm²)

Weak concrete blinding

Links (8 mm ø at 2 mc/c)

Main steel bars (typically 20 mm ø)

'L' shaped starter bars

75 mm kicker

Reinforced concrete pad foundation

Main bars in both directions

Weak concrete blinding

Figure 3.3.2 Reinforced concrete strip and pad foundations.

In poor soils the upper crust of soil (450–600 mm) is often stiffer than the lower subsoil, and to build a light raft on this crust is usually better than penetrating it with a strip foundation. It is also common to thicken the raft below any load-bearing wall with mass concrete for added strength.

Solid slab rafts are constructed of uniform thickness over the whole raft area, which can be wasteful because the design must be based on the situation existing where the heaviest load occurs. The effect of the load from columns and the ground pressure is to create areas of tension under the columns and areas of tension in the upper part of the raft between the columns, and this is where the main steel reinforcement is placed. Very often a nominal mesh of reinforcement is provided in the faces where compression occurs, to control shrinkage cracking of the concrete (see Fig. 3.3.4).

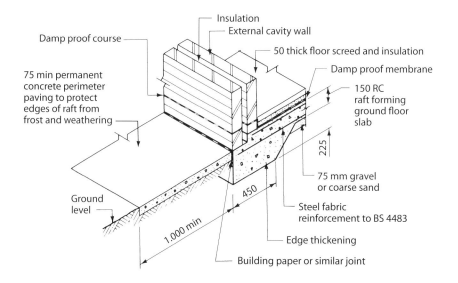

Insulation

External cavity wall

Damp proof course

50 thick floor screed and insulation

Damp proof membrane

75 min permanent concrete perimeter paving to protect edges of raft from frost and weathering

150 RC raft forming ground floor slab

225

75 mm gravel or coarse sand

Ground level

450

Steel fabric reinforcement to BS 4483

1.000 min

Edge thickening

Building paper or similar joint

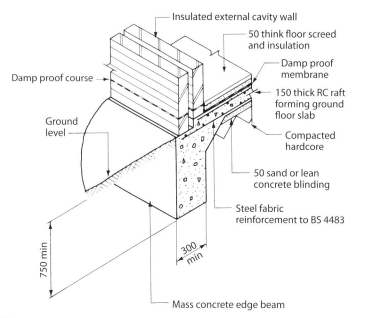

Insulated external cavity wall

50 think floor screed and insulation

Damp proof course

Damp proof membrane

150 thick RC raft forming ground floor slab

Ground level

Compacted hardcore

50 sand or lean concrete blinding

750 min

Steel fabric reinforcement to BS 4483

300 min

Mass concrete edge beam

Figure 3.3.3 **Typical raft foundations.**

Note: insulation in wall cavity omitted for clarity.

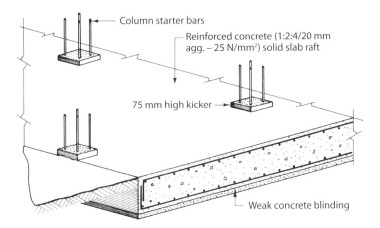

RC solid slab raft foundation

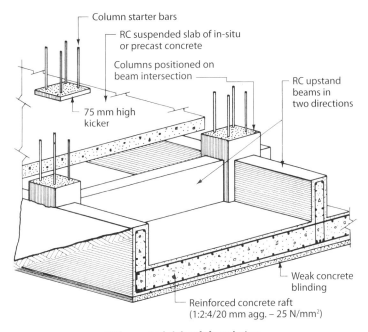

RC beam and slab raft foundation

Figure 3.3.4   Reinforced concrete raft foundations.

Beam and slab rafts are an alternative to the solid slab raft. They are used:

- where poor soils are encountered;
- to distribute column loads from framed buildings over the area of the raft (this use of these beams usually results in a reduction of the slab thickness).

The beams can be upstand or downstand depending on the bearing capacity of the soil near the surface. Downstand beams will give a saving on excavation costs, whereas upstand beams create a usable void below the ground floor if a suspended slab is used (see Fig. 3.3.4). This can be further exploited if good soils can only be found at depth because the upstand beams can form the internal walls to basement cells. This type of construction is know as 'cellular raft', details of which are illustrated in Fig. 3.3.5.

ISOLATED OR PAD FOUNDATIONS

This type of foundation is used to support and transmit the loads from piers and columns. The most economic plan shape is a square, but if the columns are close to the site boundary, it may be necessary to use a rectangular plan shape of equivalent area. The reaction of the foundation to the load and ground pressures is to require main steel in both directions. Details of typical pad foundations supporting an RC (reinforced concrete) column were shown in Fig. 3.3.2. The depth of the base will be governed by the anticipated moments and shear forces, the calculations involved being beyond the scope of this volume. Incorporated in the base will also be the L-shaped starter bars wired to the bottom reinforcement for a reinforced concrete column. If the pad foundation supports a steel column, holding-down bolts must be cast into the pad (see Part 10 on pages 521, 538).

The plan area of a pad foundation is a constant feature, being derived from:

$$\text{Area of pad} = \frac{\text{point or column load}}{\text{safe bearing capacity of subsoil}}$$

For example, if the column shown in Fig. 3.3.2 transmits a 50 kN load to subsoil of safe bearing capacity 80 kN/m², then the square column foundation dimensions are:

$$\sqrt{\frac{50\ kN}{80\ \frac{kN}{m^2}}} = 0.79 \text{ m or 790 mm (say, 800 mm} \times \text{800 mm square)}$$

Pad foundations are usually reinforced with steel mesh in the bottom of the pad, which reduces the amount of concrete required and makes them cheaper and easier to construct than equivalent mass concrete foundations. Pads are usually specified where there are point loads, as with framed buildings featuring steel or reinforced concrete columns and beams. However, they can also be used for load-bearing wall constructions such as dwellings. Here pads are located at the junctions of the load-bearing wall and around the perimeter

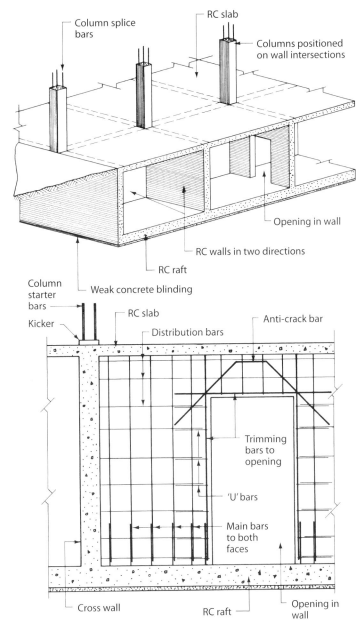

Figure 3.3.5 Typical cellular raft details.

walls typically at 3–5 m centres; spanning between the pads is a reinforced concrete ground beam that supports the walls and ground slabs.

Pads are excavated by machine within adapted trench boxes, which support the sides in sandy soils. The spoil materials are carted off site or put aside for future landscaping. The depth of the pads will depend on the strength and properties of the subsoil, and the depth of tree roots in clay soils.

PILE FOUNDATION SYSTEMS

Piling systems are generally used for medium- and high-rise structures, basement construction or where there are high building loads. In this section we will concentrate on piling systems used in low-rise, relatively lightly loaded situations. More advanced use of piling systems is shown in *Advanced Construction Technology*.

It is now common for designers of low-rise buildings to opt for a shallow piled foundation solution where previously strip foundations would have been specified, for a number of reasons:

- the increased costs and restrictions of taking excavated pile spoil (known as arisings) to landfill sites;

- the increased pressure to build on land which is not suitable for traditional strip foundations, such as land with a high water table;

- the speed of construction where the sides of the excavation are less prone to weathering and collapse;

- the availability of smaller, more mobile piling rigs and systems.

Structurally, piles support the building loads by bearing on the bottom area of the pile (end bearing) and using the frictional resistance on the vertical surface of the pile (frictional). How the pile was constructed affects whether the pile is largely end bearing or frictional. Where the soil is excavated to make the pile it is known as a replacement pile and relies on end bearing more than friction to support the building loads; where the soil is pushed aside to insert a preformed pile it is known as a displacement pile, which will have a much greater frictional component than end bearing.

Two basic types of piling system are used for low-rise applications, which are typically around 6 m deep and are known as short replacement or displacement piles.

- **Bored or augered piles (replacement)** These are formed by a variety of augers where a circular borehole is filled with mass concrete and an inserted helical steel reinforcing cage as shown in Fig. 3.3.6. These are quick to construct, particularly in clays, and require no prefabricated elements except the steel reinforcement. And although there is less spoil material generated than, say, a strip foundation, there will need to be appropriate disposal of the pile arisings.

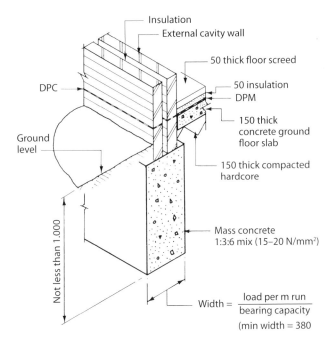

Insulation
External cavity wall
50 thick floor screed
DPC
50 insulation
DPM
150 thick concrete ground floor slab
Ground level
150 thick compacted hardcore
Not less than 1.000
Mass concrete 1:3:6 mix (15–20 N/mm²)
$$\text{Width} = \frac{\text{load per m run}}{\text{bearing capacity}}$$
(min width = 380

Deep strip foundation

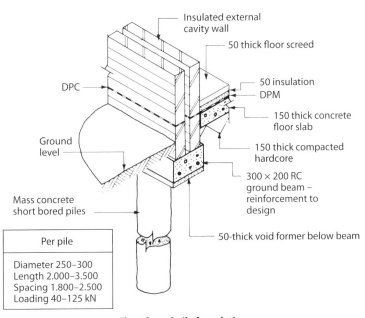

Insulated external cavity wall
50 thick floor screed
DPC
50 insulation
DPM
150 thick concrete floor slab
Ground level
150 thick compacted hardcore
300 × 200 RC ground beam – reinforcement to design
Mass concrete short bored piles
50-thick void former below beam

Per pile
Diameter 250–300 Length 2.000–3.500 Spacing 1.800–2.500 Loading 40–125 kN

Short bored pile foundation

Figure 3.3.6 Alternative foundations for clay soils.

Note: insulation in wall cavity omitted for clarity.

- **Pre-formed driven piles (displacement)** Usually steel or precast concrete. These piles can be installed traditionally using a top-driven hydraulic drop hammer. Hammer capacities can range from one to ten tonnes and can be mounted on a variety of tracked or wheeled rigs. Some smaller precast concrete piles can be jacked into the ground using a vibrationless system to reduce noise. Fig. 3.3.7 shows a typical precast concrete pile section used by Roger Bullivant Ltd. These piles generate little waste and are not weather-dependent. The segments are light and convenient to handle by forklift. A typical installation of a precast concrete pile and ground beam system is shown in Fig. 3.3.8.

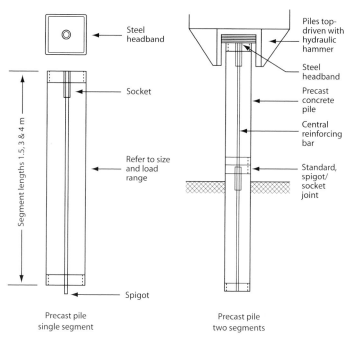

Figure 3.3.7 **Precast concrete pile.**

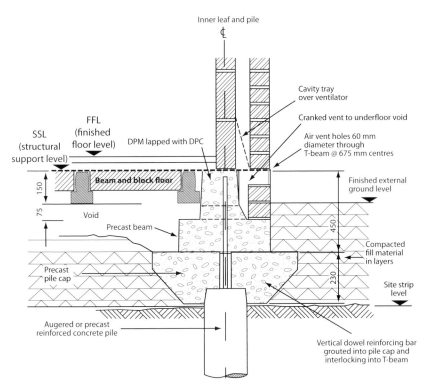

Inner leaf and pile
¢

Cavity tray
over ventilator

Cranked vent to underfloor void

Air vent holes 60 mm
diameter through
T-beam @ 675 mm centres

SSL
(structural
support level)

FFL
(finished
floor level)

DPM lapped with DPC

Beam and block floor

Finished external
ground level

150

75

Void

Precast beam

Compacted
fill material
in layers

450

Precast
pile cap

Site strip
level

230

Augered or precast
reinforced concrete pile

Vertical dowel reinforcing bar
grouted into pile cap and
interlocking into T-beam

Figure 3.3.8   Precast concrete piles, pile caps and inverted T-beams.

Trees: Effect on foundations 3.4

In the UK, shrinkable clays occur in specific areas found roughly below an imaginary line linking the Humber estuary with a point intersecting the mouth of the River Severn, as shown in Fig. 3.4.1. Generally shrinkable soils are those containing more than 35 per cent fine particles and having a Plasticity Index (PI) of 10 per cent or more, where the PI is a measure of a soil's volume change potential, determined by a specialist soil test called the Atterberg Limits test. The volume change potential is then classified as shown in Table 3.4.1.

Table 3.4.1 Volume change potential classifications.

PI	Volume change potential
40% +	High
20%– 39%	Medium
10% – 19%	Low

Clay subsoils are particularly prone to changes in moisture content, with a high degree of shrinkage occurring during periods of drought as moisture is evaporated from the clay subsoil. In these situations, ground settlement below old shallow foundation (less than 1 m deep) can occur, which is known as subsidence. If trees are located close by the building, at a distance less than the height of the mature tree, it is likely that there will be roots extending down 2–3.5 m. These will then abstract moisture and cause shrinkage of the clay at a much deeper level, where even modern deep strip trench fill foundation will fail due to subsidence. The typical metre height of UK tree species and their water demands are indicated in Table 3.4.2.

Figure 3.4.1   Area of UK with zones of shrinkable subsoils.

Clay heave is the opposite phenomenon, attributed partly to saturation from wet weather where the clay subsoil swells. However, it is more likely to be caused by removal of trees within the immediate 'footprint' of new buildings or the immediate surroundings. Here clay soil that was dry and desiccated due to the presence of a tree, which has subsequently been removed, will swell up as moisture soaks back into the clay to reach a more balanced moisture content. Fig. 3.4.2 shows cracking patterns in buildings due to these indirect sources of damage.

Thus the accepted depth for foundations in clay subsoils is at least 1,000 mm in the absence of trees, or as agreed in consultation with the local authority Building Control officer or chartered engineer. However, where trees are apparent or have been within the preceding three years, the minimum depth of foundation must be calculated. Where insurers issue structural guarantees, this depth can be determined from tables provided in procedure manuals issued by the National House Building Council (NHBC) or Zurich Municipal. Fig. 3.4.3 provides some guidance, as an indication only, of the likely depths of foundation excavation necessary where tree distance (*d*) to height (*h*) ratios

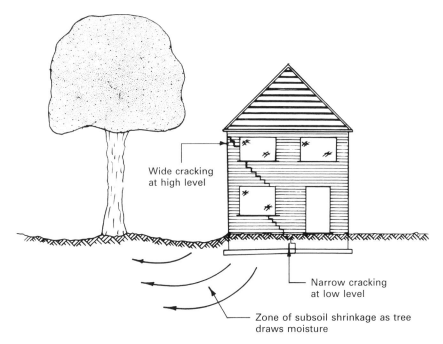

Wide cracking
at high level

Narrow cracking
at low level

Zone of subsoil shrinkage as tree
draws moisture

Subsoil dehydration and shrinkage

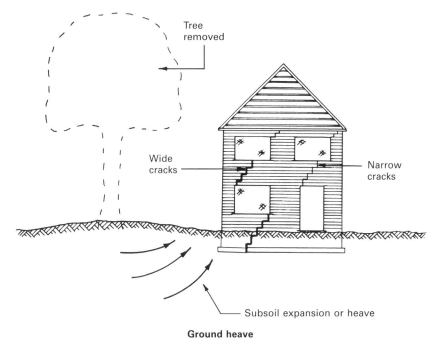

Tree
removed

Wide
cracks

Narrow
cracks

Subsoil expansion or heave

Ground heave

Figure 3.4.2   Indirect damage to buildings.

are less than unity (for more detailed tables, refer to the NHBC's Standards Guide – Part 4: Foundations) Beyond 1.5 m, depth trench-filled concrete will be appropriate but for foundations over 2.5 m, short-bored piles with ground beams are preferred.

Table 3.4.2 **Tree characteristics.**

Species	Approximate mature height (m)	Water demand: H = high, M = medium, L = low
Oak	30	H
Poplar	30	H
Ash	30	M
Cedar	30	M
Chestnut	30	M
Hemlock	30	M
Lime	30	M
Larch	30	L
Scots pine	30	L
Cypress	25	H
Willow	25	H
Maple	25	M
Sycamore	25	M
Walnut	25	M
Beech	25	L
Birch	25	L
Holly	20	L
Cherry	15	M
Hawthorn	15	M
Rowan	15	M

Note: Where groups of trees occur and the subsoil is known to be moisture responsive (i.e. shrinkable clay), the suggested building distance is 1.5 × mature height.

HEAVE PRECAUTIONS

In addition to foundation depths exceeding soil movement zones, the areas closer to the surface must be provided with additional treatment to cope with any surface-level swelling of the clay subsoil. Here, prior to concreting, the trench is prepared with a lining of low-density expanded polystyrene (or other approved proprietary compressible material). For details of the required thicknesses of void former below ground slabs and ground beams, see Table 3.4.3. The polystyrene is designed to absorb seasonal clay movement, and the polythene acts as a slip membrane preventing soil adhesion to the foundation. Fig. 3.4.4 shows application to trench fill and piled foundations.

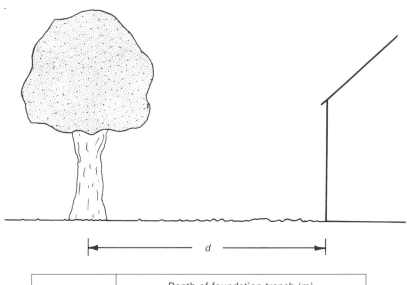

Species of tree	Depth of foundation trench (m)						
	$d/h = 0.1$	0.25	0.33	0.5	0.66	0.75	1.0
Poplar, willow and oak	Not acceptable	2.8	2.6	2.3	2.1	1.9	1.5
Others	Not acceptable	2.4	2.1	1.8	1.5	1.2	1.0

Figure 3.4.3   Tree proximity and foundation depth.

Table 3.4.3   Required thicknesses of void former below ground slabs and ground beams

Heave protection voids / Volume change potential	Thickness of void former against SIDE of foundation / ground beam (mm)	Thickness of void former UNDER ground beam / in-situ ground slab (mm)	Void dimension UNDER precast concrete floors / timber floors (mm)
High	35	150	300
Medium	25	100	250
Low	0	50	200

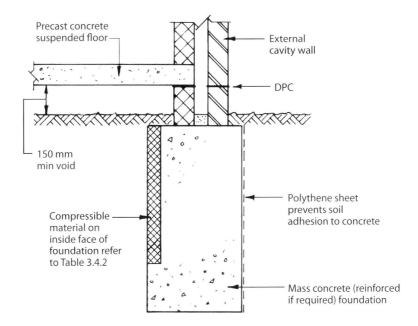

Precast concrete
suspended floor

External
cavity wall

DPC

150 mm
min void

Compressible
material on
inside face of
foundation refer
to Table 3.4.2

Polythene sheet
prevents soil
adhesion to concrete

Mass concrete (reinforced
if required) foundation

Note: Compressible material may be required to both sides of foundation

Deep strip or trench fill

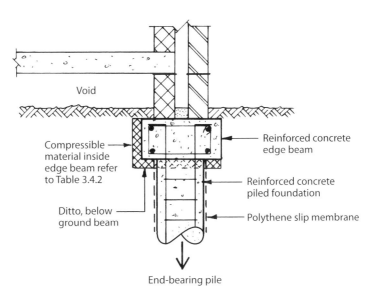

Void

Compressible
material inside
edge beam refer
to Table 3.4.2

Reinforced concrete
edge beam

Reinforced concrete
piled foundation

Ditto, below
ground beam

Polythene slip membrane

End-bearing pile

Alternative piled foundation

Figure 3.4.4   Precautionary treatment to foundations in shrinkable clay.

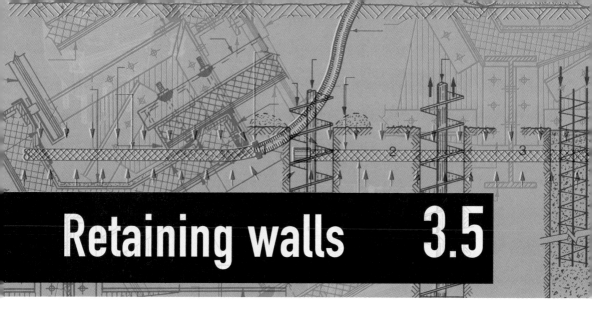

Retaining walls 3.5

The basic function of a retaining wall is to retain soil at a slope that is greater than it would naturally assume, usually at a vertical or near-vertical position. The natural slope taken up by any soil is called its **angle of repose** and is measured in relationship to the horizontal. Angles of repose for different soils range from 45° to near 0° for wet clays, but for most soils an average angle of 30° is usually taken. It is the wedge of soil resting on this upper plane of the angle of repose that a retaining wall has to support. The walls are designed to offer the necessary resistance by using their own mass to resist the thrust or relying upon the principles of leverage. The terminology used in retaining wall construction is shown in Fig. 3.5.1.

DESIGN PRINCIPLES

The design of any retaining wall is basically concerned with the lateral pressures of the retained soil and any subsoil water. The wall must be designed to ensure that:

- overturning does not occur;
- sliding does not occur;
- the soil on which the wall rests is not overloaded;
- the materials used in construction are not overstressed.

It is difficult to define accurately the properties of any soil, because they are variable materials. The calculation of pressure exerted at any point on the wall is a task for the expert, who must take into account the following factors:

- nature and type of soil;
- subsoil water movements;
- height of water table;
- type of wall;
- materials used in the construction of the wall.

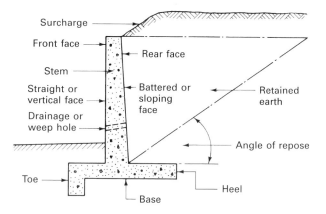

Retaining wall terminology

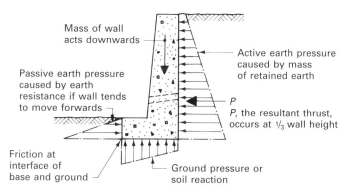

Mass retaining walls

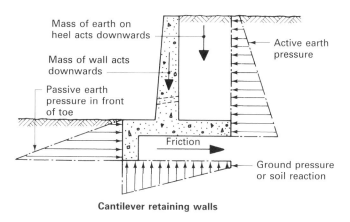

Cantilever retaining walls

Figure 3.5.1 Retaining wall terminology and pressures.

STRUCTURAL DESIGN

When working with retaining walls, knowledge of the relevant Eurocodes is vital to ensure the structural stability and safety of the project during construction and for the life of the wall. Those of relevance to retaining walls are:

- Eurocode 1: Actions on structures BS EN 1991
- Eurocode 2: Design of concrete structures BS EN 1992
- Eurocode 3: Design of steel structures BS EN 1993
- Eurocode 7: Geotechnical design BS EN 1997

You may wish to look back at pages 18–19 for more general information about Eurocodes.

EARTH PRESSURES

The designer is mainly concerned with the effect of two forms of earth pressure:

- active earth pressure;
- passive earth pressure.

Active earth pressures are those that at all times are tending to move or overturn the retaining wall, and are composed of the earth wedge being retained together with any hydrostatic pressure caused by the presence of groundwater. The latter can be reduced by the use of subsoil drainage behind the wall, or by inserting drainage openings called **weep holes** through the thickness of the stem, enabling the water to drain away.

Passive earth pressures are reactionary pressures that will react in the form of a resistance to movement of the wall. If the wall tends to move forward, the earth in front of the toe will be compressed, and a reaction in the form of passive pressure will build up in front of the toe, to counteract the forward movement. This pressure can be increased by enlarging the depth of the toe or by forming a rib on the underside of the base. Typical examples of these pressures were shown in Fig. 3.5.1.

STABILITY

The overall stability of a retaining wall is governed by the result of the action and reaction of two sorts of loads:

- **applied loads** such as soil and water pressure on the back of the wall, the mass of the wall and, in certain forms of cantilever wall, the mass of the soil acting with the mass of the wall;
- **induced loads** such as the ground pressure under the base, the passive pressure at the toe, and the friction between the underside of the base and the soil.

Effects of water

Groundwater behind a retaining wall, whether static or percolating through a subsoil, can have adverse effects upon the design and stability. It will increase the pressure on the back of the wall, and by reducing the soil shear strength it can reduce the bearing capacity of the soil. It can reduce the frictional resistance between the base and the soil, and reduce the possible passive pressure in front of the wall. It is, therefore, important to provide measures to drain away water that may collect at the rear face of the wall.

There are a number of methods to achieve this, the most common being the insertion of drainage holes (weep holes) passing from the back of the wall to the front. Weep holes are usually 50–100 mm in diameter at regular spacings of between 0.9 m and 3 m along the face of the wall. Alternatively there may be an open-jointed land drain bedded in a granular material, laid parallel behind the wall, to channel excess water to a safe outfall. This may be used in combination with a geotextile drainage blanket placed on the rear face of the wall so as to form a drained cavity to speed up the dispersal of water pressure.

Slip circle failure

This type of failure, shown in Fig. 3.5.2, is sometimes encountered with retaining walls in clay soils, particularly where there is a heavy surcharge of retained material. It takes the form of a rotational movement of the soil and

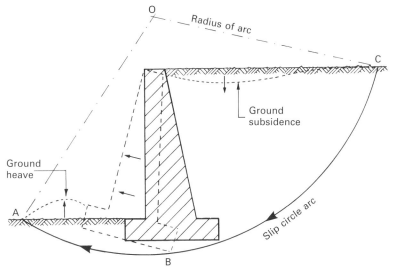

Moment due to weight of retained earth and wall above slip circle arc about O is greater than restoring moment (RM).
RM = permissible shear stress × length of arc ABC × arc radius OC.
Result: Mass above ABC rotates about O. Wall tilts forward and earth heaves in front.

Figure 3.5.2   Retaining wall failure due to rotational movement, or slip circle failure.

wall along a circular arc. The arc commences behind the wall and passes under the base, resulting in a tilting and forward movement of the wall. Further movement can be prevented by driving sheet piles into the ground in front of the toe, to a depth that will cut the slip circle arc.

Mass retaining walls

These are sometimes called **gravity walls** and rely upon their own mass together with the friction on the underside of the base to overcome the tendency to slide or overturn. They are generally economic only up to a height of 1.800 m. Mass walls can be constructed of semi-engineering quality bricks bedded in a strong cement mortar or of mass concrete.

Another common form of mass retaining walls is the gabion wall system. This is a modular containment system that enables rock, stone or other inert materials to be used as a construction material. It is often used to stabilise road cuttings in rural environments. The modules or cages, as they are known, are formed of wire mesh fabric panels, jointed to form square, rectangular or trapezoidal-shaped units. These cages are part pre-assembled then carefully filled by mechanical means, with the contractor picking the stone over by hand to reduce excessive voids. The advantages of this system are that exposed faces can also be systematically hand-packed to provide an appearance of a natural dry stone wall, and due to the innate voids in the gabions the wall drains naturally, without the need for weep holes or land drains. Also this can be a very sustainable use of inert demolition waste if present, which otherwise would need to be carted off to landfill.

Typical examples of these walls are shown in Fig. 3.5.3.

Cantilever walls

These are usually of reinforced concrete, and work on the principle of leverage. Two basic forms can be considered: a base with a large heel so that the mass of earth above can be added to the mass of the wall for design purposes, or, if this form is not practicable, a cantilever wall with a large toe (see Fig. 3.5.4). The drawings show typical sections and patterns of reinforcement encountered with these basic forms of cantilever retaining wall. (It is useful to note that these are a common form of technique for constructing basement walls, as described in the following section.) The main steel occurs on the tension face of the wall, and nominal steel is very often included in the opposite face to control the shrinkage cracking that occurs in in-situ concrete work.

Reinforced cantilever walls have an economic height range of 1.200–6.000 m; walls in excess of this height have been economically constructed using pre-stressing techniques. Any durable facing material may be applied to the surface to improve the appearance of the wall, but it must be remembered that such finishes are decorative and add nothing to the structural strength of the wall.

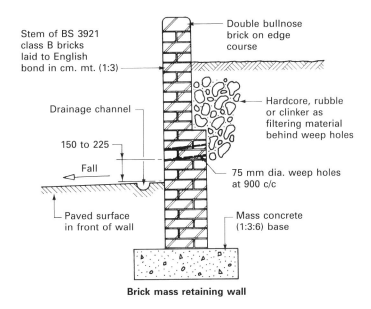

Brick mass retaining wall

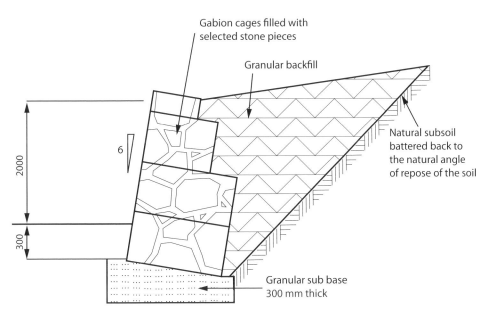

Figure 3.5.3 Typical mass retaining walls.

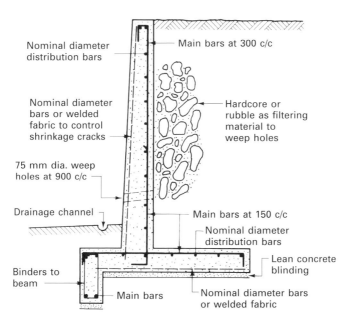

Nominal diameter distribution bars

Main bars at 300 c/c

Nominal diameter bars or welded fabric to control shrinkage cracks

Hardcore or rubble as filtering material to weep holes

75 mm dia. weep holes at 900 c/c

Drainage channel

Main bars at 150 c/c

Nominal diameter distribution bars

Lean concrete blinding

Binders to beam

Main bars

Nominal diameter bars or welded fabric

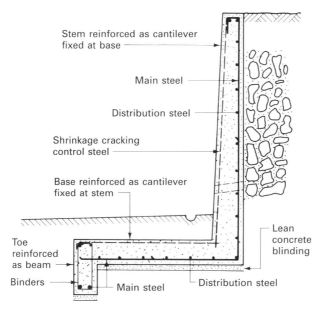

Stem reinforced as cantilever fixed at base

Main steel

Distribution steel

Shrinkage cracking control steel

Base reinforced as cantilever fixed at stem

Lean concrete blinding

Toe reinforced as beam

Binders

Main steel

Distribution steel

Figure 3.5.4 Typical reinforced concrete cantilever retaining walls.

Counterfort retaining walls

These walls can be constructed of reinforced or pre-stressed concrete, and are considered suitable if the height is over 4.500 m. The counterforts are triangular beams placed at suitable centres behind the stem and above the base to enable the stem and base to act as slabs spanning horizontally over or under the counterforts. Fig. 3.5.5 shows a typical section and pattern of reinforcement for a counterfort retaining wall.

If the counterforts are placed on the face of the stem they are termed **buttresses**, and the whole arrangement is called a **buttress retaining wall**. The design and construction principles are similar in the two formats. This type of construction is often useful in the construction of basement walls, where the counterforts become part of an internal buttressing wall.

Precast concrete retaining walls

There are now a number of different precast concrete retaining walls available on the market, all of which are manufactured from high-grade precast concrete on the cantilever principle. A typical 600 mm-wide module system is shown in Fig. 3.5.6. These walls can be erected on a foundation as a permanent retaining wall, or can be free-standing to act as a dividing wall between heaped materials, such as aggregates for concrete. In the latter situation they can increase by approximately three times the storage volume for any given area. Other advantages are a reduction in time by eliminating the curing period that is required for in-situ walls and eliminating the need for costly formwork, together with the time required to erect and dismantle the temporary forms. The units are reinforced on both faces to meet all forms of stem loading. Lifting holes are provided, into which temporary steel eyelets can be screwed and then used to sling lifting straps where required. Special units to form internal angles, external angles, junctions and curved walls are also available to provide flexible layout arrangements.

Crib retaining walls

Crib walls are designed on the principle of a mass retaining wall. They consist of a framework or crib of precast concrete or timber units within which the soil is retained. They are constructed with a face batter of between 1:6 and 1:8 unless the height is less than the width of the crib ties, in which case the face can be constructed vertical. Subsoil drainage is not required, because the open face provides for adequate drainage (see Fig. 3.5.6).

Both crib walls and gabion walls are often preferred solutions where the retained wall forms part of the visual landscaping. These methods allow a planting and wildlife habitat to be created while also providing a structural solution.

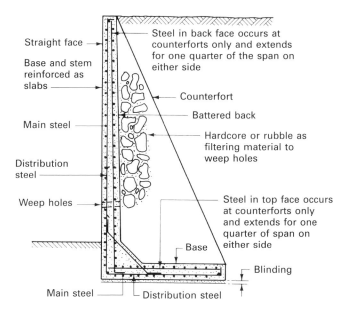

Straight face

Base and stem reinforced as slabs

Main steel

Distribution steel

Weep holes

Steel in back face occurs at counterforts only and extends for one quarter of the span on either side

Counterfort

Battered back

Hardcore or rubble as filtering material to weep holes

Steel in top face occurs at counterforts only and extends for one quarter of span on either side

Base

Blinding

Main steel

Distribution steel

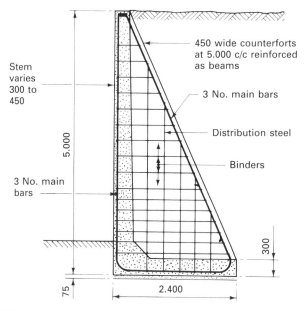

Stem varies 300 to 450

5.000

3 No. main bars

450 wide counterforts at 5.000 c/c reinforced as beams

3 No. main bars

Distribution steel

Binders

300

75

2.400

Figure 3.5.5 Typical reinforced concrete counterfort retaining wall.

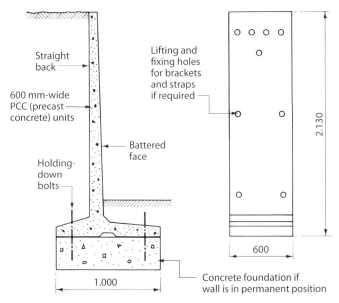

Straight back

600 mm-wide PCC (precast concrete) units

Holding-down bolts

Battered face

Lifting and fixing holes for brackets and straps if required

2.130

600

1.000

Concrete foundation if wall is in permanent position

Typical 'Marley' precast concrete retaining wall

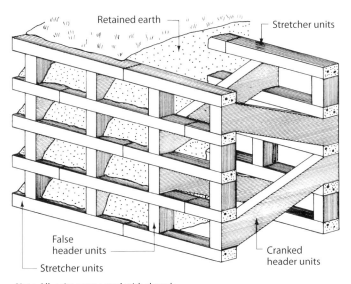

Retained earth

Stretcher units

False header units

Stretcher units

Cranked header units

Note: All units connected with dowels

'Anda-Crib' precast concrete retaining wall

Figure 3.5.6   Precast concrete retaining walls.

Reinforced masonry retaining walls

Steel reinforcement may be used in brick retaining walls to resist tensile forces and to prevent the effects of shear. Full structural design details are contained in Eurocode 6: Design of masonry structures BS EN 1996.

A traditional way of reinforcing brickwork is to use a brick-bonding arrangement known as **quetta bond**, which enables a uniform distribution of vertical voids to be created (see Part 4 – Superstructure, Fig. 4.1.9, page 150). Vertical steel reinforcement is tied to the foundation reinforcement and spaced to coincide with the purpose-made voids. When the brickwork is completed, the voids are filled with concrete to produce a series of reinforced concrete mini-columns within the wall. A simpler and slightly cheaper method is the construction of a cavity wall from dense concrete blockwork, where the cavity is used to thread through the vertical reinforcement prior to the grouting up of the cavity. It is also common practice to fix reinforcing mesh within the bed joints of the masonry for extra lateral strength.

Where appearance is not important, or the wall is to receive a surface treatment, reinforcement and in-situ concrete within hollow concrete blockwork provide for economical and functional construction. Fig. 3.5.7 shows the application of standard-profile, hollow, dense concrete blocks laid in stretcher bond as permanent formwork to continuous vertical columns.

The height potential and slenderness ratio (effective height to width) for reinforced masonry walls can be enhanced by post-tensioning the structure. The principle is explained in Part 10 of *Advanced Construction Technology*. For purposes of brick walls there are a number of construction options, including:

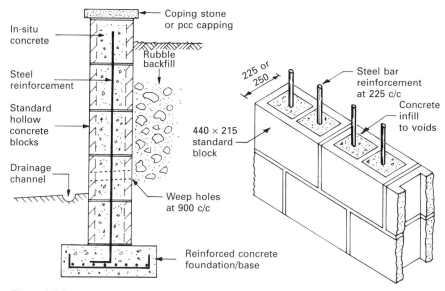

Figure 3.5.7   Reinforced concrete block retaining wall.

- quetta bond with steel bars and concrete in the voids;
- stretcher-bonded wide cavity with reinforced steel bars coated for corrosion protection;
- solid wall of perforated bricks with continuous voids containing grouted steel reinforcement bars.

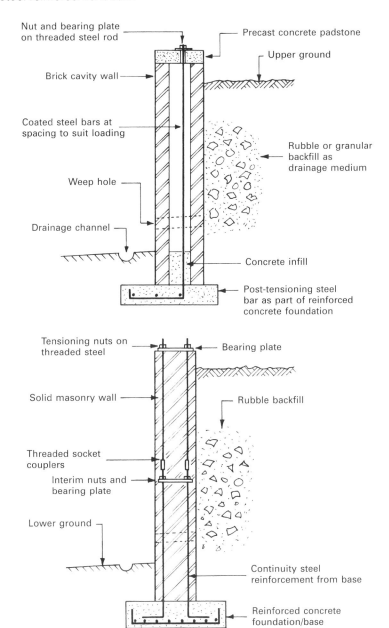

Figure 3.5.8   Post-tensioned brick retaining walls.

Basements 3.6

The demand for higher-density housing and larger living space, particularly in cities, is encouraging designers and contractors to look again at the provision of basements in dwellings. Appendix E in Approved Document B of the Building Regulations defines a basement storey as a storey with a habitable floor that at some point is more than 1.200 m below the highest level of ground adjacent to the outside walls (as opposed to a cellar, which is for storage only). This definition is given in the context of inhibiting the spread of fire within a building: generally the fire-resistance requirements for basements are more onerous than for the ground or upper storeys in the same building. This section on basements is concerned only with basement storeys that are below ground level.

The structural walls of a basement below ground level are in fact retaining walls, which have to offer resistance to the soil and groundwater pressures as well as assisting to transmit the superstructure loads to the foundations. It is possible to construct a basement free of superstructural loadings, but these techniques are beyond the scope of this book.

BASEMENTS IN NEW BUILD WORK

There are basically two approaches to the construction of basements in new-build work:

- **bottom-up construction** – basements of one storey for buildings using a range of techniques as discussed below;
- **top-down construction** – usually deep basements with two or more storeys specified in major commercial and retail projects, where intermediate floors help support the basement walls as the basement is moled or tunnelled out below floor slabs that have been previously cast. Details of this are covered in *Advanced Construction Technology*.

Bottom-up construction – the simpler method – involves building the walls and floor slab within an excavated pit with battered sides excavated to the angle of repose of the soil, which is the angle to the horizontal that the soil will remain stable (clay 25–45°, sands 15–30°, gravels 30–45°). This groundwork construction method is known as forming a batter. The advantages of this method are:

■ no specialist excavation plant is required;

■ standard proprietary precast or in-situ concrete systems can used to form the basement slab and walls;

■ there is easy access to earth or internal faces to construct waterproofing, drainage and service connections;

■ standard de-watering systems and pumps can be utillised to keep the construction area dry.

However, one limiting disadvantage is that there needs to be sufficient work space available to enable the batter to be safely constructed. This may be impossible to achieve in cramped urban sites where the basement is to be constructed close to a boundary line or an existing building; here other methods may need to be employed to provide temporary support during construction.

HEALTH AND SAFETY IN BASEMENTS

When constructing any type of basement, the management of health and safety is crucial to avoid accidents. Factors that the contractor needs to consider include:

■ extra excavation;

■ working in excavation;

■ working from height;

■ temporary stability of walls;

■ craning of large basement parts or movement of smaller units.

Each of these operations requires the hazards to be identified and suitable control measures to be put in place to ensure a safe system of work.

■ No operatives should be working in unsupported ground.

■ Safe methods of access and egress should be provided into the basement area.

■ Edge protection and vehicle stops should be provided around the excavation.

■ Any lifting operations need to be carefully planned in advance.

Various construction methods are used to form basements.

MASONRY WALL SYSTEMS

Traditionally basement walls were constructed of engineering brickwork. However, this has been largely replaced by cavity walls constructed in dense concrete blockwork, with leaves of 100 mm minimum thickness. The cavity can contain high-yield steel reinforcing bars that are tied into the basement slab reinforcement and lapped horizontally and at corners. The cavity is grouted up with a strong cement mortar (25 N/mm^2). This method will require temporary support of the surrounding soil to construct the basement, or the basement can be constructed within a battered excavation. The type of wall system has been illustrated on page 120 under Reinforced masonry retaining walls.

CAST-IN-SITU REINFORCED CONCRETE

In this method, using standard proprietary formwork and falsework systems, reinforced concrete retaining walls are constructed in a similar method to normal above-ground reinforced concrete walls. These systems are designed to withstand the loads of wet concrete and reinforcement as it is fixed, placed and compacted. Care must be taken at cold or construction joints to ensure that no water can seep through at these weak points, by the detailing and installation of appropriate water bars. This method will require temporary support of the surrounding soil to construct the basement, such as the use of temporarily installed steel-sheet interlocking piles, known as a coffer dam; this is removed once the reinforced concrete walls have reached sufficient strength to support the lateral forces of the retained soil.

PRECAST REINFORCED CONCRETE MODULAR UNITS

This is a fairly recent development which utilises factory-constructed precast concrete units that have pre-insulated voids filled with polystyrene. The precast units can incorporate door and window openings. These units slot together on site on a carefully prepared in-situ base slab. External drainage is provided by a geocomposite drain membrane, 600 mm of clean stone backfill and a geotextile membrane placed against the external face of the precast concrete unit. This method will require temporary support of the surrounding soil to construct the basement or can be constructed within a battered excavation. See Fig. 3.6.1 on the next page.

STEEL-SHEET PILING

This is a common method in marine and civil engineering works that has now been applied to domestic basement construction. With this method,

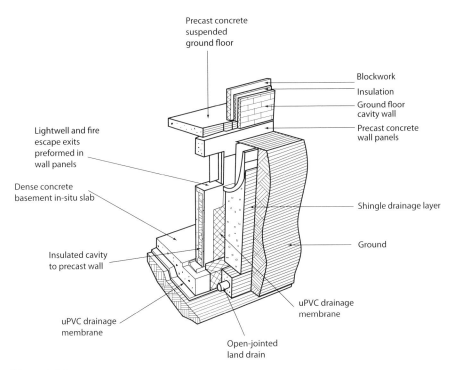

Precast concrete
suspended
ground floor

Blockwork

Insulation

Ground floor
cavity wall

Precast concrete
wall panels

Lightwell and fire
escape exits
preformed in
wall panels

Dense concrete
basement in-situ slab

Shingle drainage layer

Ground

Insulated cavity
to precast wall

uPVC drainage
membrane

uPVC drainage
membrane

Open-jointed
land drain

Figure 3.6.1   Precast concrete basement method.

interlocking steel-sheet piles are driven or vibrated into the ground prior to excavation to provide temporary or permanent basement walls. They interlock by the use of simple mechanical sliding grooves (known as clutches) along the vertical edges of the sheet piles. The piles are driven into the ground before any excavation work is done. The interlocking sheet piles can either be toed into the subsoil or propped, as in Fig. 3.6.2. It is quite common to shore up the piles temporarily with heavy steel props prior to the construction of the permanent ground floor slab, which then acts as a permanent support to the sheet piles. Alternatively the steel-sheet piles can be driven at least a third deeper than the formation level of the basement slab, forming a vertical cantilever. These cantilevered sheet piles do not require any temporary cross-bracing which would interfere with the internal construction of the permanent basement walls or internal features. In areas with a high water table, the steel-sheet piles can be permanently installed with welded seals at their clutches, to form a watertight box as well as supporting the permanent building loads.

In situations like this where the steel-sheet piling becomes a permanent part of the wall, it is usual to build an internal masonry skin to form a drained cavity for waterproofing purposes (see page 128).

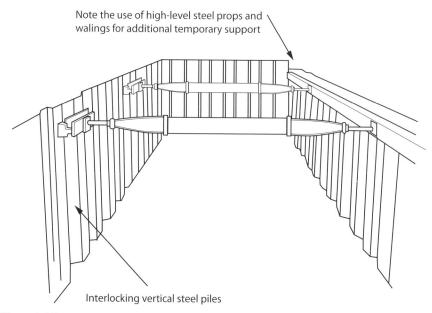

Note the use of high-level steel props and walings for additional temporary support

Interlocking vertical steel piles

Figure 3.6.2 Interlocking steel-sheet piles used as permanent basement retaining wall.

CONTIGUOUS AND SECANT PILING METHODS

This is an increasingly popular method of providing a structural wall. A series of augered reinforced concrete piles is constructed at close centres to form a continuous wall, again prior to excavation of the basement box. In dry soils, only close-bored contiguous piles need to be used, but an internal drained cavity may need to be constructed to catch any groundwater that may be present (see page 128). Where a watertight self-supporting solution is needed, secant piles should be specified as these are bored so that the pile diameters overlap, providing an impermeable barrier to groundwater flow. This method is also self-supporting, provided that the piles are sunk as cantilevered piles below the formation level, although it is quite common to shore up the piles temporarily with heavy steel props prior to the construction of the permanent ground floor slab, which then acts as a permanent prop.

DIAPHRAGM WALLS

Diaphragm walls are similar to contiguous piles in that they are constructed of reinforced concrete and constructed ahead of the main basement excavation; however, they comprise of panels 2–4 m in width and up to 20 m deep. The construction method is quite involved and requires specialist excavation plant and temporary infilling with a clay-bentonite slurry to prevent the panels collapsing prior to concrete being placed. This method is covered in detail in *Advanced Construction Technology*.

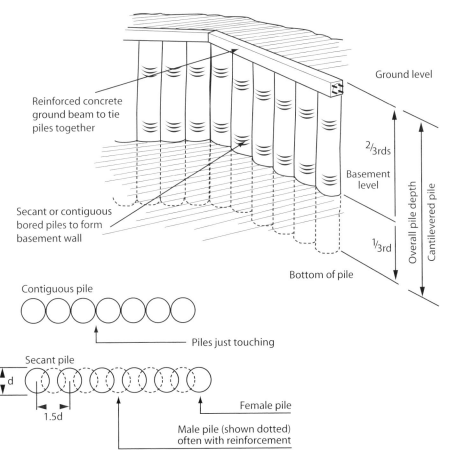

Reinforced concrete
ground beam to tie
piles together

Secant or contiguous
bored piles to form
basement wall

Ground level

²/₃rds

Basement
level

¹/₃rd

Overall pile depth

Cantilevered pile

Bottom of pile

Contiguous pile

Piles just touching

Secant pile

d

1.5d

Female pile

Male pile (shown dotted)
often with reinforcement

Figure 3.6.3 **Modern method of contiguous and secant piled walls.**

WATERPROOFING

Apart from the structural design of the basement walls and floor,
waterproofing presents the greatest problem in basement construction.
Building Regulation C2 requires such walls to be constructed so that they will
not transmit moisture from the ground to the inside of the building or to any
material used in the construction that would be affected adversely by
moisture. Building Regulation C2 also imposes similar conditions on the
construction of floors. BS 8102:2009 *Code of practice for protection of below
ground structures against water from the ground* considers three main
methods for waterproofing basements:

■ Type A – Barrier protection;

■ Type B – Structurally integral protection;

■ Type C – Drained protection.

These three systems are illustrated in Fig. 3.6.4. The system to be used depends on:

- whether it is a sloping site or a full excavation;
- the depth and extent of the basement;
- the potential level of the water table;
- the level of tanking required to prevent water ingress;
- the specialism of the contractors employed to construct the basement.

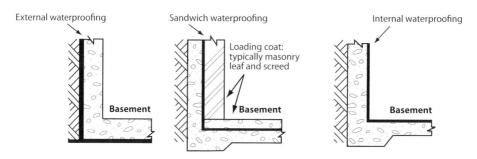

Type A structure: Tank protection

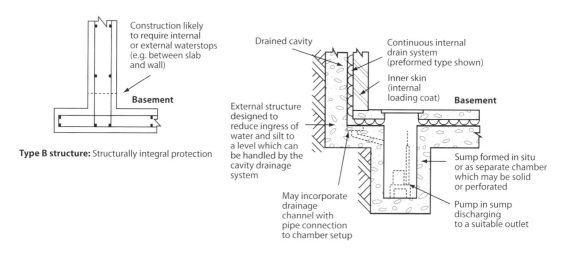

Type C structure: Drained cavity

Figure 3.6.4   The three waterproofing methods used for basement construction.

TYPE A – BARRIER PROTECTION

A membrane is a relatively thin material placed on either the external or internal face of a basement wall or floor to provide the resistance to the passage of moisture to the inside of the basement.

If the membrane is applied externally, protection is also given to the structural elements and the hydrostatic pressure will keep it firmly in place, but a reasonable working space must be allowed around the perimeter of the basement. This working space will entail extra excavation and subsequent backfilling after the membrane has been applied. If adequate protection is not given to the membrane, it can easily be damaged during the backfilling operation. A typical use of this method is in the waterproofing of a reinforced concrete basement wall and slab using mastic asphalt as the barrier membrane, as shown in Fig. 3.6.5.

An internally applied membrane gives no protection to the structural elements, and there is the danger that the membrane may be forced away from the surfaces by water pressure unless it is adequately loaded, i.e. sandwiched within the bulk of the wall by adding an internal masonry skin to the wall. These additional loading coats, although they can keep the membrane in place by resisting the water pressure, will reduce the usable volume within the basement.

Suitable materials that can be used for forming membranes are:

- mastic asphalt tanking to BS EN 12970 (a naturally occurring bitumen-based mineral);
- bonded sheet membranes to BS 8747 (such as fibre-reinforced bituminous felt, polyisobutylene plastic or polythene sheet);
- liquid-applied membranes to BS 8102 (such as epoxy resin compounds or bituminous compounds);
- geosynthetic clay liners (such as bentonite-impregnated matting);
- cementitious slurries and layer coats.

Mastic asphalt tanking

Asphalt is a natural or manufactured mixture of bitumen with a substantial proportion of inert mineral matter. When heated, asphalt becomes plastic and can be moulded by hand pressure into any shape. The basic principle of asphalt tanking is to provide a continuous waterproof membrane to the base and walls of the basement. Continuity between the vertical and horizontal membranes is of the utmost importance. As asphalt will set rapidly once removed from the heat source used to melt the blocks, it is applied in layers over small areas; again continuity is the key to a successful operation. Joints in successive coats should be staggered by at least 150 mm in horizontal work and at least 75 mm in vertical work.

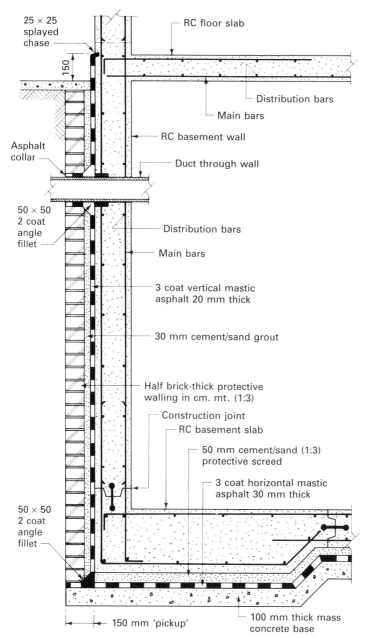

25 × 25 splayed chase

150

RC floor slab

Distribution bars

Main bars

RC basement wall

Asphalt collar

Duct through wall

50 × 50 2 coat angle fillet

Distribution bars

Main bars

3 coat vertical mastic asphalt 20 mm thick

30 mm cement/sand grout

Half brick-thick protective walling in cm. mt. (1:3)

Construction joint

RC basement slab

50 mm cement/sand (1:3) protective screed

3 coat horizontal mastic asphalt 30 mm thick

50 × 50 2 coat angle fillet

150 mm 'pickup'

100 mm thick mass concrete base

Figure 3.6.5   In-situ reinforced concrete basement – including waterproofing methods Type A and Type B.

During the construction period, the asphalt tanking must be protected against damage from impact, following trades and the adverse effects of petrol and oil. Horizontal asphalt tanking coats should be covered with a fine concrete screed at least 50 mm thick as soon as practicable after laying. Vertical asphalt tanking coats should be protected by building a half-brick or block wall 30 mm clear of the asphalt; the cavity so formed should be filled with a mortar grout as the work proceeds to ensure perfect interface contact. In the case of internal tanking, this protective wall will also act as the loading coat.

Any openings for the passages of pipes or ducts may allow moisture to penetrate unless adequate precautions are taken. The pipe or duct should be primed and coated with three coats of asphalt so that the sleeve formed extends at least 75 mm on either side of the tanking membrane before being placed in the wall or floor. The pipe or duct is connected to the tanking by a two-coat angle fillet.

Mastic asphalt has a number of advantages as a waterproof membrane.

- It is a thermoplastic material and can therefore be heated and reheated if necessary to make it pliable for moulding with a hand float to any desired shape or contour.

- It is durable: bituminous materials have been used in the construction of buildings for over 5,000 years and have remained intact to this day, as shown by excavations in Babylonia.

- It is impervious to both water and water vapour.

- It is non-toxic, vermin- and rot-proof, and is odourless after laying.

- It is unaffected by sulphates in the soil, which, if placed externally, will greatly improve the durability of a concrete structure.

The application of mastic asphalt is recognised as a specialist trade in the building industry. Most asphalt work is placed in the hands of specialist subcontractors, most of whom are members of the Mastic Asphalt Council and Employers Federation Limited.

Bonded sheet membranes

Plastic and bitumen sheeting materials are suited to shallow basements and can be applied externally or sandwiched within the structural wall or floor of the basement. The base structure of concrete or masonry is prepared with a primer of bituminous solution before sheeting, and is hot-bitumen bonded with 100 mm side and 150 mm end lapping in at least two layers. Fig. 3.6.6 shows application to wall and floor.

Liquid-applied membranes

These are one- or two-part systems of bitumen, elastomeric urethane or modified epoxy, where they are applied wet by painting on to a trowel-finished

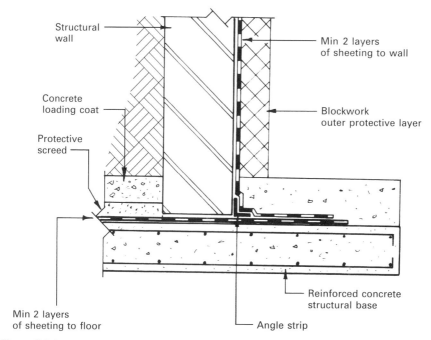

Figure 3.6.6 Impervious membrane tanking to shallow basement.

concrete, screeds or renderings. In basement wall construction the membrane can applied internally, with a masonry loading coat internally to resist hydrostatic pressure, or externally applied in a minimum of three coats, although protection will be needed to prevent danger of damage from digging and backfilling. Brick and block masonry can also be treated but joints should be flush-pointed or finished with a sand/cement render. In all cases defects in existing surfaces should be made good. The final coat should be blinded with clean, sharp sand while still tacky. There are a number of proprietary brands on the market.

Geosynthetic clay liners

This product consists of a thin layer of naturally occurring bentonite clay sandwiched between two layers of extremely strong open-weave man-made geotextile material that are needle-punched together. This geosynthetic layer is placed on the external side of the basement wall or on the underside of the basement slab where it will become damp. When the bentonite becomes moist, it can swell up to 15 times its dry volume and form a dense, impervious waterproofing membrane. It can be fitted to the inside face of reinforced concrete wall formwork or pinned against the outside face of either a masonry or precast concrete wall prior to backfilling. The geosynthetic material comes in rolls and must be lapped 150 mm at horizontal and vertical junctions.

Cementitious slurries and layer coats

There are a variety of cementitious waterproof materials on the market. Some like slurries are used on building materials containing free lime, where they react to block cracks and capillaries. They remain active and self-seal any future hairline cracks. These can be applied internally or externally. Others containing additional waterproofing agents are supplied as a pre-mixed slurry and are applied as a thin single or multi-layer coats depending on the degree of protection required.

TYPE B–STRUCTURALLY INTEGRAL PROTECTION

This is achieved in concrete basement wall and floor construction by forming dense reinforced concrete from impervious aggregates and well-compacted concrete to form a barrier to water penetration. If a satisfactory water barrier is to be achieved, great care must be taken with the design of the mix, and the actual mixing and placing, together with careful selection and construction of the formwork. The design of reinforced concrete basements falls within the suite of Eurocodes, the most relevant to basement construction being:

- Eurocode 2: Design of concrete structures BS EN 1992;
- Eurocode 3: Design of steel structures BS EN 1993;
- Eurocode 7: Geotechnical design BS EN 1997.

Shrinkage cracking can be largely controlled by increasing the reinforcement or forming construction crack joints at regular intervals. These construction joints should provide continuity of reinforcement and, by the incorporation of a PVC or rubber variety of water bars, a barrier to the passage of water; typical examples are shown in Fig. 3.6.7.

- In Fig. 3.6.7 (a) there is a traditional uPVC or rubber bar placed externally where the construction joint is formed.
- In Fig. 3.6.7 (b) an internal water bar is fixed within the steel reinforcement cage; this can be made of metal or uPVC.
- In Fig. 3.6.7 (c) there can be either a hose system that allows the injection of acrylic resins into the cast joint, or sodium bentonite expansion seals, which swell by up to 350 per cent on contact with penetrating groundwater.

TYPE C – DRAINED PROTECTION

This method provides an excellent barrier to moisture penetration of basements by allowing any moisture that has passed through the structural wall to drain down within a cavity formed between the inner face of the structural wall and an inner non-loadbearing wall. This internal wall is built of a floor covering of special triangular precast concrete tiles, which allows the

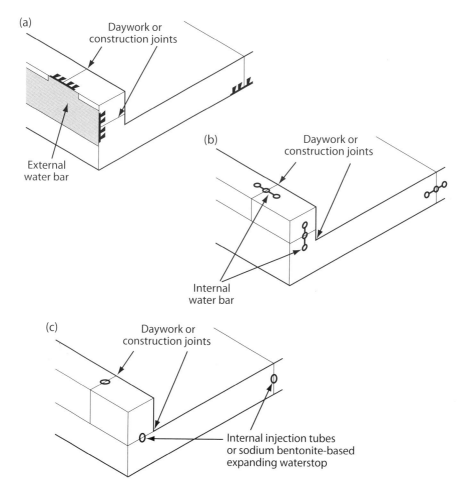

(a) Daywork or construction joints

External water bar

(b) Daywork or construction joints

Internal water bar

(c) Daywork or construction joints

Internal injection tubes or sodium bentonite-based expanding waterstop

Figure 3.6.7 Modern types of water bar used in RC basement construction.

moisture from the cavity to flow away under the tiles to a sump, where it is discharged into a drainage system either by gravity or by pumping. A simple illustration of this method is included in Fig. 3.6.4. This method of waterproofing is usually studied in detail during advanced courses in construction technology.

4

Superstructure

Brickwork 4.1

Brickmaking is a longstanding craft, traditionally used for making a hard-wearing, weather-resistant and durable material that can be economically produced in large quantities to a consistent quality. Bricks were introduced into the UK by the Romans, who fired clay bricks to produce a harder-wearing product. Large brick manufacturers still produce bricks in the UK today using just such methods.

The European CEN (European Committee for Standardisation) Standard Specification for clay masonry units, BS EN 771-1 (replacing the old standard, BS 3921) specifically relates to bricks manufactured from various materials for use in walling. Other standards exist for units produced from calcium silicate (BS 187) and concrete (BS 6073). A brick can be defined as a walling unit with co-ordinating or format size. They are normally produced in the UK to a standard size of 215 mm length, 102.5 mm width and 65 mm height; introducing a 10 mm joint gives the 'format size' that bricks are known by: 225 mm length, 112.5 mm width and 75 mm height. These are metric brick sizes, but a range of alternative sizes is available to match the old imperial bricks, for use when undertaking extensions or refurbishment work where the depth of the bricks is greater than 65 mm. The terms used for bricks and brickwork are shown in Figs. 4.1.1, 4.1.2 and 4.1.3.

Brickwork is used primarily in the construction of walls by the bedding and jointing of bricks into established brick bonding arrangements. The majority of the bricks used today are made from clay or shale conforming to the requirements of BS EN 771-1.

MANUFACTURE OF CLAY BRICKS

The basic raw materials are clay or shale, both of which are in plentiful supply in this country. The raw material is excavated, and then prepared either by

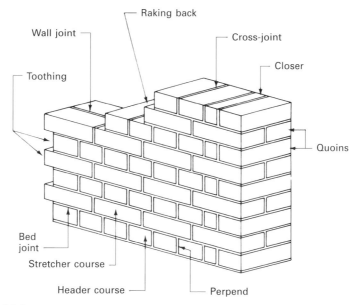

Figure 4.1.1   Brickwork terminology.

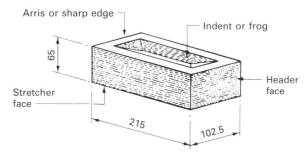

Figure 4.1.2   Standard brick.

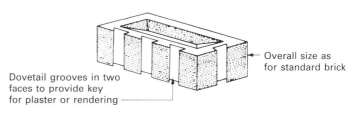

Figure 4.1.3   Keyed brick.

weathering or by grinding before being mixed with water to the right plastic condition. It is then formed into the required brick shape before being dried and fired in a kiln.

The use of different clays and surface textures creates a product that has a great deal of variety in its composition, colour and properties. This then lends the product to be used in a variety of applications, from heat resistance in kiln lining through to high-strength engineering applications.

PRESSED BRICKS

This is the type of brick most commonly used, accounting for nearly 60 per cent of the two billion produced on average in this country each year. There are two processes of pressed brick manufacture: the semi-dry method and the stiff plastic method.

By far the greatest number of bricks is made by the **semi-dry pressed process** and the bricks are called **flettons**; these form over 40 per cent of the total brick production in Britain. The name originates from the village of Fletton outside Peterborough, where the bricks were first made. This process is used for the manufacture of bricks from the Oxford clays, which have a low natural plasticity. The clay is ground, screened and pressed directly into the moulds.

The **stiff plastic process** is used mainly in Scotland, the north of England and south Wales. The clays in these areas require more grinding than the Oxford clays, and the clay dust needs tempering (mixing with water) before being pressed into the mould.

All pressed bricks contain **frogs**, which reduce the weight of the brick and make it more economical to produce. In general, pressed bricks are more accurate in shape than other clay bricks, with sharp arrises and plain faces.

WIRE-CUT BRICKS

Approximately 28 per cent of bricks produced in Britain are made by this process. The clay, which is usually fairly soft and of a fine texture, is extruded as a continuous ribbon and is cut into brick units by tightly stretched wires spaced at the height or depth for the required brick. Allowance is made during the extrusion and cutting for the shrinkage that will occur during firing. Wire-cut bricks do not have frogs, and on many the wire-cutting marks can be clearly seen.

SOFT MUD PROCESS BRICKS

This process is confined mainly to the south-eastern counties of England, where suitable soft clays are found. The manufacture can be carried out by machine or by hand, either with the natural clay or with a mixture of clay and lime or chalk. In both methods the brick is usually frogged, and it is less accurate in shape than other forms of brick. Sand is generally used in the moulds to enable the bricks to be easily removed, and this causes an uneven patterning or creasing on the face.

Bricks are a diverse and varied product: therefore they are difficult to standardise for classification. Under the EN classification EN 771-1, they can be specified in terms of:

- density (low and high);
- compressive strength;
- thermal properties;
- durability;
- water absorption and movement;
- salt content.

The following sections give some common terminology.

VARIETIES

- **Common** Suitable for general building work where appearance is not a critical element.
- **Facing** Specially made or selected to have an attractive appearance; used as the outer skin of external walls.
- **Engineering** Having a dense and strong semi-vitreous body conforming to defined limits for absorption and strength; used in applications of high moisture.

TYPES

- **Solid** A solid mass unit with no frogs or holes.
- **Frogged** In which frogs do not exceed 20 per cent of the brick's volume.
- **Perforated** With holes passing through it that are vertically perforated.

OTHER CLASSIFICATIONS

Bricks may also be classified by one or more of the following:

- place of origin – for example, London Brick Company (LBC);
- raw material – for example, clay;
- manufacture – for example, wire-cut;
- use – for example, foundation;
- colour – for example, blue;
- surface texture – for example, sand-faced.

CALCIUM SILICATE BRICKS

These bricks are also called sandlime or sometimes flintlime bricks, and are covered by BS 187 EN 771-2, which classifies the brick by compressive strength or density. They are made from carefully selected clean sand and/or crushed flint mixed with controlled quantities of lime and water. At this stage, colouring pigments can be added if required to provide variation in colour; the relatively dry mix is then fed into presses to be formed into the required shape. The moulded bricks are then hardened in sealed and steam-pressurised autoclaves. This process, which takes 7–10 hours, causes a reaction between the sand and the lime, resulting in a strong homogeneous brick that is ready for immediate delivery and laying. The bricks are very accurate in size and shape, but do not have the individual characteristics of clay bricks.

CONCRETE BRICKS

These are made from a mixture of inert aggregate and cement in a similar fashion to calcium silicate bricks, and are cured either by natural weathering or in an autoclave. Details of the types and properties available as standard concrete bricks are given in BS EN 771-3. They are sometimes used in high-strength areas where cuts are required to make up levels: for example, beneath precast concrete floor beams.

MORTARS FOR BRICKWORK

The mortar used in brickwork transfers the tensile, compressive and shear stresses uniformly between adjacent bricks, thus spreading the loads. To do this it must satisfy certain requirements; please see the list below.

- It must have adequate consistent strength, but not greater than that required for the design strength.
- It must have good workability to aid bricklaying.
- It must retain plasticity long enough for the bricks to be laid and should not set immediately but cure over a period.
- It must be durable over a long period to prevent uneven weathering of the brick face.
- It must bond well to the bricks, maintaining strength.
- It must be able to be produced at an economic cost.

If the mortar is weaker than the bricks, shrinkage cracks will tend to follow the joints of the brickwork, and these are reasonably easy to make good. If the mortar is stronger than the bricks, shrinkage cracks will tend to be vertical through the joints and the bricks, thus weakening the fabric of the structure.

MORTAR MIXES

Mortar is a mixture of sand and lime or a mixture of sand and cement with or without lime. Nowadays lime is only used where conservation is required within a graded and listed structure. Proportioning of the materials prior to mixing it is best undertaken by weight as this is more accurate. The effect of the lime is to make the mix more workable, but as the lime content increases, the mortar's resistance to damage by frost action decreases.

Plasticisers, which are additives that trap small bubbles of air in the mix and cause the breaking down of surface tension, will also increase the workability of a mortar.

Mortars should never be re-tempered, and should be used within two hours of mixing or be discarded.

The specification of mortars is now governed by BS EN 998-2.

Mortar can be classified in two ways: by its properties and use, or by its mode of manufacture.

Properties and uses

- General purpose masonry mortar (G) – a mortar that does not have any specified characteristics.
- Thin layer masonry mortar (T) – mortar with a specified aggregate size.
- Lightweight masonry mortar (L) – a low-density mortar.

Mode of manufacture

- Factory-made masonry mortar.
- Semi-finished factory-made masonry mortar.
- Pre-batched masonry mortar.
- Pre-mixed lime-sand masonry mortar.
- Site-made masonry mortar.

DAMPNESS PENETRATION

It is possible for moisture to penetrate into a building through single-skin brick walls by one or more of three ways:

1. by the rain penetrating the top of the wall and soaking down into the building below the roof level;
2. by ground moisture entering the wall at or near the base, moving up the wall by capillary action, and entering the building above the ground floor level;
3. by the rain landing on the external wall and soaking through the fabric and joints into the building.

The first two ways can be overcome by the insertion of a suitable damp-proof course across the thickness of the wall. The other can be overcome by one of two methods:

- applying a barrier to the exposed face of the wall a barrier, such as cement rendering, or some suitable cladding, such as vertical tile hanging;
- constructing a cavity wall, whereby only the external skin becomes damp, the cavity providing a suitable barrier to the passage of moisture through the wall.

DAMP-PROOF COURSES

A damp-proof course is a physical barrier that prevents moisture crossing its path and entering a building, causing damage and disturbance to its occupants. Damp-proof courses may be inserted either horizontally or vertically, and can generally be divided into three two types:

- those below ground level to prevent the entry of moisture from the soil;
- those placed 150 mm above ground level to prevent moisture moving up the wall by capillary action (sometimes called **rising damp**);
- those placed at openings, parapets and similar locations to exclude the entry of the rainwater that falls directly onto the fabric of the structure.

Materials for damp-proof courses

BS 743: *Specification for materials for damp-proof courses* (partially superseded) indicates suitable materials for the construction of damp-proof courses, all of which should be:

- completely impervious;
- durable, having a longer life than the other components in the building and therefore not needing replacing during its lifetime;
- in comparatively thin sheets so as to prevent disfigurement of the building;
- strong enough to support the loads placed upon it without exuding from the face of the wall;
- flexible enough to give with any settlement of the building without fracturing.

The following also apply:

- BS 6398: *Specification for bitumen damp-proof courses for masonry;*
- BS 6515: *Specification for polyethylene damp-proof courses for masonry;*
- BS 8215: *Code of practice for design and installation of damp-proof courses in masonry construction.*

The following sections look at the most common materials that are specified for a damp-proof course.

Polyethylene

Black low-density polythene sheet of single thickness not less than 0.5 mm thick should be used; it is easily laid but can be torn and punctured easily. Polythene DPC is often specified as 1000, 1200, 1500 or 2000 gauge, with 1200 gauge being standard.

Engineering bricks

These should comply with the requirements of EN 771 (specification for clay masonry units) 'engineering' classification. They are laid in two courses in cement mortar, and may contrast with the general appearance of other brickwork in the same wall. However, a damp-proof course should be used in conjunction with the engineering brick.

BRICKWORK BONDING

When building with bricks, it is necessary to lay the bricks to some recognised pattern or bond in order to ensure stability of the structure and to produce a pleasing appearance. All the various bonds are designed so that no vertical joint in any one course is directly above or below a vertical joint in the adjoining course. To simplify this requirement, special bricks are produced to BS 4729: *Specification for dimensions of bricks of special shapes and sizes*. Alternatively, the bricklayer can cut from whole bricks on site, using a water-table saw or petrol-driven hand saw. Application of some of these special bricks is shown in Fig. 4.1.4, with some additions in Fig. 4.1.5. The various bonds are also planned to give the greatest practical amount of lap to all the bricks, and this should not be less than a quarter of a brick length. Properly bonded brickwork distributes the load over as large an area of brickwork as possible, so that the angle of spread of the load through the bonded brickwork is 45°.

COMMON BONDS

- **Stretcher bond** Consists of all stretchers in every course and is used for half-brick walls and the half-brick skins of hollow or cavity walls (see Fig. 4.1.6).

- **English bond** A very strong bond consisting of alternate courses of headers and stretchers (see Fig. 4.1.7).

- **Flemish bond** Each course consists of alternate headers and stretchers; its appearance is considered to be better than English bond, but it is not quite so strong. This bond requires fewer facing bricks than English bond, needing only 79 bricks per square metre as opposed to 89 facing bricks per square metre for English bond. This bond is sometimes referred to as **double Flemish bond** (see Fig. 4.1.8).

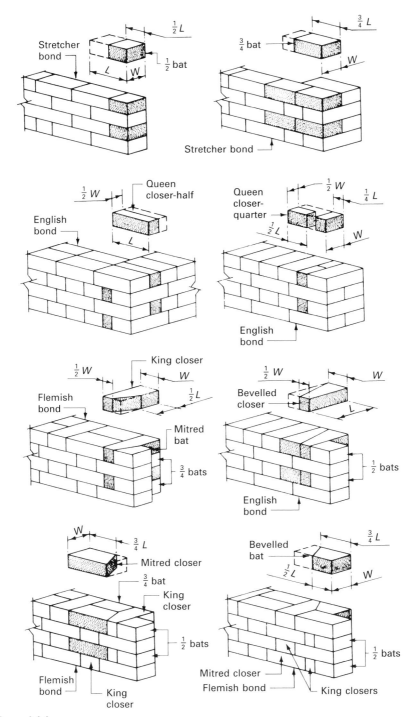

Figure 4.1.4   Special bricks.

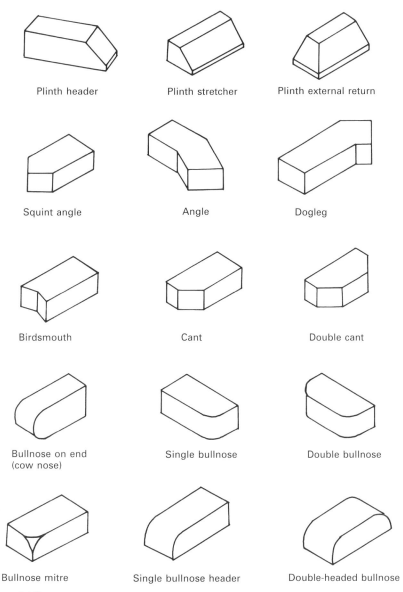

Plinth header	Plinth stretcher	Plinth external return
Squint angle	Angle	Dogleg
Birdsmouth	Cant	Double cant
Bullnose on end (cow nose)	Single bullnose	Double bullnose
Bullnose mitre	Single bullnose header	Double-headed bullnose

Figure 4.1.5 Purpose-made bricks.

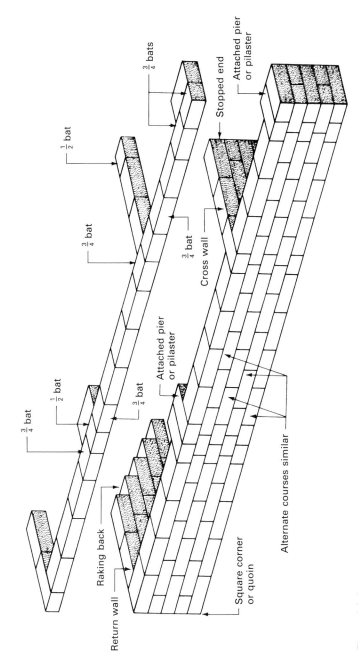

Figure 4.1.6 Typical stretcher bond details.

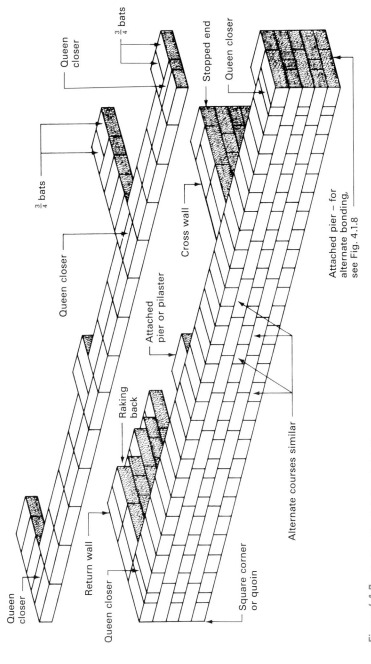

Queen closer

Queen closer

Return wall

Queen closer

Square corner
or quoin

Raking
back

Attached
pier or pilaster

$\frac{3}{4}$ bats

Queen closer

Cross wall

Alternate courses similar

Queen
closer

$\frac{3}{4}$ bats

Stopped end

Queen closer

Attached pier – for
alternate bonding,
see Fig. 4.1.8

Figure 4.1.7 Typical English bond details.

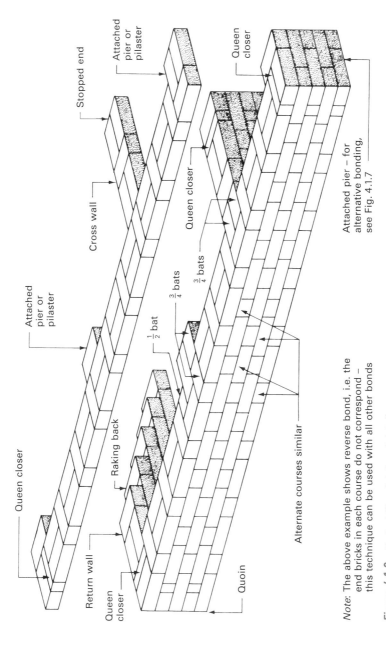

Queen closer

Attached pier or pilaster

Stopped end

Attached pier or pilaster

Queen closer

Cross wall

Queen closer

Return wall

Raking back

$\frac{3}{4}$ bats

Queen closer

$\frac{1}{2}$ bat

$\frac{3}{4}$ bats

Attached pier – for alternative bonding, see Fig. 4.1.7

Quoin

Alternate courses similar

Note: The above example shows reverse bond, i.e. the end bricks in each course do not correspond – this technique can be used with all other bonds

Figure 4.1.8 Typical Flemish bond details.

- **Single Flemish bond** A combination of English and Flemish bonds, having Flemish bond on the front face with a backing of English bond. It is considered to be slightly stronger than Flemish bond. The thinnest wall that can be built using this bond is a one-and-a-half brick wall.

- **English garden wall bond** Consists of three courses of stretchers to one course of headers.

- **Flemish garden wall bond** Consists of one header to every three stretchers in every course; this bond is fairly economical in facing bricks and has a pleasing appearance.

SPECIAL BONDS

- **Rat-trap bond** A brick-on-edge bond that gives a saving on materials and lodings; suitable as a backing wall to a cladding such as tile hanging (see Fig. 4.1.9).

- **Quetta bond** Used on one-and-a-half brick walls for added strength; suitable for retaining walls (see Fig. 4.1.9).

- **Stack bond** A brickwork feature used for partitions and infill panels. Bricks may be laid on bed or end with continuous vertical joints. Fig. 4.1.10 shows steel-mesh-reinforced horizontal bed joints to compensate for the lack of conventional bond.

BOUNDARY WALLS

These are subjected to exposed weather conditions and therefore require correct design and construction. If these walls are also acting as a retaining wall structure, the conditions will be even more extreme, but the main design principle of the exclusion of water remains the same. The presence of water in brickwork can lead to frost damage, mortar failure and efflorescence. The incorporation of adequate damp-proof courses and overhanging throated copings is of the utmost importance in this form of structure to throw and direct water off the brickwork (see Fig. 4.1.11).

Efflorescence

This is a white stain appearing on the face of brickwork, caused by deposits of soluble salts formed on or near the surface of the brickwork as a result of evaporation of the water in which they have been dissolved. It is usually harmless, and disappears within a short period of time; brushing dry or with clean water may be done to remove the salt deposit.

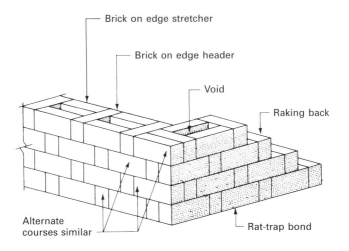

Brick on edge stretcher

Brick on edge header

Void

Raking back

Alternate
courses similar

Rat-trap bond

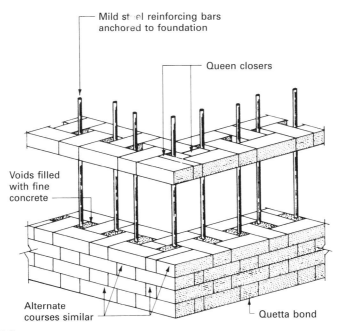

Mild steel reinforcing bars
anchored to foundation

Queen closers

Voids filled
with fine
concrete

Alternate
courses similar

Quetta bond

Figure 4.1.9 Special bonds.

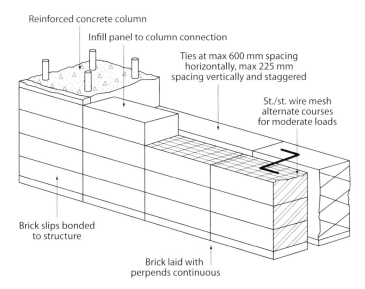

Figure 4.1.10   Reinforced stack bond.

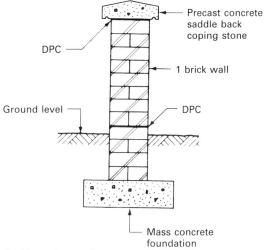

Figure 4.1.11   Typical boundary wall.

FEATURE BRICKWORK

Repetitive brickwork can be visually very monotonous. Some relief can be provided with coloured mortars and the method of pointing the mortar joint while wet. Greater opportunity for artistic expression is achieved by varying the colours and textures of the bricks to produce patterns and images on a wall. A further dimension is obtained by using cut and purpose-made bricks as projecting features. Some possible applications are shown in Figs. 4.1.12–4.1.16.

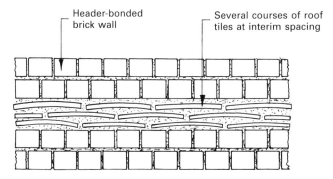

Header-bonded brick wall

Several courses of roof tiles at interim spacing

Figure 4.1.12 Lacing course (see also Fig. 4.1.7).

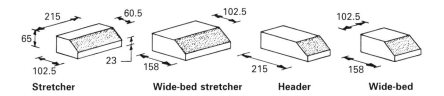

215 60.5 102.5 102.5

65 23 158 215 158

102.5

Stretcher **Wide-bed stretcher** **Header** **Wide-bed**

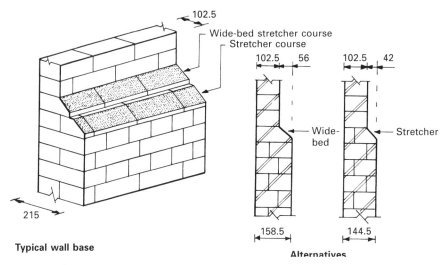

102.5

Wide-bed stretcher course
Stretcher course

102.5 56 102.5 42

Wide-bed

Stretcher

215

158.5 144.5

Typical wall base

Alternatives

Figure 4.1.13 Plinth brickwork.

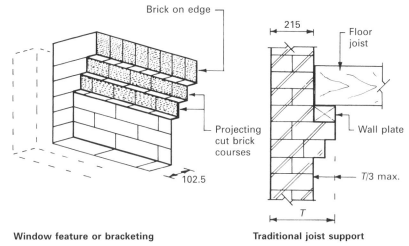

Window feature or bracketing

Traditional joist support

Figure 4.1.14 Corbelling.

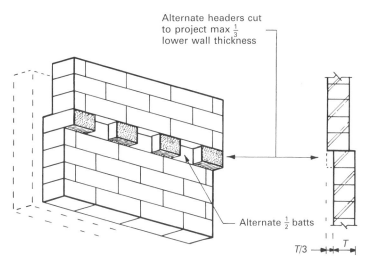

Figure 4.1.15 Dentil coursing.

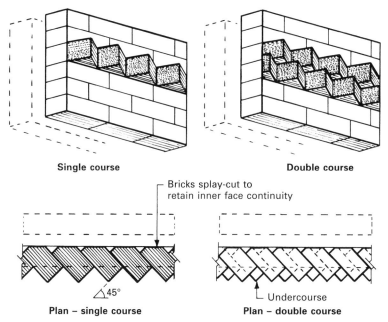

Single course

Double course

Bricks splay-cut to
retain inner face continuity

△45°

Plan – single course

Undercourse

Plan – double course

Figure 4.1.16   Dogtooth coursing.

Blockwork   4.2

A block is essentially a unit that is generally six times larger than a brick's dimensions. It is manufactured from a thermally efficient concrete mix that entrains air bubbles into the process; for higher load-bearing applications, blocks are manufactured from stronger concrete mixes. Blocks are a more economical alternative to brickwork and are used in locations where their appearance will be covered up by a secondary finish.

PRECAST CONCRETE BLOCKS

The specification and manufacture of precast concrete blocks or masonry units is covered in BS EN 771-3:2003 and BS EN 772-2:1998 BS 6073. In the UK, classification is often also by strength (for example, 3.5 and 7 N/mm^2) and by size, which is typically 440 mm in length x 215 mm in height. The width of blocks varies with their application; standard commercial thicknesses are 100, 150 and 200 mm.

The density of a precast concrete block gives an indication of its compressive strength – the greater the density, the stronger the block. Similarly the lower the density, the lower the block's capacity for thermal transfer: trapped air reduces the transfer of heat and increases the thermal properties of a block. The reduction of sound through a block relies on its density: the greater the density, the better its sound-reduction qualities.

The actual properties of different types of precast concrete block can be obtained from manufacturers' literature together with their appearance classification, such as plain, facing or paint grade.

Aerated concrete for blocks is produced by introducing air or gas into the mix so that, when set, a uniform cellular block is formed. The usual method employed is to introduce a controlled amount of fine aluminium powder into the mix; this reacts with the free lime in the cement to give off hydrogen, which is quickly replaced by air and so provides the aeration.

Typical details are shown in Fig. 4.2.1. The use of proprietary cavity closers aids thermal efficiency.

Concrete blocks are laid in what is essentially stretcher bond, and joined to other walls by block bonding or by leaving metal ties or strips projecting from

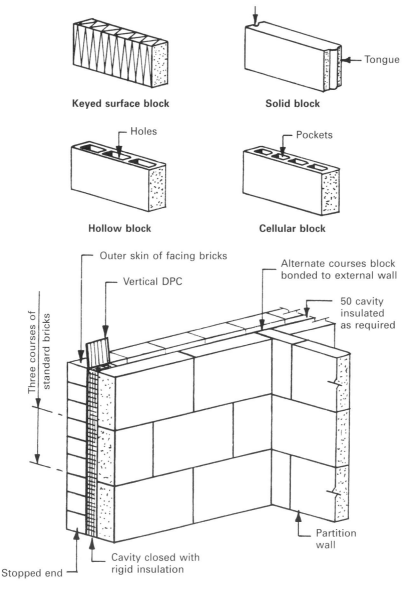

Figure 4.2.1 Precast concrete blocks and brickwork.

suitable bed courses. As with brickwork, the mortar used in blockwork should be weaker than the material of the walling unit; generally a 1:6 cement/sand mix will be suitable for work above ground level.

Concrete blocks shrink on drying out: therefore they should not be laid until the initial drying shrinkage has taken place (this is about 14 days under normal drying conditions), and should be protected on site to prevent them from becoming wet, expanding and causing subsequent shrinkage, possibly resulting in cracking of the blocks and any applied finishes such as plaster. To aid this, blocks are normally delivered shrink-wrapped on pallets.

Where the length of a wall exceeds 6 m or thereabouts, it is necessary to incorporate vertical movement joints. These can comprise a stainless steel or galvanised steel former or a flexible compressible board. Ties must be incorporated across the joint, with one end debonding to allow movement. Fig. 4.2.2 shows the installation with bonded profiled or perforated ties to one side and plastic-sleeved ties to the other to maintain continuity and facilitate movement. If movement joint filler strips are used, the joint would need pointing with a polysulphide sealant that can expand and contract.

The main advantages of blockwork over brickwork are:

- labour-saving – easy to cut, larger units;
- easier fixings – most take direct fixing of screws and nails;
- higher thermal insulation properties;
- lower density;
- provides a suitable key for plaster and cement rendering.

The main disadvantages are:

- lower strength;
- less resistance to rain penetration;
- load-bearing properties less (one- or two-storey application);
- lower sound insulation properties.

THIN JOINT MASONRY

Thin joint masonry blocks are high-quality dimensioned blocks that are normally 610 mm long and available in a range of heights and widths. They enable large areas of blockwork to be installed efficiently using a cement-based bonding mortar 2 mm thick. They offer several advantages:

- faster construction;
- set time of approximately two hours;
- thermal efficiency;
- reduction in programme times;
- less waste in mortar;
- easily cut.

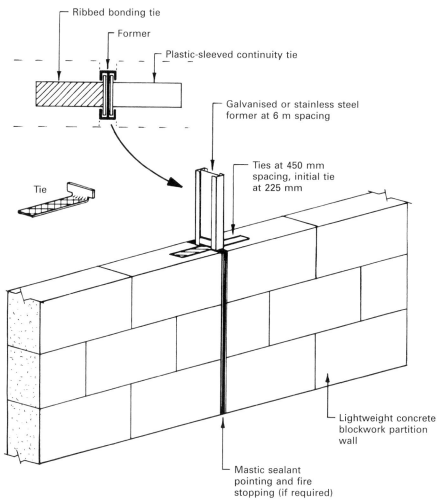

Ribbed bonding tie

Former

Plastic-sleeved continuity tie

Galvanised or stainless steel former at 6 m spacing

Tie

Ties at 450 mm spacing, initial tie at 225 mm

Lightweight concrete blockwork partition wall

Mastic sealant pointing and fire stopping (if required)

Figure 4.2.2   Blockwork movement joint (BS 5628: Part 3: *Use of masonry*).

Cavity walls 4.3

A wall constructed in two leaves or skins with a space or cavity between them is called a **cavity wall**, and it is the most common form of external wall used in domestic building today. A cavity wall is a traditional method of construction. The main purpose of constructing a cavity wall is to prevent the penetration of rain to the internal surface of the wall. It is essential that the cavity is not bridged or blocked in any way, as this would provide a passage for the moisture to migrate across.

Modern cavity wall construction now often requires that the cavity is fully filled with insulation in order to meet the requirements of the Building Regulations Part L. Traditionally air bricks have been used to ventilate the cavity. However, modern insulation sheets and fibres now provide a solution to include within the cavity wall, clipped to the wall ties or allowed to fully expand and fill a cavity.

The main consideration in the construction of a cavity wall above ground-level damp-proof course is the choice of a brick or block that will give the required durability, strength and aesthetic appearance, and also conform to Building Regulations requirements for loading and stability. The main function of the wall below ground-level damp-proof course is to transmit the load safely to the foundations; in this context, the two half-brick or block leaves forming the wall act as retaining walls. To aid stability and strength, the cavity below ground level is filled with a weak-mix concrete that is sloped to the outside at the top, so any moisture can be drained towards the outside through weep holes left in the brickwork (see Fig. 4.3.1).

A modern approach is to use trench blocks. These are concrete blocks constructed to the full width of the cavity wall that are laid in stretcher bond up to 150 mm below ground level. From this point up to the DPC, the external wall is constructed using engineering brickwork. The use of this material at the junction of the wall and the ground provides a suitably durable, dense and weather-resistant function at a point where rainwater will bounce back against the outside of the wall.

Parapets, whether solid or of cavity construction, are exposed to the elements on three sides and need careful design and construction. These are elements that you would find on traditional Victorian properties. Because the parapet is exposed on three sides, adequate barriers must be included to prevent the ingress of moisture down the wall and into the building. A DPC is required to prevent moisture capillary action down the wall under the force of gravity, along with a parapet cap and an integral gutter that is tucked into the DPC. A solid parapet wall should not be less than 150 mm thick and not less than the thickness of the wall on which it is carried, and its height should not exceed four times its thickness. The recommended maximum heights of cavity wall parapets are shown in Fig. 4.3.2.

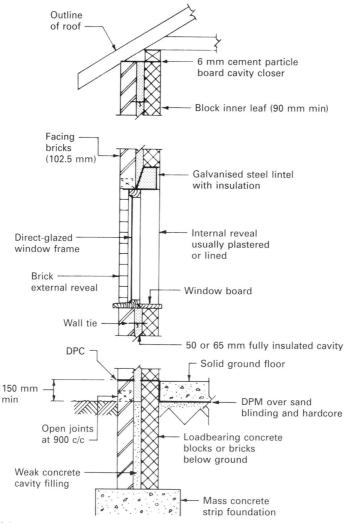

Figure 4.3.1 Typical cavity wall details.

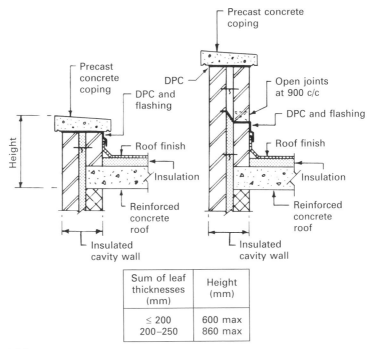

Sum of leaf thicknesses (mm)	Height (mm)
≤ 200	600 max
200–250	860 max

Figure 4.3.2 Parapets.

BUILDING REGULATIONS

Regulation A1 requires that a building shall be so constructed that the combined dead, imposed and wind loads are sustained and transmitted to the ground safely and without causing any movement that will impair the stability of any part of another building. Guidance to meet these requirements for cavity walls is given in Approved Document A. Section 2C of this document considers full storey-height cavity walls for residential buildings of up to three storeys and makes a number of requirements.

1. The compressive strengths of bricks and blocks should be not less than 7.5 N/mm² (Group 1) and 11 N/mm² (Group 2) respectively for buildings up to two storeys. Greater strength classifications are required for three-storey buildings (see Approved Document A, Table 7, Diagram 9). Group 1 masonry units have not more than 25 per cent formed voids (20 per cent for frogged bricks) and Group 2 masonry units have formed voids greater than 25 per cent but not exceeding 55 per cent.

2. Cavities (gaps between masonry faces) should be at least 50 mm, but may be up to 300 mm. Both leaves in cavity walls should have ties embedded at least 50 mm into adjacent masonry.

3. Wall ties should comply with BS EN 845-1: *Specification for ancillary components for masonry. Ties, tension straps, hangers and brackets.* Plastic and galvanised steel ties have been used, but now the preferred material is austenitic stainless steel. Maximum spacings are shown in Fig. 4.3.3.

4. Cavity walls normally have skins that are standard with the brick and block manufacturer's dimensions, which are 102 mm for a brick and 100 mm for a block.

5. The combined thickness of the two leaves of a cavity wall should be not less than 190 mm for a maximum wall height of 3.5 m and length not exceeding 12 m, and also for heights between 3.5 m and 9 m in wall lengths not exceeding 9 m. Wall lengths and heights up to 12 m require a minimum thickness of 290 mm.

Table 4.3.1 **Minimum thicknesses of certain external walls, separating and compartment walls.**

Wall height	Wall length	Wall thickness
12 m	9 m	290 mm for one storey
		190 mm for remainder
12 m	12 m	290 mm for two storeys
		190 mm for remainder

The current width of a wall for a new building equates to 300 mm wide, which will satisfy the requirements of Part A of the Building Regulations.

6. Mortar should be specified using BS EN 998-1:2003 BS 5628-1: *Structural use of unreinforced masonry*, which covers the specification for mortar for masonry, rendering and plastering mortar or a gauged mortar mix of 1:1:6 by volume, or its equivalent.

7. Cavity walls of any length need to be provided with roof lateral support. Walls over 3 m length will also require floor lateral support at every floor forming a junction with the supported wall. If roof lateral support is not provided by type of covering (tiles or slates), a pitch of 15° or more plus a minimum wall plate bearing of 75 mm, durable metal straps with a minimum cross-section of 30 mm × 5 mm will be needed at not more than 2 m centres (see Fig. 6.1.9, page 263). If the floor does not have at least a 90 mm bearing on the supported wall, lateral support should be provided by similar straps at not more than 2 m spacing, or the joists should be fixed using restraint-type joist hangers (see Fig. 4.4.1, page 169).

PREVENTION OF DAMP IN CAVITY WALLS

Approved Document C recommends a cavity to be carried down at least 150 mm below the lowest damp-proof course, and that any bridging of the cavity, other than a wall tie or closing course protected by the roof, is to have

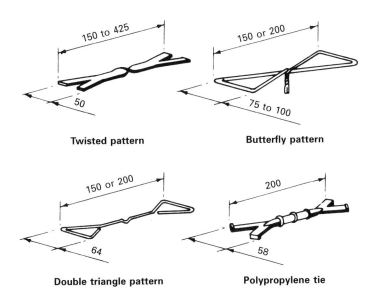

Twisted pattern

Butterfly pattern

Double triangle pattern

Polypropylene tie

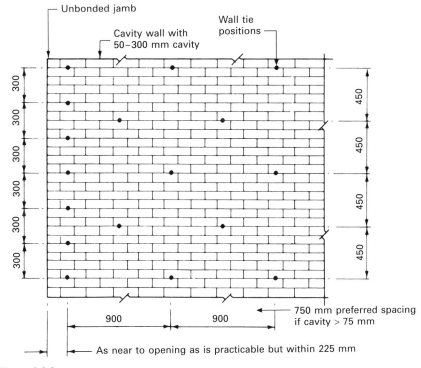

Figure 4.3.3   Wall ties.

a suitable damp-proof course to prevent the passage of moisture across the cavity. Where the cavity is closed at the jambs of openings a vertical damp-proof course should be inserted. The use of proprietary cavity closers that incorporate a DPC and an insulated cavity closer are a current modern method of closing the cavity at an opening point.

Approved Document C recommends a damp-proof course to be inserted in all external walls at least 150 mm above the highest adjoining ground or paving to prevent the passage of moisture rising up the wall and into the building, unless the design is such that the wall is protected or sheltered.

ADVANTAGES AND DISADVANTAGES

ADVANTAGES OF CAVITY WALL CONSTRUCTION

These can be listed as follows:

- able to prevent a driving rain in all situations from penetrating to the inner wall surface;
- gives good thermal insulation, keeping the building warm in winter and cool in the summer;
- enables the use of cheaper and alternative materials for the inner construction;
- increased sound insulation;
- width provides good lateral stability.

DISADVANTAGES OF CAVITY WALL CONSTRUCTION

These can be listed as follows:

- requires a high standard of design and workmanship to produce a soundly constructed wall;
- the need to include a vertical damp-proof course to all openings;
- longer construction period than modern methods of construction (MMC);
- weather-dependent during its construction;
- scaffolding required to internal skin.

JOINTING AND POINTING

These terms are used for the finish given to both the vertical and the horizontal joints in brickwork irrespective of whether the wall is of brick, block, solid or cavity construction.

Jointing is the method of completing the mortar joint on the external face of the brick or block.

Pointing is the finish given to the joints by raking out to a depth of approximately 20 mm and filling in on the face with a hard-setting cement mortar, which could have a colour additive. This process can be applied to both new and old buildings. Typical examples of jointing and pointing are shown in Fig. 4.3.4.

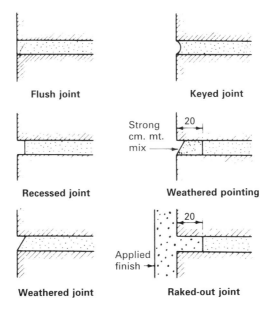

Figure 4.3.4   Brick joints.

THERMAL BRIDGING

Thermal bridging in a cavity wall can lead to the introduction of condensation and pattern staining on the internal decorated surfaces of the building. It occurs where there is an inconsistency of construction of the cavity wall. This can occur through several areas such as:

- bridging of the cavity by mortar dropping onto and setting on the wall ties;
- using metal lintels which are not insulated internally;
- weaknesses in construction around jambs of openings;
- gaps in the coverage of cavity insulation;
- using dense concrete lintels;
- any failure in an element that crosses the cavity.

This results in variations in thermal transmittance values as one material transmits heat differently to another. Where this does occur, the thermal or cold bridge will incur a cooler surface temperature than the remainder of the wall, greater heat loss at this point and a possible dew-point temperature manifesting in water droplets or condensation. The moisture will attract dust and dirt to contrast with adjacent clean areas.

Apart from examples of bad practice, such as using bricks to make up coursing in lightweight insulating concrete block inner leaf walls, the most common areas for thermal bridging are around door and window openings, at junctions between ground floor and wall, and the junction between wall and roof. These are shown in Fig. 4.3.5.

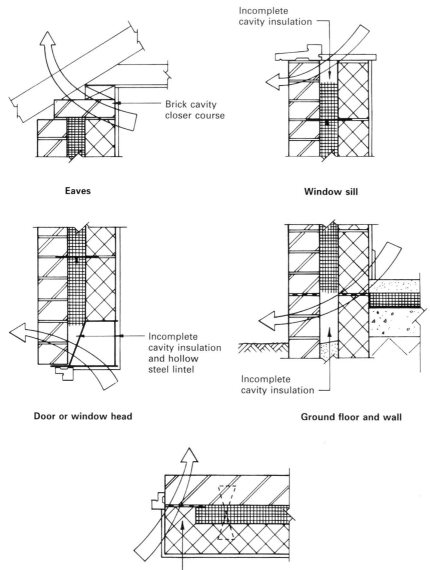

Figure 4.3.5 Thermal bridging – areas for concern.

Attention to detail at construction interfaces is the key to preventing thermal bridging. The details shown in Fig. 4.3.6 incorporate the necessary features to promote good practice.

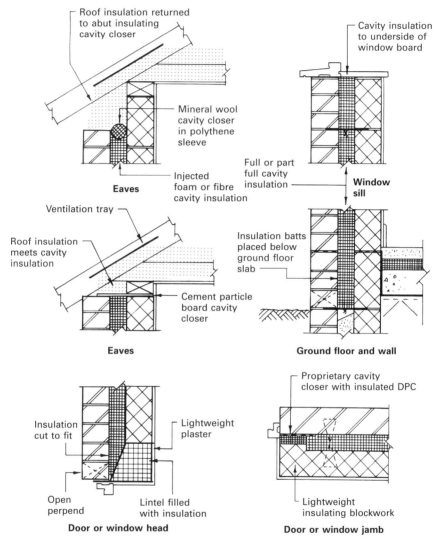

Figure 4.3.6 Thermal bridging – treatment at openings and junctions.

Lateral restraint 4.4

Walls have been designed such that their mass, together with the loads carried from floors and roofs, provides adequate resistance to buckling. Buckling can be controlled using buttresses or piers, but these may encroach on space and be visually unattractive. Contemporary design favours more slender construction: therefore stability and resistance to tensile lateral forces such as wind, and possibly impact, is achieved by using support from the floor and roof (see Part 6 on page 263).

Floor structures function as a stable horizontal platform supported by perimeter and intermediate walls. They complement that support by providing restraint and stability for the wall against lateral overturning forces, in conjunction with adequate structural connections. Connections to external, compartment (fire), separating (party) and internal loadbearing walls in houses where walls exceed 3 m length may be achieved by:

- 90 mm minimum direct bearing of the floor by timber joists (see Fig. 5.3.4, page 230), or from a continuous concrete slab;

- BS EN 845-1 approved galvanised steel restraint-type joist hangers with a minimum 100 mm bearing, spaced at no more than 2 m centres;

- galvanised steel purpose-made straps of minimum cross-section 30 mm × 5 mm, spaced at no more than 2 m, secured to the inner leaf of masonry and over at least three joists;

- galvanised steel straps, as above, applied to adjacent floors; continuity of contact between the floor and wall is also acceptable.

Note: The first two items assume that joist spacing does not exceed 1.2 m and the house is not over two storeys. If over two storeys, supplementary strapping should be provided to joists in the longitudinal direction, as shown in Fig. 4.4.1.

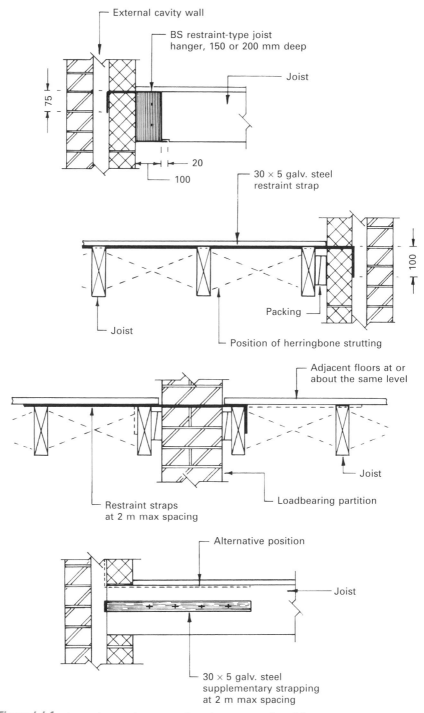

External cavity wall

BS restraint-type joist
hanger, 150 or 200 mm deep

Joist

75

20

100

30 × 5 galv. steel
restraint strap

100

Packing

Joist

Position of herringbone strutting

Adjacent floors at or
about the same level

Joist

Restraint straps
at 2 m max spacing

Loadbearing partition

Alternative position

Joist

30 × 5 galv. steel
supplementary strapping
at 2 m max spacing

Figure 4.4.1   Lateral restraint connections between wall and floor.

As shown in Fig. 4.4.1, lateral restraint from floors is necessary to stabilise adjacent walls (see Fig. 6.1.9, page 263). Many older buildings (mainly housing) have been constructed without strapping. Consequently, as these buildings age, there can be a tendency for their walls to show signs of outward bulging. Walls rarely bulge inwards, because the effect of thermal movement from the sun causes the outer surface to expand. Also, floor loads will tend to push outwards rather than inwards, particularly if there is an inner leaf carrying a greater load. Wall bulging can also be caused where wall ties have failed because of corrosion.

Remedial measures

- Traditional through-tie (Fig. 4.4.2).
- Retro-strap (Fig. 4.4.3).
- Retro-stud (Fig. 4.4.4).
- Driven resin-fixed screw (Fig. 4.4.5).
- Self-tapping helix (Fig. 4.4.6).

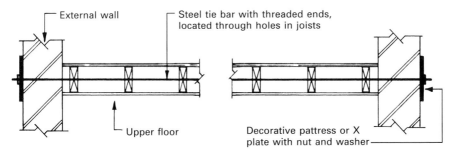

Figure 4.4.2   Traditional through-tie.

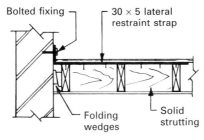

Figure 4.4.3   Retro-strap.

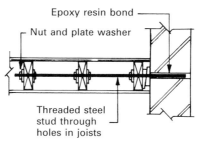

Figure 4.4.4   Retro-stud.

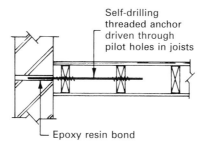

Figure 4.4.5   Driven resin-fixed screw.

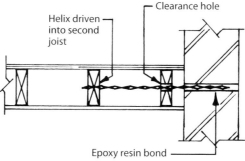

Figure 4.4.6   Self-tapping helix.

WALL TIE DECAY

Wall ties should be specified as manufactured from corrosion-resistant stainless steel. Historically, they have been produced from iron or steel, at best galvanised with a coating of zinc. Iron and steel ties can corrode and break, particularly where the zinc coating to steel is of poor quality, or it has been damaged during construction. Ties will have greater exposure to corrosion where pointing has broken down, and where old lime mortars have softened by decomposition due to atmospheric pollutants and frost damage.

Inevitably, there will be loss of stability between the two leaves of masonry wall, creating areas of outward bulging from the outer leaf.

A solution is to demolish the outer leaf, one small area at a time, and rebuild it complete with new ties. A simpler, less dramatic approach is to apply chemical cavity stabilisation. The material used is structural polyurethane foam of closed cell characteristic. The polyurethane is created by on-site mixing of an isocyanate with a resin to produce a liquid for injection by hand gun through 12 mm-diameter holes. Holes are pre-drilled in the outer leaf at intervals as shown in Fig. 4.4.7. Within the cavity the fluid mix reacts and expands to produce a rigid foam mass. This foam combines as an adhesive agent and as an insulant.

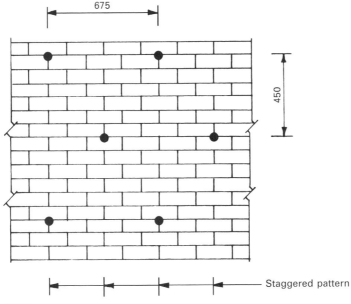

Staggered pattern

Notes:
1. At door and window openings, hole spacing is 300 mm vertically (200 mm horizontally from openings) and 400 mm horizontally above the opening.
2. At party walls, vertical holes are placed at 300 mm spacing, 300 mm from the party wall line.
3. Any unsleeved air bricks are replaced with sleeved, to prevent foam from leakage to where it is not required, e.g. under suspended floors.
4. At upper floor level supplementary stainless steel anchors are positioned at 1,000 mm horizontal intervals.

Figure 4.4.7   Standard spacing of polyurethane injection holes.

Forming openings in walls 4.5

Openings have to be formed within external walls for the inclusion of doors and windows. Doors allow access and windows let light and ventilation into a building. The weight of the wall above any opening has to be supported by a suitable structural support, which is termed a lintel. The amount of support will depend on the self-weight of the wall materials and any load that is carried by the wall above, such as floor joists. Modern lintels are available to undertake this support and are tested for light and heavy duty loading conditions. Openings are an aesthetic feature of a building that can be used to great effect in designing a living or working space.

HEAD

The function of a head is to carry the load of brickwork over the opening and transmit this load to the jambs at the sides. In order to accomplish this it must transmit the loads without any settlement or deflection. A variety of materials and methods is available for producing a lintel or beam, such as mild steel, precast concrete, stone and brick.

- **Mild steel** A variety of different preformed lintels is available for different applications and uses, from box section beams to full-width lintels that carry both skins of the cavity. For large spans, a universal beam section to design calculations will be needed so it can carry the load above with a minimum of deflection. Steel lintels and beams that are exposed to the elements should be either galvanised or painted with several coats of anticorrosive paint to give them protection against corrosion of any exposed steel. The provision of fire protection to the steelwork must also be considered in certain applications.

- **Precast concrete** Modern factory-manufactured precast concrete beams or lintels can be used for moderate spans. Pre-stressing wires in the lintels enables greater loads to be carried.

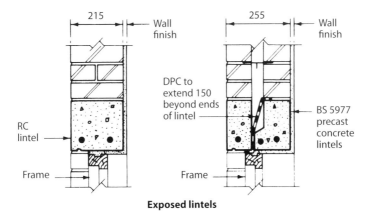

Exposed lintels

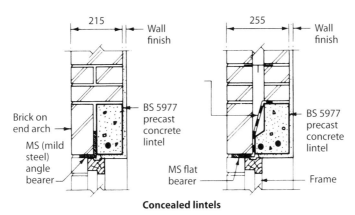

Concealed lintels

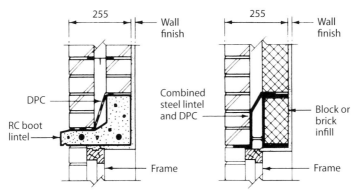

Figure 4.5.1 Typical head treatments to openings (you will still find this construction on refurbishment work).

- **Stone** These can be natural, artificial or reconstructed stone but are generally used as a facing to a steel or concrete lintel. Stone lintels are used for conservation and graded building work where replacements have to be like for like.

- **Brick** This is often used as a facing material in a soldier course which is supported by a plate that extends from the main body of a steel lintel.

Fig. 4.5.1 illustrates some of the typical concrete lintel arrangements that would have been used pre-1985 Building Regulations.

To carry a load safely a lintel requires suitable end bearing to support the loads down onto the jambs of an opening. Part A of the Building Regulations states that where the lintel span is less than 1200 mm the end bearing can be reduced to 100 mm. BS EN 845 states that the end bearing should not normally be less than 100 mm. It is generally accepted that it is good practice to ensure that the end bearing is 150 mm.

Open joints or proprietary cavity weep inserts are sometimes used to act as weep holes; these are placed at 900 mm centres in the outer leaf immediately above the damp-proof course. This ensures that any trapped moisture retained by the DPC over the lintel is allowed to exit via the weep holes. Typical examples of head treatments to openings are shown in Figs. 4.5.1 and 4.5.2.

JAMBS

In traditional solid wall constructions, jambs are bonded to give the required profile and strength; examples of bonded jambs are shown in Fig. 4.5.3. In cavity walls the cavity is now closed at the opening by using a suitable frame,

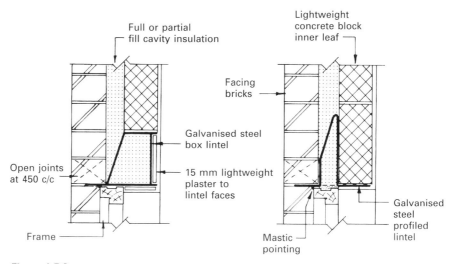

Figure 4.5.2 Typical head treatments to openings.

or by the insertion of an insulated cavity closer that prevents a cold bridge as required by the Building Regulations. Typical examples of jamb treatments to openings are shown in Figs. 4.5.3 and Fig 4.5.4.

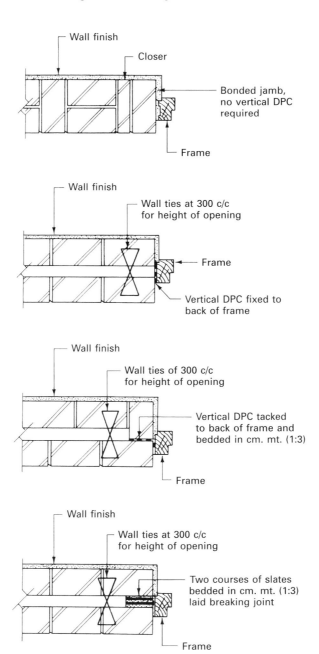

Figure 4.5.3 Typical jamb treatments to openings (you will still find this construction on refurbishment work).

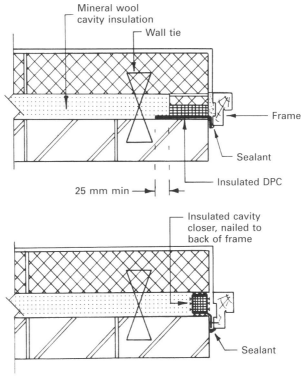

Figure 4.5.4   Typical jamb treatments to openings.

The function of a sill is to divert water away from the face of the building so it does not stain the brickwork below if used in a window position. Sills can be formed in facing brickwork, quarry tiles or a proprietary cast stone or precast concrete. Appearance and durability are the main requirements, as a sill is not a member that is needed to carry heavy loads. Sills, unlike lintels, do not require a bearing at each end, but architecturally often run past the jamb. Typical examples of sill treatments to openings are shown in Fig. 4.5.5 and 4.5.6.

CHECKED OPENINGS

REVEAL

This is the part of an opening returning at right angles from the front face of a wall. The traditional construction for door and window reveals includes a check or recess to accommodate the frame. This is a method used in areas of severe, driving and penetrating rain as described in the Building Regulations Part C. It also provides a barrier against draughts. Examples of this dated (but nevertheless existing) construction are shown in Fig. 4.5.7.

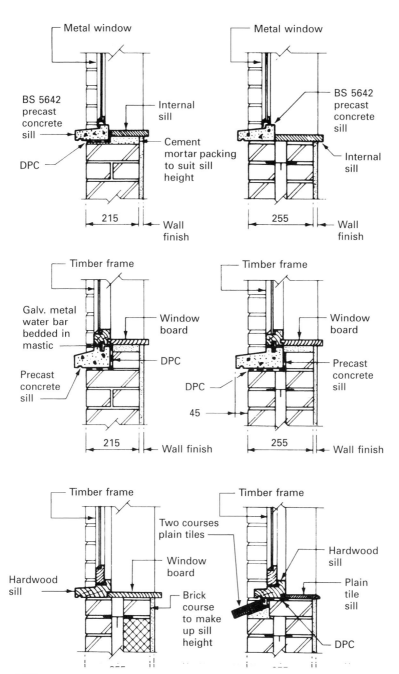

Figure 4.5.5 Typical sill treatments to openings (pre-1985 Building Regulations).

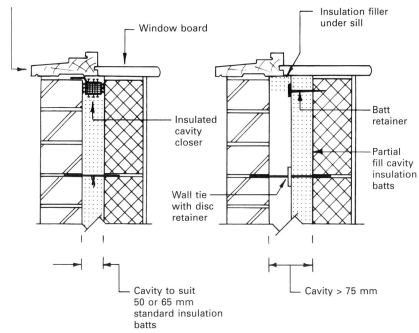

Hardwood sill

Window board

Insulation filler under sill

Insulated cavity closer

Batt retainer

Partial fill cavity insulation batts

Wall tie with disc retainer

Cavity to suit 50 or 65 mm standard insulation batts

Cavity > 75 mm

Figure 4.5.6 Typical sill treatments to openings.

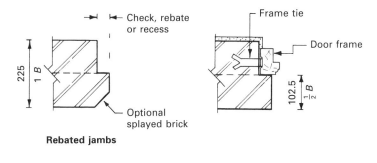

Check, rebate or recess

Frame tie

Door frame

Optional splayed brick

225

1 B

102.5

$\frac{1}{2}$ B

Rebated jambs

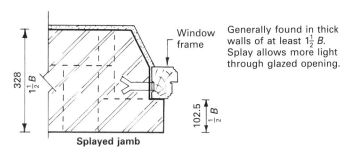

Window frame

Generally found in thick walls of at least $1\frac{1}{2}$ B. Splay allows more light through glazed opening.

328

$1\frac{1}{2}$ B

102.5

$\frac{1}{2}$ B

Splayed jamb

Figure 4.5.7 Traditional checked openings (sectional plans).

BUILDING REGULATIONS

Approved Document C: *Site preparation and resistance to contaminants and moisture* identifies the need to incorporate checked rebates at openings exposed to severe and penetrating driving rain. One leaf of the cavity wall, normally the facing brickwork, projects past the inner leaf, forming a check that the window or door frame sits into. In the UK, driving rain exposure zones are mainly to the west and south-west coasts, as shown in Fig. 4.5.8. Isolated situations also occur, and locations of these can be obtained from the local authority building control department. In keeping with the traditional construction shown, the frame is set behind the outer leaf of masonry. Contemporary practice also requires an insulated cavity, vertical DPC and cavity closer, as shown in Fig. 4.5.9.

Figure 4.5.8 Principal areas exposed to driving rain.

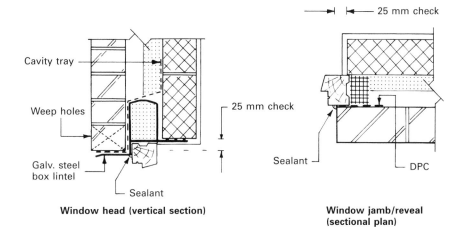

Window head (vertical section)

Cavity tray

Weep holes

Galv. steel
box lintel

Sealant

25 mm check

25 mm check

**Window jamb/reveal
(sectional plan)**

Sealant

DPC

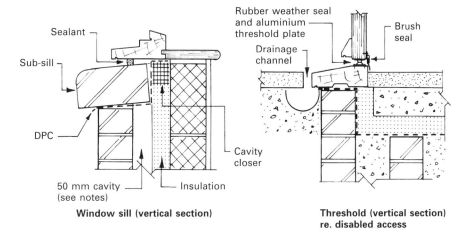

Window sill (vertical section)

Sealant

Sub-sill

DPC

50 mm cavity
(see notes)

Insulation

Cavity
closer

Rubber weather seal
and aluminium
threshold plate

Drainage
channel

Brush
seal

**Threshold (vertical section)
re. disabled access**

Notes:
1. In areas most exposed to driving rain, a 50 mm air gap in the cavity between inner and outer leaf is necessary to prevent dampness bridging the insulation.
2. The detail shown has a 100 mm cavity part insulated and a 100 mm insulating block inner leaf. This will provide a *U* value of about 0.35. A 100 mm fully insulated cavity in a brick and insulating block wall has a *U* value of about 0.27.
3. To achieve the lesser *U* value and still retain a 50 mm air gap, the thickness of the inner leaf of insulating blockwork may be increased and/or additional insulation applied to the inside of the inner leaf behind a protective plasterboard lining.

Figure 4.5.9 Contemporary checked openings.

Arches 4.6

Arched features are features of specially manufactured or cut wedge-shaped bricks called **voussoirs**. These bricks are designed to support each other and transmit the load over and around the opening, following a curved profile, to abutments on either side. When constructing an arch it must be given temporary support until the mortar joints have set and the arch has gained sufficient strength to support itself and transmit the load over and around the opening. These temporary supports are called **centres** and are usually made of timber; their design is governed by the size of the span, the weight of the load to be transmitted and the thickness of the arch to be constructed. Lintel manufacturers now make special arch formed lintels that can be permanently installed into the cavity wall construction, providing permanent support.

SOLDIER ARCHES

This is an exception to the usual arch type as it has a flat span with no camber or curve over an opening. This type of arch consists of a row of bricks showing on its face either the end or the edge of the bricks to form a decorative feature. Soldier arches have no real strength, and if the span is over 1,000 mm they will require some form of permanent support, such as a metal flat or angle (see Fig. 4.6.4, page 187). Modern pressed-steel lintels now carry both leaves of a cavity wall and allow the outer soldier course to be carried on a flat plate. If permanent support is not given, the load will be transferred to the head of the frame in the opening instead of the jambs on either side. This can cause problems when, for example, uPVC windows or doors are installed as replacement upgrades. Relatively small spans can have an arch of bonded brickwork by inserting some form of joint reinforcement into the horizontal joints immediately above the opening, such as expanded metal or bricktor, which is a woven strip of high-tensile steel wires designed for the reinforcement of brick walls. It is also possible to construct a soldier arch by inserting metal cramps in the vertical joints or by using an angle, and fixing these into an in-situ backing lintel of reinforced or precast concrete.

ROUGH ARCHES

These arches are constructed of ordinary uncut bricks; being rectangular in shape, they give rise to wedge-shaped joints, where the width of the joint tapers. To prevent the thick end of the joint from becoming too wide, rough arches are usually constructed in header courses, which break up the length of the joint. The rough arch is used mainly as a backing or relieving arch to a gauged brick or stone arch, but they are sometimes used in facework, for the cheaper form of building or where appearance is of little importance.

GAUGED ARCHES

These are the true arches and are dimensionally accurate cut bricks, called voussoirs. The purpose of voussoirs is to produce a uniform, thin joint that converges onto the centre point or points of the arch. This produces an arch that is aesthetically pleasing to look at within the brick structure. Traditionally there are two methods of cutting the bricks to the required wedge shape: **axed** and **rubbed**. If the brick is of a hard nature, it is first marked with a tin saw, to produce a sharp arris, and then it is axed to the required profile. For rubbed brick arches, a soft brick called a **rubber** is used; the bricks are first cut to the approximate shape with a saw, and are then finished off with an abrasive stone or file to produce the sharp arris. In both cases, a template of plywood or hardboard to the required shape will be necessary for marking out the voussoirs. Modern methods of brick production now include the formation of arch kits, with all the bricks moulded to the required shape ready for installation by the bricklayer. The terminology and setting out of simple brick arches is shown in Fig. 4.6.1.

CENTRES

These are the temporary structures, usually manufactured from plywood and timber, which are strong enough to fulfil their function of supporting the arches of brick or stone while they are being built, and until they are sufficiently set to support themselves and the load over the opening.

Centres can be an expensive item for a contractor as they have to be accurate and strong enough to support the load. Therefore their design should be simple and adaptable so that as many uses as possible can be obtained from any one centre. A centre is always smaller in width than the arch to allow for plumbing – that is, alignment and verticality of the brick face with a level or rule.

The type of centre to be used will depend upon:

■ the weight to be supported;   ■ the span;   ■ the width of the soffit.

Typical examples of centres for brick arches are shown in Figs. 4.6.2, 4.6.3 and 4.6.4.

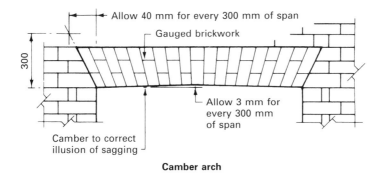

Allow 40 mm for every 300 mm of span

Gauged brickwork

300

Allow 3 mm for every 300 mm of span

Camber to correct illusion of sagging

Camber arch

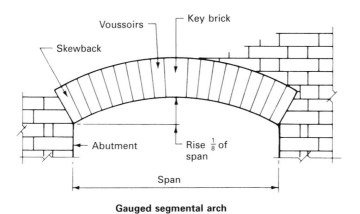

Key brick

Voussoirs

Skewback

Abutment

Rise $\frac{1}{8}$ of span

Span

Gauged segmental arch

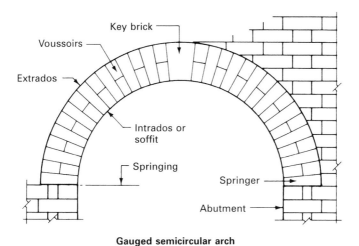

Key brick

Voussoirs

Extrados

Intrados or soffit

Springing

Springer

Abutment

Gauged semicircular arch

Figure 4.6.1 Typical examples of brick arches.

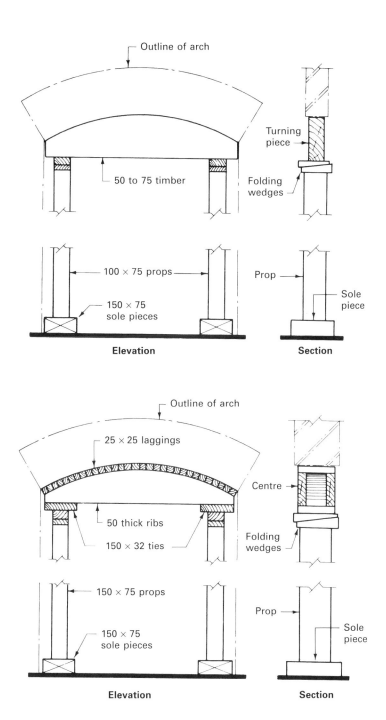

Figure 4.6.2 Centres for small-span arches.

Diagram labels (upper figure):
- Outline of arch
- 50 to 75 timber
- Turning piece
- Folding wedges
- 100 × 75 props
- 150 × 75 sole pieces
- Prop
- Sole piece
- **Elevation**
- **Section**

Diagram labels (lower figure):
- Outline of arch
- 25 × 25 laggings
- 50 thick ribs
- 150 × 32 ties
- Centre
- Folding wedges
- 150 × 75 props
- 150 × 75 sole pieces
- Prop
- Sole piece
- **Elevation**
- **Section**

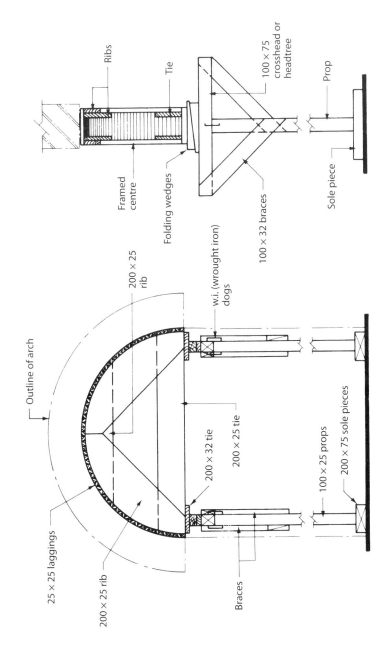

Figure 4.6.3 Typical framed centre for spans up to 1.500 mm.

Ribs

Tie

100 × 75 crosshead or headtree

Prop

Sole piece

Framed centre

Folding wedges

100 × 32 braces

Outline of arch

200 × 25 rib

w.i. (wrought iron) dogs

25 × 25 laggings

200 × 25 rib

200 × 32 tie

200 × 25 tie

Braces

100 × 25 props

200 × 75 sole pieces

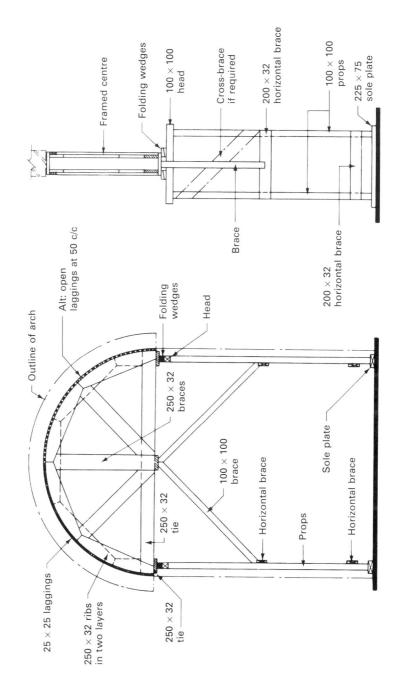

Figure 4.6.4 Typical framed centre for spans up to 4.000 mm.

Framed centre

Folding wedges

100 × 100 head

Cross-brace if required

200 × 32 horizontal brace

100 × 100 props

225 × 75 sole plate

Brace

200 × 32 horizontal brace

Outline of arch

Alt: open laggings at 50 c/c

Folding wedges

Head

250 × 32 braces

25 × 25 laggings

250 × 32 ribs in two layers

250 × 32 tie

100 × 100 brace

250 × 32 tie

Horizontal brace

Props

Sole plate

Horizontal brace

ARCH FORMERS

Modern methods of construction now allow the use of manufactured arch formers, made of steel, which are permanent supports acting as a lintel. The arch former or lintel has the same end bearing as a standard lintel and sits upon the jamb of the opening. The arch is then constructed around the former and the brickwork then continued.

Using an arch former is a quicker method than using a centre: the work can proceed while the mortar sets, and there is no need for any temporary supports to be placed beneath the arch while it is being constructed. The former can be galvanised or painted with a protective coating to ensure that the exposed edge remains free of corrosion. Arch formers or lintels are available to order in a variety of sizes and configurations.

Fig. 4.6.5 illustrates an application of the arch support. The cavity tray (see below) unifies the structural support with the damp-proof course and remains an integral part of the structure. Arch profiles are limited to tray manufacturers' designs and spans, but most will offer semicircular and segmental as standard, with gothic, elliptical, ogee and triangular styles available to order. Clear spans up to about 2 m are possible.

CAVITY TRAYS

A cavity tray must be provided above every opening to capture and drain away any moisture that has managed to enter the cavity through the external leaf. This should direct the water to weep holes that are left at every 900 mm along the opening. A weep hole allows any trapped moisture to leave the cavity; this can be an open joint or a proprietary manufactured weep hole (see Fig. 4.6.6). A cavity tray must be fitted with stop ends to prevent moisture running off the ends and down the cavity.

The cavity tray is normally constructed from damp-proof course materials and must be impervious to water. It saves considerable construction time as temporary supports are not required.

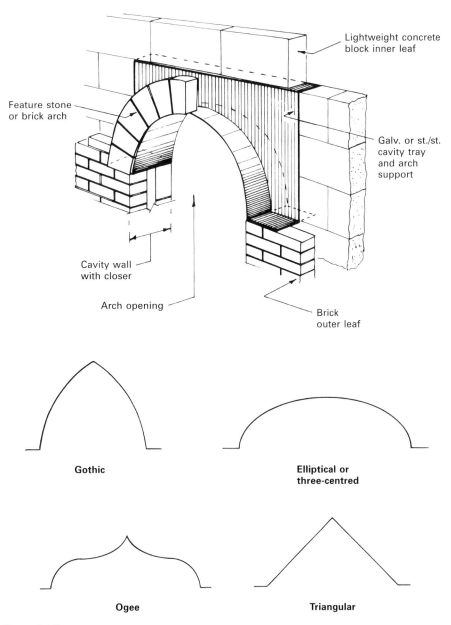

Lightweight concrete
block inner leaf

Feature stone
or brick arch

Galv. or st./st.
cavity tray
and arch
support

Cavity wall
with closer

Arch opening

Brick
outer leaf

Gothic

**Elliptical or
three-centred**

Ogee

Triangular

Figure 4.6.5   Arch former and arch profiles.

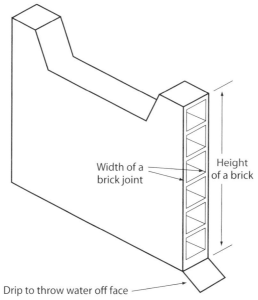

Width of a
brick joint

Height
of a brick

Drip to throw water off face

Figure 4.6.6   Proprietary manufactured weep hole.

Timber-framed housing

4.7

Timber framing is a current sustainable method of domestic housing construction. It uses frames manufactured from processed softwood timber faced with plywood sheet materials. Loadbearing timber walls are constructed by creating a framework of relatively small timbers (up to ex. 150 mm × 50 mm – the 'ex' means 'cut out of') spaced vertically at 400 or 600 mm centres. These are known as **studs** and have short struts of timber or noggins placed between them at 1 m maximum spacing to prevent distortion. Some panel prefabricators provide thin galvanised-steel diagonal bracing as an alternative to struts. Head and sole plates complete the framing.

With the benefit of factory quality-control procedures and efficient industrial manufacturing processes, the high cost of imported timber in the UK is offset by efficient factory prefabrication of complete wall, floor and roof units. Furthermore, efficient site assembly methods using a mobile crane and semi-skilled operatives provides for a considerable saving in construction time and resources.

TIMBER FRAMING

ADVANTAGES

- Timber by its nature is light compared to a traditional inner leaf of a cavity wall. It is therefore easier to handle and is less of a structural dead weight factor than traditional masonry. This will produce savings on the foundation design.

- It has a high strength-to-weight ratio and is also very stiff when related to its strength.

- Prefabrication of structural units eliminates the need for skilled carpentry as all joints are simply nailed together.

- It is highly thermally efficient, reducing heat loss from the structure.
- Savings can be made by using dry-lining methods to the inside of the frame, saving time waiting for drying out.
- The method creates a service void for running cables and pipes.
- The process is not weather-dependent.
- It is a sustainable and environmentally friendly method of construction.

DISADVANTAGES

- Poor quality control, as damp-proof membranes and breather membranes can lead to damp penetration.
- Poor resistance to fire in sizes under 150 mm x 100 mm, as there is insufficient substance to char-protect the inner structure; the effect of fire is limited by cladding with non-combustible material such as plasterboard internally and facing brick externally.
- Its hygroscopic nature, which can support decay in damp situations; however, provided the timber is installed dry and assembled with correct protection (damp-proof course, vapour-control layers, etc.), dampness penetration will not be an issue.
- The fixing of fittings and furniture to the dry-lined wall may present problems.
- The transmission of sound across the structure may be greater than with traditional methods.

Manufacture and construction techniques derive from the applications of **platform** or **balloon** framing. Balloon framing uses full-height studs from ground level to the roof; platform framing uses storey-height studs – that is, from ground to first floor, etc. Platform framing is the more common method adopted in the UK today. Fig. 4.7.1 shows the principle of storey-height platform-framed walls made up of ex. 100 mm x 50 mm timber studding at 400 or 600 mm centres. These are supported at intermediate floor levels or platforms by nailing through the sole plates. Head plates are similarly fixed to a head binder, which links each prefabricated panel. The sill plate must be level and true to prevent any distortion in the frames when erected.

Fig. 4.7.2 shows the ground floor to eaves full-height balloon-framed wall. This is independent of the intermediate floor except for complementary bracing where the floor joists fix to studs and a **ribbon** or **ledger** let into the stud framing.

Storey-height platform framing is simpler to handle and transport, but less rapidly constructed on site than the one lift to roof level that balloon framing permits. The effects of timber movement are more easily contained and restrained in the smaller units of a platform frame.

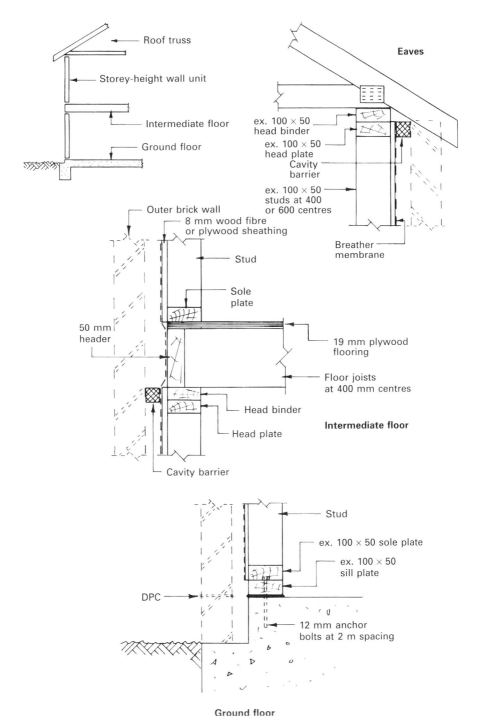

Roof truss

Storey-height wall unit

Intermediate floor

Ground floor

Eaves

ex. 100 × 50 head binder

ex. 100 × 50 head plate

Cavity barrier

ex. 100 × 50 studs at 400 or 600 centres

Breather membrane

Outer brick wall

8 mm wood fibre or plywood sheathing

Stud

Sole plate

50 mm header

19 mm plywood flooring

Floor joists at 400 mm centres

Head binder

Head plate

Intermediate floor

Cavity barrier

Stud

ex. 100 × 50 sole plate

ex. 100 × 50 sill plate

DPC

12 mm anchor bolts at 2 m spacing

Ground floor

Figure 4.7.1   Timber-framed construction – platform frame.

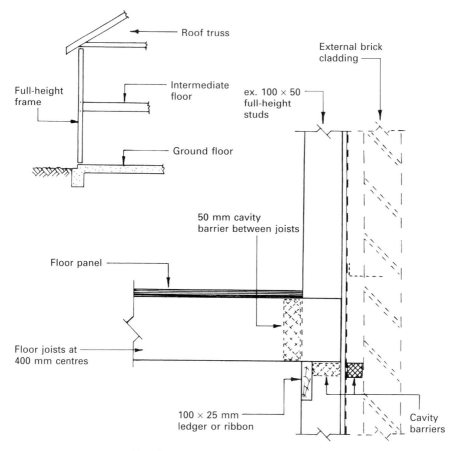

Note: Eaves and ground floor as Fig. 4.7.1

Figure 4.7.2 Timber-framed construction – balloon frame.

TIMBER-FRAMED CLADDING

A variety of methods are available to clad the outside of the timber-framed construction.

- An outer cladding of brickwork, which provides a traditional appearance, with stainless steel angle wall ties spanning a cavity to secure the masonry outer leaf to the timber framing.
- A blockwork skin, provided it is covered with a waterproof rendered finish.
- Timber cladding.
- uPVC cladding.
- Clay tiles hung vertically.
- Specialist insulation systems with a rendered application.

The possibility of fire spread through the cavity is restrained by the use of cavity barriers at strategic intervals. These normally occur at intersections of elements of construction: for example, wall to roof and party wall to external wall. Fire resistance of cavity barriers in dwellings should be constructed to provide at least 30 minutes of fire resistance (see Building Regulations Approved Document B: (Tables A1 and A2 of appendix A of Part B1 Approved Document)).

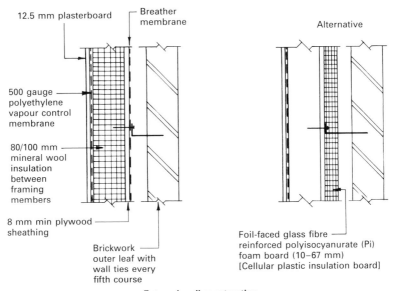

External wall construction

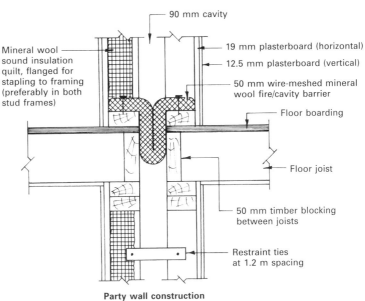

Party wall construction

Figure 4.7.3 Timber-framed construction – further details.

Timber technology 4.8

The use of timber as sustainable superstructure material is now more common in homes that have to meet carbon emission regulations. Softwood timber is a natural product that is grown in several European countries and imported into the UK where it is processed into structural carcassing for timber-framed construction. Each piece of timber is unique and its strength properties may differ by the number of defects within its structure. This section examines the properties of timber and the factors that might lead to its deterioration over time.

TIMBER PROPERTIES

Timber can be derived from thousands of different tree species, but the botanical classification divides timber into **hardwood** and **softwood**. Hardwoods are from broad-leaved trees, most of which are deciduous, and these woods are used for furniture and fixtures. Softwoods are from coniferous trees, and this is the type of timber most commonly used for construction elements and framing.

For structural purposes, strength characteristics and durability are the more important means for classification, as the timber must be able to withstand the loading and stress that is exerted upon it. This is particularly important with softwood, which is used predominantly for structural timbers in the UK, as this is the most economical timber to use for structural applications.

When specifying softwood structural timbers, the following should be stated:

- any preservative treatment;
- cross-sectional size and surface finish;
- the moisture content;
- the strength class.

PRESERVATIVE TREATMENT

The majority of softwood contains a high proportion of sapwood, which is essentially the growing layer of a tree. This is more prone to attack than the denser structural heartwood found closer to the core of a timber trunk. Timber can be specified to be chemically treated to prevent possible fungal decay (wet or dry rot) and potential infestation from wood-boring insects.

The requirement for timber preservation will be determined by the position of the timbers within a building and the geographical location. In parts of southern England there is a perceived risk of softwood timber infestation by the house longhorn beetle. The Building Regulations Approved Document to support Regulation 7, Approved Document A defines the particular boroughs where softwoods must be treated for this beetle if applied in any aspect of roof construction. See also BS 8417: *Preservation of timber. Recommendations, for clarification of acceptable chemical treatment.*

CROSS-SECTIONAL SIZE AND SURFACE FINISH

Softwood cross-sectional size has previously been specified under three classifications:

1. sawn;

2. planed (machined);

3. regularised.

Sawn is the result of converting a timber log into commercially accepted sizes: for example, 50 mm x 100 mm, 75 mm x 225 mm. Application of these sections is largely for unseen structural use, such as joists and rafters, as the sawing process leaves a rough finish to the timber surface.

Planed involves machining about 1.5 mm off each surface to provide a smooth finish: for example, 50 mm x 100 mm sawn becomes 47 mm x 97 mm, otherwise known as ex. 50 x 100. This application is for exposed timbers that would be seen, such as door linings and frames.

Regularised timber has a uniform width to provide a more even final product. The timber may be sawn or planed to, say, 47 mm for uniformity in timber studding. The corners of the studs produced are often slightly rounded in the process and are not square.

The above terms are common terms used in the specification but BS EN 336: *Structural timber* details the sizes and deviations that are permitted. This standard specifies target sizes in order to simplify the previous confusion over the finished dimensions. Target size is in only two tolerance classes:

- T1, applicable to sawn surfaces;

- T2, applicable to planed timber.

For example, if a section of timber is required 50 mm wide sawn and 197 mm in depth planed, it will be specified as 50 mm (T1) × 197 mm (T2). The tolerances are detailed within the BS EN 336 in millimetres for the different classes T1 and T2, and thicknesses and widths.

MOISTURE CONTENT

When timber is felled it is saturated with moisture and unusable. In this state it is unworkable and, if allowed to remain so, will be prone to shrinkage, distortion and the effects of fungal growths. The felled timber has to have its moisture content reduced so it can be processed into further products. Natural drying or seasoning is possible in some climates, but this is usually done commercially. Most commercial organisations use kilns to reduce the moisture content of rough-sawn softwood to between 12 per cent and 20 per cent. The moisture content is expressed as the weight of water in the timber as a percentage of the weight of the timber dry. Here is the formula:

$$\text{Moisture content \%} = \frac{(\text{wet weight}) - (\text{dry weight})}{\text{dry weight}} \times 100$$

After seasoning it is essential that the product remains in a well-ventilated, atmospherically stable environment, so its moisture content can be controlled. Timber is hygroscopic, which means it is made up of cells that will absorb moisture readily. Therefore significant changes in temperature and humidity may cause it to react by expanding, contracting, deforming or twisting. Storage should be under covered and well-ventilated sheds. Moisture content of stored timber can be checked regularly with a surveyor's moisture meter, which measures the conduction of electricity across two probes to give a reading.

STRENGTH CLASSES

The grading of timber where each piece has different characteristics is not an easy task. In an attempt to rationalise this, various European standards have been introduced. Grading of timber can be undertaken visually or by the more efficient use of computerised grading machines. Individual pieces are assessed against permissible defect limits and marked accordingly. Some examples of grade markings are shown in Fig. 4.8.1.

Grading occurs in this country or the country of origin. BS EN 14081-1:2005, BS EN 518 and BS EN 14081-3:2005 519 (European standards for visual and machine grading, respectively) effectively rationalise much of the timber graded in Europe and many surrounding countries, such as Scandinavian countries. However, confusion may occur with timber graded elsewhere in the absence of an international standard.

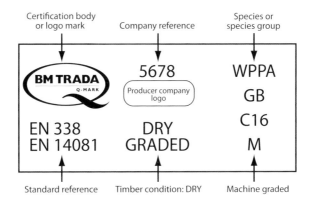

Figure 4.8.1 Examples of the format of grading stamps on softwood timber.

Visual grading in the UK may be in accordance with BS 4978 2007. This provides two quality grades: GS (general structural) and SS (special structural) grades will meet the requirements of the aforementioned European standards.

Timber imported from North America is independently graded to rules applicable in Canada and the USA. BS EN 1912 also assigns strength classes for softwoods graded to the following standards of Canada and the USA:

- national grading rules for dimension lumber. NLGA (National Lumber Grades Authority), Canada;

- national grading rules for softwood dimension lumber. NGRDL (National Grading Rules for Dimension Lumber), USA;

- North American export standard for machine stress-rated (MSR) lumber (NAMSR-North American machine stress-rated lumber) and EU Export Annexes.

For more details, see Table 4.8.1.

Table 4.8.1 North American timber grades and markings.

Timber mark	Grade	Application	Range of nominal widths
Selstr No. 1 No. 2 No. 3	Select structural Number 1 Number 2 Number 3	Structural joists and planks	150 mm and over
Selstr No. 1 No. 2 No. 3	Select structural Number 1 Number 2 Number 3	Structural light framing	50 mm to 100 mm
Const Std Util Stud	Construction Standard Utility Stud	Light framing	50 mm to 100 mm

Comparisons can be drawn, and Table 4.8.2 provides an indication of the similarity of product grades and strength classes to BS EN 338 2009: *Structural timber. Strength classes.*

Table 4.8.2 Timber grading and strength class comparisons.

Species	Standard		C14	C16	C18	C22	C24	C27/TR26	C30
	BS EN 338		C14	C16	C18	C22	C24	C27/TR26	C30
	Bldg Regs			SC3			SC4	SC5	
Whitewood or redwood	BS 4978			GS			SS		
British spruce	BS 4978		GS		SS				
British pine	BS 4978		GS			SS			
Canadian S-P-F or hem-fir	NLGA or NGRLD (J&P/SLF)			No. 1			Sel		
				No. 2					
	NLGA or NGRLD (LF)		Const						
	NLGA or NGRLD (Stud)		Stud						

Key: S-P-F = Spruce, pine, fir.
hem = Hemlock.
Others: see Table 4.8.1.

Notes
North American MSR grades are also accepted in the UK and correspond across the BS EN 338 range given above.

For more detailed information, refer to: BS EN 1995-1-1:2004; UK National Annex to Eurocode 5: *Design of timber structures.*

As can be seen from Table 4.8.2, using GS- or SS-graded timber from different sources may not result in the same strength characteristics, as these can vary between species. The full listings of BS EN 338 2009 strength classes are shown in Table 4.8.3, with kind permission of the British Standards Institution. It provides a comprehensive comparison of timber strength properties and characteristics, which may be used as a basis for structural design and timber selection.

	Softwood species												Hardwood species							
	C14	C16	C18	C20	C22	C24	C27	C30	C35	C40	C45	C50	D18	D24	D30	D35	D40	D50	D60	D70
Strength properties (in N/mm²)																				
Bending $f_{m,k}$	14	16	18	20	22	24	27	30	35	40	45	50	18	24	30	35	40	50	60	70
Tension parallel $f_{t,0,k}$	8	10	11	12	13	14	16	18	21	24	27	30	11	14	18	21	24	30	36	42
Tension perpendicular $f_{t,90,k}$	0,4	0,4	0,4	0,4	0,4	0,4	0,4	0,4	0,4	0,4	0,4	0,4	0,6	0,6	0,6	0,6	0,6	0,6	0,6	0,6
Compression parallel $f_{c,0,k}$	16	17	18	19	20	21	22	23	25	26	27	29	18	21	23	25	26	29	32	34
Compression perpendicular $f_{c,90,k}$	2,0	2,2	2,2	2,3	2,4	2,5	2,6	2,7	2,8	2,9	3,1	3,2	7,5	7,8	8,0	8,1	8,3	9,3	10,5	13,5
Shear $f_{v,k}$	3,0	3,2	3,4	3,6	3,8	4,0	4,0	4,0	4,0	4,0	4,0	4,0	3,4	4,0	4,0	4,0	4,0	4,0	4,5	5,0
Stiffness properties (in kN/mm²)																				
Mean modulus of elasticity parallel $E_{0,mean}$	7	8	9	9,5	10	11	11,5	12	13	14	15	16	9,5	10	11	12	13	14	17	20
5% modulus of elasticity parallel $E_{0,05}$	4,7	5,4	6,0	6,4	6,7	7,4	7,7	8,0	8,7	9,4	10,0	10,7	8	8,5	9,2	10,1	10,9	11,8	14,3	16,8
Mean modulus of elasticity perpendicular $E_{90,mean}$	0,23	0,27	0,30	0,32	0,33	0,37	0,38	0,40	0,43	0,47	0,50	0,53	0,63	0,67	0,73	0,80	0,86	0,93	1,13	1,33
Mean shear modulus G_{mean}	0,44	0,5	0,56	0,59	0,63	0,69	0,72	0,75	0,81	0,88	0,94	1,00	0,59	0,62	0,69	0,75	0,81	0,88	1,06	1,25
Density (in kg/m³)																				
Density ρ_x	290	310	320	330	340	350	370	380	400	420	440	460	475	485	530	540	550	620	700	900
Mean density ρ_{mean}	350	370	380	390	410	420	450	460	480	500	520	550	570	580	640	650	660	750	840	1080

Note 1 Values given above for tension strength, compression strength, shear strength, 5% modulus of elasticity mean modulus of elasticity perpendicular to grain and mean shear modulus, have been calculated using the equations given in Annex A.

Note 2 The tabulated properties are compatible with timber at a moisture content consistent with a temerature of 20 °C and a relative humidity of 65%.

Note 3 Timber conforming to classes C45 and C50 may not be readily available.

Note 4 Characteristic values for shear strength are given for timber without fissures, according to EN 408. The effect of fissures should be covered in design codes.

Table 4.8.3 Strength classes for different types of timber
Source: British Standard BS EN 338: 2009 Structural timber. strength classes.

Timber deterioration 4.9

Timber has been used for thousands of years as a building material. It is a very resilient and robust material under normal atmospheric conditions. It can even retain its structure in a permanently wet environment, as evidenced by many archaeological finds. As well as being very durable, timber will outlast many other building materials succumbing to the effects of rain, frost and chemicals. However, the weakness of both hardwoods and softwoods is their source as a food for plant growths in the form of fungi and to certain species of insect, which weaken their structure and durability.

FUNGAL ATTACK

Fungi do not rely on photosynthesis in order to grow and reproduce, unlike other plant growths, but when conditions are right they will readily consume organic material. A secondary element needs to be in place for this to occur, which is a high moisture content in excess of 20 per cent; this can occur due to moisture penetrating into a structure, or in an unventilated space with high humidity. Generally, softwoods for internal structural use will be commercially seasoned to a moisture content of about 12–16 per cent. Internal softwood for joinery and hardwoods will be seasoned to a lower figure of around 10 per cent. As long as these values are maintained, fungal growth cannot start.

CATEGORIES OF ROT ATTRIBUTED TO FUNGI

There are generally two types of fungal attack on timber: dry and wet rot. The initial indication of rot infestation is usually a stale or musty smell emanating from a damp source below floors, in cellars or within a roof space. Closer investigation often reveals whitish/grey fungal plant growths and possibly coloured fruiting bodies, depending on how far the fungus has advanced.

DRY ROT

Dry rot is known biologically by the Latin name *Serpula lacrymans* or *Merulius lacrymans*. It is a 'brown rot', leaving the timber dry and friable, so named because the wood becomes dark in colour. It is further characterised by the timber drying and cracking to produce small surface squares, manifesting as cubes within the depth of the wood. This shrinking is caused by destruction of the cellulose tissues.

Fungal growth is caused when red/rusty-coloured spores from an established fungus drift through the air to settle and germinate on damp timber. The spores develop into white strands (hyphae) appearing as cottonwool-like patches (mycelium) growing flesh-textured fruiting bodies (sporophores), in turn producing more spores.

Early detection is essential to eradicate the problem. Any source of moisture ingress must be removed or stopped to prevent further damage. Possible sources of moisture are leaking plumbing, or condensation due to inadequate underfloor or roof space ventilation. Removing the source will not necessarily kill off the dry rot as it can still thrive by developing small vein-like tubing of 2–3 mm diameter (rhizomorphs) to extract dampness from areas adjacent to the timber food source. This can include moisture in brickwork, render, plaster or concrete. Moisture from the air can be sufficient for dry rot to live, particularly in areas of high humidity.

Optimum growth conditions are a combination of dampness and warmth. Temperatures between 13 °C and 24 °C are ideal. At freezing temperatures the fungus becomes dormant, but it will die in temperatures above 40 °C.

Treatment of dry rot

- Eliminate the source of moisture ingress.
- Dry out the building area affected by ventilation.
- Cut out all affected timber at least 500 mm beyond the decay.
- Remove all affected plaster and other finishes within the vicinity of attack.
- Sterilise the affected area: where the attack is severe, drill close-spaced inclined holes of about 12 mm diameter into the adjacent masonry and liberally feed the holes with patent fungicide to saturate the structure.
- Make good jointing and pointing to masonry and plasterwork with a zinc oxychloride additive or similar fungicide in the mix. Specialist paints containing fungicide are also available.
- Replace all affected timber with well-seasoned, preservative-treated wood.

Following the remedial work, regular monitoring is essential for several months to ensure successful eradication of dry rot. Even though the source of dampness may be removed, dry rot has an irritating habit of reappearing, somehow managing to thrive on only a nominal amount of moisture.

WET ROT

The most common types of wet rot fungus in the UK are known biologically by the Latin names of *Coniophora puteana* or *Coniophora cerebella* and *Phellinus contiguus* or *Poria contigua*. The growth and development of both fungi are similar to those of dry rot, but fruiting bodies are rare.

Coniophora, otherwise known as cellar fungus, prefers very damp conditions. It is a 'brown rot', splitting the internal structure of the host timber longitudinally and laterally into small cube shapes, but leaving the surface largely intact. The timber crumbles when prodded, and it is dull dark brown or black in colour.

Phellinus has become more associated with decay in external joinery. It is a 'white rot', attacking poor-quality sapwood where used in door and window frames, fascia boards, cladding, etc. Inadequate treatment of the timber, poor jointing techniques, inappropriate adhesives and lack of decorative maintenance will all lead to rainwater penetration. Wood in these situations is often painted, and the first sign of fungal decay is usually surface irregularities. Closer inspection will reveal splitting within the body of wood, breaking into soft strands.

Optimum growth conditions are as for dry rot, but wet rots are more easily controlled as they require greater exposure to dampness to thrive. Hence, once the moisture source is removed, the decayed timber can be replaced. Adjacent timber, brickwork, plaster, etc. are unlikely to be affected, but as a precaution the whole area should be brush-treated with a fungicide.

INSECT ATTACK

SPECIES

There are several species of wood-boring insect that are capable of seriously damaging structural timber and joinery in buildings. During their relatively short time in adult form they are classified as beetles. Most of their life is in the larval stage, gnawing and burrowing through timber – hence the general descriptive term of woodborer or woodworm. Most species of larvae have a preference for the less dense sapwood growth areas as their source of food.

The most common species in the UK are:

- common furniture beetle (*Anobium punctatum*);
- death watch beetle (*Xestobium rufovillosum*);
- powder post beetle (*Lyctus* family);
- house longhorn beetle (*Hylotrupes bajulus*).

Fig. 4.9.1 compares the species proportionally. The largest is the adult house longhorn beetle at about 25 mm overall.

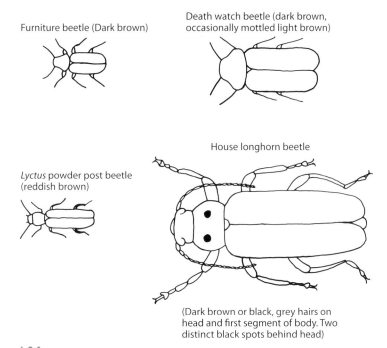

Furniture beetle (Dark brown)

Death watch beetle (dark brown, occasionally mottled light brown)

House longhorn beetle

Lyctus powder post beetle (reddish brown)

(Dark brown or black, grey hairs on head and first segment of body. Two distinct black spots behind head)

Figure 4.9.1 Common wood-boring beetles.

Habitat

Most species prefer a slightly damp, draught-free environment. Unventilated roof spaces and recesses behind eaves cupboards are ideal areas in which to thrive. Other areas suitable to the lifestyle of the woodworm include understair cupboards and voids within enclosed baths and other sanitary fittings. Timber of intermediate floors in housing is also vulnerable to attack, but raised timber ground floors are less vulnerable if peripheral air vents remain clear for air circulation.

Lifecycle

Wood-boring insects are most active during the warmer spring and summer months. The lifecycle shown in Fig. 4.9.2 is similar for all species, progressing from egg, through larva (grub) and pupa (chrysalis), to adult (beetle). Adult female beetles seek rough crevices of sawn timber or former borehole exits to deposit their eggs. As the larvae hatch they bore into the wood, using it as food and shelter. The tunnelling effect of hundreds of larvae from each batch of eggs can be extremely damaging. The larval stage is the predominant part of an insect's life, extending for several years before maturing to a chrysalis. The chrysalis develops into an adult beetle just beneath the timber surface, from where it emerges to reproduce, lay eggs and generate more damage. Table 4.9.1 provides some comparison of the behavioural characteristics of species.

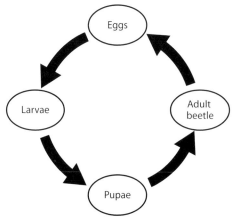

Figure 4.9.2 Woodborers' lifecycle.

Table 4.9.1 Common woodborers.

Name	Egg quantity	Egg maturity	Laval period	Pupal stage	Emergence as a beetle	Location	Exit holes
Furniture beetle	20–50	4–5 weeks	2–3 years	4–8 weeks	May–Sept	Sapwood of soft or hard wood	1–2 mm dia.
Death watch beetle	40–80	2–8 weeks	4–6 years	2–4 weeks	Mar–June	Hardwoods, preferably decayed oak	3 mm dia.
Powder post beetle	70–200	2–3 weeks	1–2 years	3–4 weeks	May–Sept	Sapwood of new hardwood	1–2 mm dia.
Longhorn beetle	< 200	2–3 weeks	3–10 years	About 3 weeks	July–Sept	Sapwood of softwood	5 mm × 10 mm oval

Note: Periods given will vary relative to ambient temperatures.

RECOGNITION AND TREATMENT

Woodworm is usually first recognised by the distinctive exit flight holes and powdery deposits (**frass**) on the surface of timber. By this stage the internal structure of the timber may have suffered considerable damage. The extent of damage may be established by chiselling away the timber surface to examine the borings or galleries produced by the larvae.

If the damage is slight and there are only a few exit holes, remedial spraying or brushing with liberal applications of a proprietary insecticide will control the problem. Structural timbers seriously damaged must be removed and burned. All new timber must be pre-treated, preferably with a vacuum/pressure-impregnated insecticide sourced from a commercial timber supplier.

In new construction, prevention of woodworm infestation by using pre-treated timber is the only effective method of control. The Building Regulations Approved Document A: *Structure* (Section 2B) provides specific reference and guidance for the use of treated timber for roof construction in parts of Surrey and adjacent areas.

Reference sources for acceptable methods of timber preservation include:

- The Wood Protection Association's Manual;
- BS 8417:2003 *Preservation of timber – Recommendations*;
- EN 1995 (Design of timber structures);
- EN 335, EN 351 and EN 599 – European standards for wood preservatives and treated wood.

Volumetric construction 4.10

With the introduction of modern methods of construction (MCC) comes the use of volumetric construction techniques, which can utilise several different methods of construction:

- light steel frames;
- timber-framed construction;
- structurally insulated panels (SIPS).

DEVELOPMENT

Sustainability in construction has developed over recent years by the use of modular construction techniques, which have replaced the traditional cavity wall of brick and block construction. Modular construction allows the quick and efficient production of multiple units that can be assembled to form many applications, such as student accommodation and hotels. The application of this type of construction has been tested in housing but has not been formally adopted.

LIGHT STEEL FRAMES

Steel-framed housing is similar to timber frame in that standard sections are used to make up a structural frame. The frame consists of a volumetric unit, with walls, floor and ceiling panels, all factory-finished, transported and craned into position. An outer leaf of brick cladding retains a traditional image and protection from the elements.

Advantages

- structural strength;
- fast method of construction;

- will not rot;
- fire-resistant finishes;
- savings in time and resources;
- requires only semi-skilled labour to assemble;
- components factory-manufactured to quality-controlled standards;
- high level of prefabrication of finishes possible.

Disadvantages

- limited design aesthetics;
- fixings to hollow walls difficult;
- risk of corrosion;
- steel requires rust protection.

STRUCTURALLY INSULATED PANELS (SIPS)

The use of an insulated core that is protected by a layer of plywood on each side forms the basis of a SIP. The insulation is left short within the panel so a connecting timber can be used to hold two panels together and maintain structural integrity. The panel is similar to timber-framed construction, but faster in the time it takes to assemble because the insulation is integral to the panel. The panels acts as the inside skin of a wall (see Fig. 4.10).

Advantages

- fast method of construction;
- direct finish can be applied externally and internally;
- highly thermally efficient;
- minimum wastage on site;
- airtight structure achieved;
- lighter construction method;
- green technology.

Disadvantages

- fireproofing;
- fixings to hollow walls difficult;
- sound transmittance.

The use of prefabrication in the production of domestic dwellings is starting to become a common method especially with the SIPS panel system and the timber framed construction. The larger the panel that can be produced and transported the better this is for the environment and the quicker a home can be produced on site.

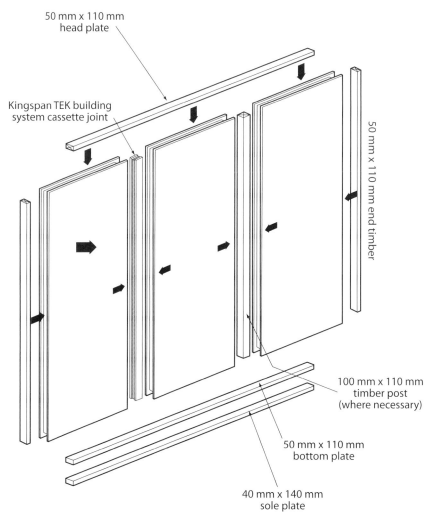

50 mm x 110 mm
head plate

Kingspan TEK building
system cassette joint

50 mm x 110 mm end timber

100 mm x 110 mm
timber post
(where necessary)

50 mm x 110 mm
bottom plate

40 mm x 140 mm
sole plate

Figure 4.10 System wall panel.

5

Floors

Solid concrete ground floor construction 5.1

Solid concrete ground floor construction is the traditional method for solid ground floors on good subsoil, where the slab construction rests directly on the ground. It comprises the following elements:

- hardcore;
- blinding;
- damp-proof membrane;
- insulation;
- concrete slab.

HARDCORE

Hardcore is clean, broken brick or similar inert material. Its purpose is to provide a firm base on which to place a concrete ground slab and to help spread any point loads over a greater area. Often the hardcore material is supplied from crushing existing demolished masonry that is already on site. It also acts against capillary action of moisture within the soil. Hardcore is usually laid in 150 mm layers to the required depth, and it is important that each layer is well compacted, using a vibrating whacker plate if necessary, to prevent any unacceptable settlement beneath the solid floor. The maximum depth allowed by Part C: Section 4 of the Building Regulations is 600 mm; any depth greater than this requires a suspended ground floor to be specified.

Approved Document C recommends that no hardcore laid under a solid ground floor should contain water-soluble sulphates or other harmful matter in such quantities as to be liable to cause damage to any part of the floor. This recommendation prevents the use of any material that may swell upon becoming moist, such as colliery shale. Furthermore it is necessary to ascertain that brick rubble from demolition works and clinker furnace waste intended for use as hardcore does not have any harmful water-soluble sulphate content.

BLINDING

This is used to even off the surface of hardcore if a damp-proof membrane is to be placed under the concrete slab or if a reinforced concrete slab is specified. It will prevent the damp-proof membrane being punctured by the hardcore and will provide a true surface from which any reinforcement can be positioned. Blinding generally consists of a layer of sand 25–50 mm thick or a 50–75 mm layer of weak concrete if a true surface for reinforced concrete is required.

DAMP-PROOF MEMBRANE

Part C: Section 4 also requires the ground slab to resist the passage of ground moisture to the upper surface of the floor, and to resist the passage of ground gases, such as methane and radon. To resist moisture, a damp-proof membrane is usually formed using 0.3 mm thick (1200 gauge) polyethylene sheet laid on the blinding material with minimum laps of 300 mm. The membrane should be turned up at the edges to meet and lap with the damp-proof course in the walls, to prevent any penetration of moisture by capillary action at edges of the bed. The membrane also acts as a temporary barrier to stop cement grout from the placed wet concrete from draining away and preventing the satisfactory curing of the concrete slab.

However, there are alternative methods of creating a suitable DPM in a solid ground floor slab. These are:

- hot-poured bitumen, which should be at least 3 mm thick cold-applied bitumen/rubber emulsions, which should be applied in not less than three coats;

- asphalt/pitchmastic, which could be dual-purpose by acting as a finish and a damp-proof membrane.

Note: Prevention of the ingress of radon and/or methane gases from the ground will require a wire- or fibre-reinforced polythene membrane of up to 1 mm thickness. Radon is a naturally occurring radioactive gas originating from uranium and radium deposits in certain rock subsoils found primarily in parts of the West Country, northern England and areas of Scotland. Methane is an explosive gas, which can build up in the ground as a result of deposited decaying organic materials. Where either of these gases is prevalent, it is now standard practice to construct suspended precast concrete floors with natural draught underfloor ventilation (see Part 5.2, Fig. 5.2.1, on page 219)

Typical details of solid floor construction are shown in Figs. 5.1.1 and 5.1.2.

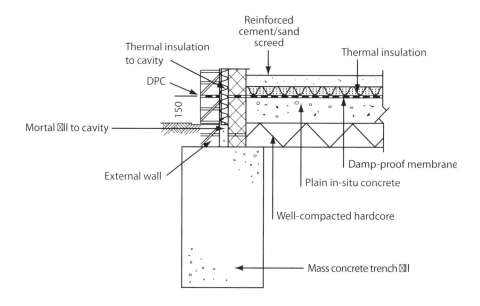

Reinforced
cement/sand
screed

Thermal insulation
to cavity

Thermal insulation

DPC

150

Mortal ⊠ll to cavity

External wall

Damp-proof membrane

Plain in-situ concrete

Well-compacted hardcore

Mass concrete trench ⊠ll

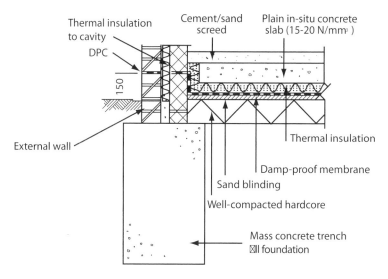

Thermal insulation
to cavity

Cement/sand
screed

Plain in-situ concrete
slab (15-20 N/mm³)

DPC

150

External wall

Thermal insulation

Damp-proof membrane

Sand blinding

Well-compacted hardcore

Mass concrete trench
⊠ll foundation

Figure 5.1.1 Typical solid floor details at external walls.

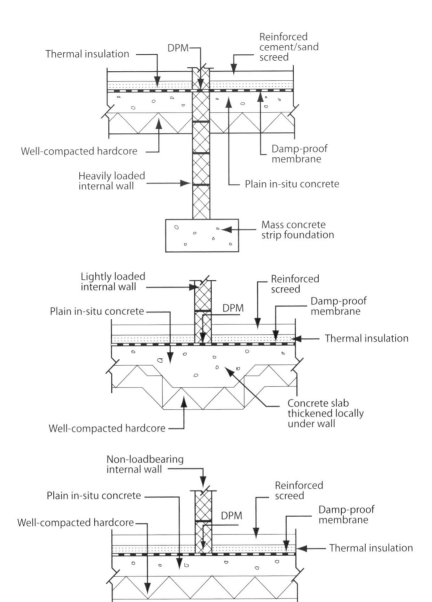

Figure 5.1.2 Typical solid floor details at internal walls.

INSULATION

It is current practice to ensure that all ground floors are suitably insulated. They need to comply with the Building Regulations Part L in that the limiting fabric heat loss or U-value must not be greater than 0.25 W/m²K. The choice of insulation materials and their location within the floor should suit the local conditions in terms of compressive strength, water absorption and durability. For ground floor construction, plastic insulation products in solid or board form, such as expanded polystyrene or polyurethane, are generally specified because of their strength and durability. The required thicknesses of the insulation are shown in Table 5.1.1.

Table 5.1.1 **Insulation thickness (mm) required to achieve U-value (W/m²K).**

U-value (W/m²K)	Thermal conductivity of insulation product (W/m²K)			
	0.02	0.03	0.04	0.05
0.25	43	64	86	107
0.3	30	46	61	76
0.35	21	32	43	54

Source: BRE (Building Research Establishment) Good Building Guide 45 *Insulating Ground Floors*, C Stirling.

The insulation can be placed either above or below the concrete slab. When placed below the slab, the concrete retains heat, which can then be re-radiated back into the dwelling when the air temperature drops; this is known as a warm slab. The insulation product used in this type of slab will need to have a high compressive strength to support the weight of the concrete slab, such as a polyurethane board. Where insulation is placed above the concrete, known as a cold slab, the dwelling will heat up more quickly and will not need to be so strong. In both cases, it is important that the insulation in the floor and wall overlap so as to avoid cold spots or bridges where heat may escape.

CONCRETE SLAB

It is common to construct ground-bearing concrete slabs for dwellings on good stable subsoil of plain in-situ concrete of thickness 150 mm. In weaker ground, the concrete slab may be reinforced with steel-mesh fabric, particularly in the top of the slab to prevent surface cracking. Suspended in-situ reinforced concrete slabs are seldom used now, as suspended precast concrete beam and block floors are preferred due to their ease of construction and cost; however, they are still used in industrial and heavily loaded situations.

Suitable concrete mixes are produced to BS EN 206-1: *Concrete. Specification, performance, production and conformity*. A typical concrete mix for a plain in-situ concrete slab would be 50 kg cement:200 kg fine aggregate:400 kg coarse aggregate, or mix specification ST2, or grade GEN 1.

The reinforcement used in concrete slabs for domestic work is usually in the form of a welded steel fabric to BS 4483. Sometimes a light square-mesh fabric is placed 25 mm from the upper surface of the concrete slab to prevent surface crazing and limit the size of any cracking.

In domestic work the areas of concrete are defined by the room sizes, and it is not usually necessary to include expansion or contraction joints in the construction of the slab. The structural slab in dwellings is then covered by 50–65 mm thick concrete screed made from fine aggregate and cement. This can be trowelled smooth with a steel float to provide a flat surface to receive the floor coverings. In non-domestic dwellings, such as industrial units, a screed is not used but the surface of the structural slab is power-floated to provide a smooth dense surface.

Suspended concrete ground floor construction 5.2

Suspended concrete ground floor construction is increasingly the most common specified in domestic construction today. For a typical medium-sized detached house, with its prefabricated off-site components, it can be delivered and assembled in less than a day without the need for any specialist labour, plant or equipment.

BEAM AND BLOCK FLOOR

This type of domestic floor system is derived from the principles of the precast hollow and composite floor systems used for commercial buildings and apartments, as detailed in Part 5.5. It has developed into a cost- and time-effective means of constructing domestic ground and upper floors, by incorporating precast concrete beams with lightweight concrete blocks as an infilling. The benefits of quality-controlled factory manufacture of components and simple site assembly with the aid of a mobile crane to hoist the beams add to the following advantages:

- potential to span over unsound infilling, common to sloping sites;
- application over movable subsoils such as shrinkable clay (see Part 3, pages 104–105);
- suitability where ventilation under the ground floor is required to dilute intrusive gases;
- reduced site waste of concrete and temporary formwork materials.

The Building Regulations Approved Document C, requires a minimum clear void depth of 75 mm below these floors, but it is usual practice to leave at least 150 mm. Ventilation of the void is advisable to dilute and prevent concentration of gases from the ground (radon and/or methane) and possible leakage from piped services. Fig. 5.2.1 shows typical construction of a domestic suspended beam and block floor, where the stripped topsoil leaves the underfloor surface lower than adjacent ground. This is acceptable only if the soil is free-draining. Also, ground differentials should be minimal; otherwise the external wall becomes a retaining wall and will require specific design calculations. All organic material should be removed from the void, and the surface should be treated with weedkiller. Void depth may need to be as much as 225 mm in the presence of heavy clay subsoil and nearby trees (see Part 3.4, Table 3.4.3, page 108).

Upper floors follow the same principles of assembly, with purpose-made trimmer shoes providing support to concrete beams around stair openings. Span potential is only about 5 m: therefore intermediate support from a loadbearing partition or a steel beam is acceptable. Direct flange bearing is possible, but if a deep section is required the floor structure may be accommodated on ledger angles. Typical upper floor details are shown in Fig. 5.2.2 on the next page.

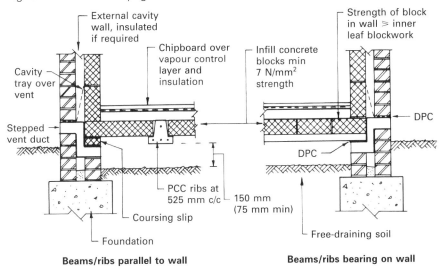

Beams/ribs parallel to wall   **Beams/ribs bearing on wall**

Figure 5.2.1 Beam and block domestic ground floor.

BEAM AND EXPANDED POLYSTYRENE (EPS) BLOCK

Precast concrete beams with EPS (expanded polystyrene) block infill units have developed from the beam and block principles applied to domestic floor construction. As a construction technique it has the advantages of speed and simplicity with exceptional thermal performance and, with each EPS block weighing less than 2 kg, the manual handling health and safety issues are minimal. Thermal insulation U-values for the floor as a whole are about 0.20 W/m²K, depending on the thickness and amount of EPS relative to rib spacing. The system is in effect a structurally adequate floor, with integral insulation.

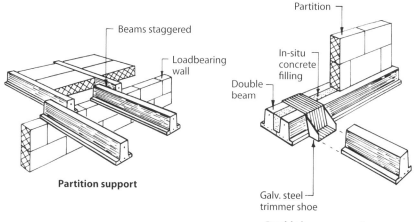

Partition support

Double beam support

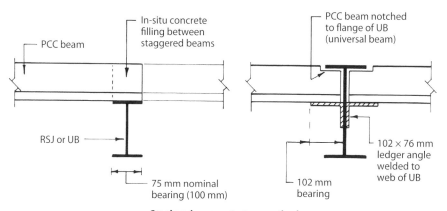

Steelwork support – two methods

Figure 5.2.2   Beam and block – upper floor intermediate support.

The construction principles are the same as described for suspended beam (rib) and block, with some variation on rib spacing to suit EPS block width. Figs. 5.2.3 and 5.2.4 show different block forms and applications. Fig. 5.2.3 shows EPS units functioning as both insulation and permanent shuttering to an in-situ reinforced concrete diaphragm suspended ground floor. Fig. 4.2.4 shows typical dry construction, using a moisture-resistant chipboard surface finish.

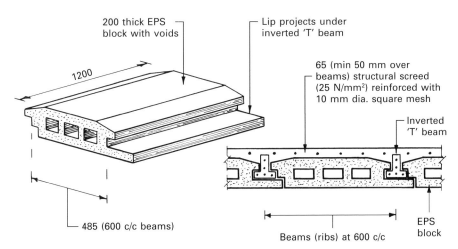

Figure 5.2.3   Typical EPS block floor for structural screed topping.

200 thick EPS block with voids

Lip projects under inverted 'T' beam

1200

65 (min 50 mm over beams) structural screed (25 N/mm²) reinforced with 10 mm dia. square mesh

Inverted 'T' beam

485 (600 c/c beams)

Beams (ribs) at 600 c/c

EPS block

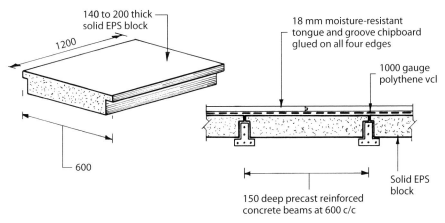

Figure 5.2.4   Typical EPS block floor for a chipboard finish.

140 to 200 thick solid EPS block

1200

18 mm moisture-resistant tongue and groove chipboard glued on all four edges

1000 gauge polythene vcl

600

150 deep precast reinforced concrete beams at 600 c/c

Solid EPS block

Suspended timber floors 5.3

Timber ground floors in houses are just as popular today as in previous generations; however, the only difference is that today the timber is usually an applied interior finish such as a laminate or timber block floor covering. The days when a master carpenter would be brought in to craft together an internal floor for a new dwelling are long past – that is, except in the small number of specialist contracts involving the maintenance of heritage or historic properties where those skills are still in demand.

HISTORICAL SUSPENDED TIMBER GROUND FLOORS

The ground floors on most UK housing stock built before 1939 comprise timber planks fixed to joists spanning over short honey-combed sleeper walls (a typical detail of this type of floor is shown in Fig. 5.3.1). After the Second World War the restricted availability of suitable timber made it difficult to continue in this way, so concrete floors replaced suspended timber floors. Although suspended timber floors are not constructed in this way anymore, you should be aware that this form of construction still needs to be maintained and adapted when the need arises. It is a more expensive form of construction than a concrete floor and can only be justified to match existing construction where a building is extended. It could be used on sloping sites that require a great deal of filling to make up the ground to the specified floor level, although a precast concrete flooring system could also be used, and this is likely to be much cheaper. Another disadvantage to the use of suspended timber floors is the possibility that the timber may get damp and hence be prone to dry rot, which is a form of fungal attack (see page 203). This can be overcome by adequate ventilation under the floor and the correct positioning of damp-proof courses to keep the underfloor area and timber dry. Through-ventilation is essential to keep the moisture content of the timber below that

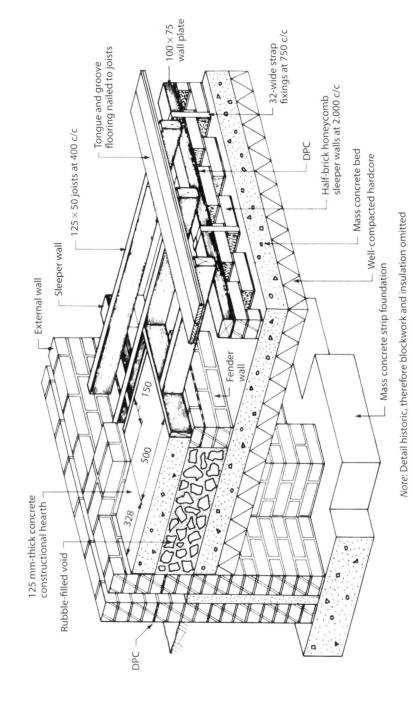

Tongue and groove flooring nailed to joists

100 × 75 wall plate

125 × 50 joists at 400 c/c

32-wide strap fixings at 750 c/c

DPC

Half-brick honeycomb sleeper walls at 2.000 c/c

Mass concrete bed

Well-compacted hardcore

Mass concrete strip foundation

External wall

Sleeper wall

Fender wall

150

500

328

125 mm-thick concrete constructional hearth

Rubble-filled void

DPC

Note: Detail historic, therefore blockwork and insulation omitted

Figure 5.3.1 **Typical details of suspended timber floor – joists at right angles to fireplace.**

which would allow fungal growth to take place: that is, 20 per cent of its oven-dry weight. The usual method is to allow a free flow of air under the floor covering by providing airbricks in the external walls. These are sited near the corners and at approximately 2 m centres around the perimeter of the building. They must have an equivalent of 1,500 mm²/m run of wall. If an old suspended timber floor is adjacent to a new solid ground floor as part of an extension, pipes of 100 mm diameter are used under the solid floor to convey air to and from the external walls to the old suspended timber floor.

Suspended timber ground floors are susceptible to draughts, and are said to be colder than other forms of flooring, and as such the retrofitting of insulation systems is a now common task for small or general builders. This can be achieved in a number of ways: for example, by the use of rigid insulation board fixed below the floor boarding on battens between the floor joists, or high-density mineral wool held on nylon netting pinned between the joists.

SUSPENDED TIMBER UPPER FLOORS

Timber, being a combustible material, is restricted by Part B of the Building Regulations to small domestic buildings as a structural flooring material. Its popularity in this context is due to its low cost in relation to other structural flooring methods and materials. Structural softwood is readily available at a reasonable cost, is easily worked and has a good strength-to-weight ratio, and is therefore suitable for domestic loadings. The most economic layout of floor joists is to span the joists across the shortest distance of the room.

Simple terminology used in upper timber floors

- **Common joist** A joist spanning from support to support.

- **Trimming joist** Span as for common joist, but it is usually 25 mm thicker and supports a trimmer joist.

- **Trimmer joist** A joist at right angles to the main span supporting the trimmed joists; it is usually 25 mm thicker than a common joist.

- **Trimmed joist** A joist cut short to form an opening, and supported by a trimmer joist; it spans in the same direction as common joists and is of the same section size.

Joist sizing

There are three ways of selecting a suitable joist size for supporting a domestic type floor.

1. Simple 'rule of thumb' for joists of 50 mm width, spaced at 400 mm centres

$$\frac{\text{span in mm}}{24} + 50 \text{ mm} = \text{depth in mm}$$

For example, for a 3.6 m span:

$$\frac{3600}{24} + 50 = 170 \text{ mm}$$

Therefore commercial size of joist chosen is 175 mm × 50 mm @ 400 mm c/c.

Some comparisons can be made with the methods shown in 2 and 3. However, the 'rule' is limited as a simple means of guidance for joist spacing at 400 mm only and should not be used to prove the structural efficacy of a timber joist.

2. Calculation

The current British Standard for the design of timber is BS5268:2-2002. The main section is Part 2, which covers design methods using traditional permissible stresses. The other part provides specific information for various forms of timber construction including fire resistance, a code of practice for the design of trussed rafter roofs and the design of timber-framed structures.

An illustrative example of the design of a timber floor joist using permissible design methods is:

$$BM = \frac{fbd^2}{6} = \frac{WL}{8}$$

where   BM = bending moment
             f = maximum fibre stress in N/mm² (see Part 4.8)
             b = breadth in mm
             d = depth in mm
            W = total load on joist in Newtons (N)
             L = clear length or span of joist in mm

Loading on each joist:
    Imposed load = 4.5 m × 0.45 m × 1.5 kN = 3.04 kN
Dead loads:
      Floorboards (4.5 m × 0.45 m × 10 kg)
  + Plasterboard (4.5 m × 0.45 m × 11 kg)
    = 42.53 kg × 9.81 × 10⁻³ = 0.42 kN

Total loading on each joist = 3.04 kN + 0.42 kN = 3.46 kN (3460 N)

Note: The self-weight of the joist is usually omitted, as this is small compared with the load carried. Moreover, the commercial joist size selected is normally greater than the calculated size, and this will cover the self-weight.

$$\frac{fbd^2}{6} = \frac{WL}{8}$$

From BS EN 338: 2009 the GS (general structural) grade timber of strength class C16 has the fibre stress (f) of 5.3 N/mm².

Transposing the bending moment formula to make d the subject and taking b, the breadth, as 50 mm:

$$d^2 = \frac{6WL}{8fb} = \frac{6 \times 3460 \times 4500}{8 \times 5.3 \times 50} = 44066$$

$$d^2 = \sqrt{44066} = 210 \text{ mm}$$

220 mm is the nearest commercial timber size above 210 mm. Therefore 220 mm × 50 mm @ 450 mm spacing will be selected.

3. Building Regulations, Approved Document A: *Eurocode 5 Span Tables for solid timber members in floors, ceilings and roofs for dwellings*, published by TRADA. An example of a span table is given in Table 5.3.1. For the above calculation, a comparable result with a maximum clear span of 4.52 m is 220 mm x 50 mm joists at 450 mm spacings with timber of strength class C16.

Table 5.3.1 provides some guidance where selection of timber is strength class C16.

Joists

If the floor is constructed with structural softwood joists of a size not less than that required by the Approved Document, the usual width is taken as 50 mm. The joists are spaced at 400–600 mm centre to centre, depending on the width of the ceiling boards that are to be fixed on the underside. Maximum economy of joist size is obtained by spanning in the direction of the shortest distance to keep within the deflection limitations allowed. The maximum economic span for joists is between 3,500 and 4,500 mm; for spans over this a double floor could be used.

Support

The ends of the joists must be supported by loadbearing walls. The common methods are to build in the ends and treat with preservative, or to use special metal fixings called **joist hangers**. Support on internal loadbearing walls can be by joist hangers or direct bearing when the joists are generally lapped (see Fig. 5.3.2).

Trimming

This is a term used to describe the framing of joists around an opening or projection. In properties built before the Second World War various carpentry joints were used to connect the members together, such as housing, bevelled and tusked tenon joints. These are often encountered in renovation of older properties but not used in modern construction, as they have been superseded by the use of metal straps and joist hangers as shown on

Fig. 5.3.2. Trimming around flues should comply with the recommendations of Approved Document J. It should be noted that the provision of upper-floor fireplaces is seldom included in modern designs. However, they are frequently encountered in renovation and maintenance work to older properties. Typical trimming joints and arrangements are shown in Fig. 5.3.3.

Table 5.3.1 Permissible clear spans for domestic floor joists for strength class C16.

Permissible clear spans for domestic floor joists
Imposed load not exceeding qk = 1.5 kN/m³ or Qk = 0.90 kN
Strength class C16

Size of joist		Dead loads gk (kNm²) excluding self-weight of joist								
		gk not more than 0.25			gk not more than 0.5			gk not more than 1.25		
		Spacing of joints (mm)								
Breadth (mm)	Depth (mm)	400	450	600	400	450	600	400	450	600
		Maximum clear span (m)								
38	97	1.75*	1.66*	1.43	1.64*	1.55*	1.35	1.43	1.35	0.71
38	120	2.36*	2.23*	1.94	2.15*	2.07*	1.80	1.66	1.77	1.55
38	145	2.85*	2.74*	2.48	2.68*	2.58*	2.32	2.33	2.22	1.96
38	170	3.33*	3.20*	2.90	3.14*	3.02*	2.73	2.74	2.63	2.37
38	195	3.81*	3.67*	3.32	3.59*	3.45*	3.12	3.14	3.01	2.71
38	220	4.29*	4.13*	3.74	4.05*	3.89*	3.52	3.53	3.39	3.06
50	97	2.00*	1.89*	1.65	1.87*	1.77*	1.54	1.61	1.53	1.34
50	120	2.59*	2.49*	2.22	2.44*	2.34*	2.05	2.09	1.99	1.75
50	145	3.12*	3.00*	2.72	2.94*	2.83*	2.56	2.57	2.47	2.21
50	170	3.65*	3.51*	3.19	3.44*	3.31*	3.00	3.01	2.89	2.61
50	195	4.17*	4.02*	3.65	3.94*	3.79*	3.44	3.45	3.31	3.00
50	220	4.70*	4.52*	4.11	4.43*	4.26*	3.87	3.85	3.73	3.38
63	97	2.23*	2.11*	1.84	2.07*	1.97*	1.72	1.78	1.70	1.50
63	120	2.80*	2.69*	2.44	2.64*	2.54*	2.28	2.30	2.19	1.94
63	145	3.37*	3.24*	2.95	3.18*	3.06*	2.78	2.79	2.68	2.42
63	170	3.3.94*	3.79*	3.45	3.72*	3.58*	3.25	3.26	3.13	2.84
63	195	4.50*	4.33*	3.94	4.25*	4.09*	3.72	3.73	3.58	3.25
63	220	5.06*	4.87*	4.44	4.78*	4.60*	4.18	4.20	4.04	3.66
75	120	2.96*	2.85*	2.59	2.79*	2.69*	2.44	2.45	2.35	2.09
75	145	3.56*	3.42*	3.12	3.37*	3.24*	2.94	2.95	2.84	2.57
75	170	4.16*	4.01*	3.65	3.93*	3.79*	3.44	3.45	3.32	3.01
75	195	4.75*	4.58*	4.17	4.49*	4.33*	3.94	3.95	3.80	3.45
75	220	5.34*	5.15*	4.70	5.05*	4.87*	4.43	4.45	4.26	3.88

*Two additional joists required – when supporting non-loadbearing partition walls.
Nominal bearing of 40 mm to be doubled.

Notes
1. The table allows for up to 1.5 kN/m² of imposed loading due to furniture and people. For imposed loading greater than this, calculations must be used.
2. For dead loading in excess of 1.25 kN/m², calculations must also be used.
3. Softwood floorboards should be at least 16 mm finished thickness for joist spacing up to 450 mm. 19 mm min thickness will be required for joist spacing of 600 mm.
4. Joists should be duplicated below a bath.
5. Dead loading is usually summated in kg. To convert to a force in newtons (N), multiply by the gravitational factor 9.81 (say 10). For example: 25 kg × 10 = 250 N, or 0.25 kN.
6. See BS 648: *Schedule of weights of building materials* for material dead loads.

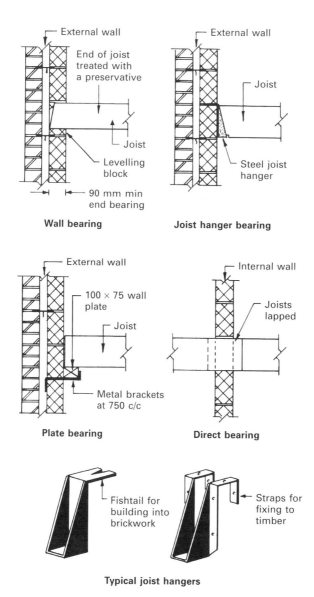

Wall bearing

Joist hanger bearing

Plate bearing

Direct bearing

Typical joist hangers

Note: Insulation in external walls had been ommitted for clarity.

Figure 5.3.2 Typical joist support details.

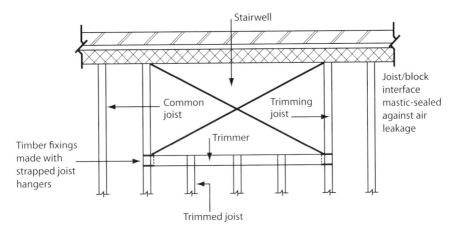

Trimming to stairwell

Stairwell

Common joist

Trimming joist

Joist/block interface mastic-sealed against air leakage

Trimmer

Timber fixings made with strapped joist hangers

Trimmed joist

Trimming around flues

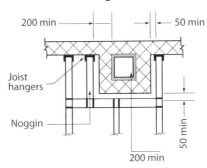

200 min

50 min

Joist hangers

Noggin

50 min

200 min

50 min

Figure 5.3.3   Typical trimming arrangements.

Strutting

Shrinkage in timber joists will cause twisting to occur: this will result in movement of the ceiling below, and could cause the finishes to crack. To prevent this, strutting is used between the joists if the total span exceeds 2,400 mm, the strutting being placed at mid-span (see Fig. 5.3.4).

Floor decking

The standard flooring provided for dwellings is moisture-resistant flooring grade particleboard, such as chipboard or oriented strand board (OSB) to BS EN 322:1993. The thickness of board required is 18 mm for joist centres not greater than 450 mm and 22 mm board for 600 mm centres. Boards can be either square-edged or tongued and grooved and come in modular sizes of either 0.6 x 1.2 m or 2.4 x 1.2 m. The boards are more expensive but provide a strong lateral bracing element, particularly when the joints are also glued as

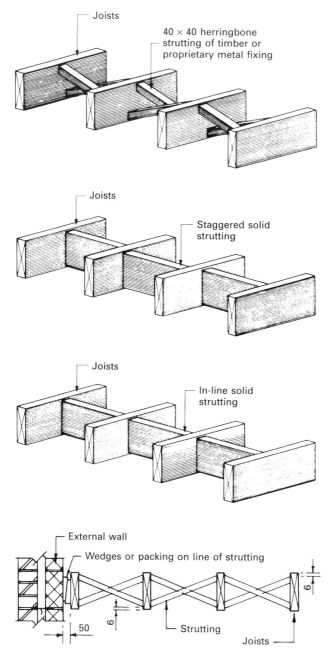

Joists

40 × 40 herringbone
strutting of timber or
proprietary metal fixing

Joists

Staggered solid
strutting

Joists

In-line solid
strutting

External wall

Wedges or packing on line of strutting

6

50

6

Strutting

Joists

Figure 5.3.4 Strutting arrangements.

well as nailed. An expansion of joint of no less than 10 mm should be left around the edge of the floor decking to allow for expansion. Access to services needs to be provided by the insertion of screw-down hatches supported on edge noggins.

Acoustic performance

Timber floors need careful detailing to prevent the passage of sound through the fabric of the structure in mixed-occupancy dwellings to comply with the Building Regulations Approved Document: Part E. Traditionally, to improve the sound insulation between ground and first-floor accommodation, a layer of dry sand known as pugging was placed within the floor void; this increased the mass of the floor and made it less prone to vibrate and transmit noise to the surrounding air. In 2004 the 'Robust Details' scheme was set up by Robust Details Limited, a UKAS (United Kingdom Accreditation Service) accredited product certification body. This scheme helps designers and developers meet the requirements of Part E by providing practical solutions to unwanted transmission of sound. Using this scheme, a number of modern methods can be used to add mass in the form of additional layers of plasterboard to the ceiling or sound-absorbent quilt materials within the floor void, or enabling the timber floor boarding to 'float' above the structural floor deck by supporting it on proprietary resilient battens or pads that absorb vibration. It is also critical to consider the edge detailing around the floor to reduce sound leakage at points of contact between floor decks, skirtings and walls. Fig. 5.3.5 shows a typical arrangement for sound-proofing a timber floor.

Double floors

These can be used on spans over 4,500 mm to give a lower floor area free of internal walls. They traditionally consisted of a timber binder spanning the shortest distance, which supports common joists spanning at right angles. However, in modern construction the timber binder is usually replaced by a steel beam as it is generally considered to be more economic and quicker to construct. The beam reduces the span of the common joists to a distance that is less than the shortest span to allow an economic joist section to be used. Typical details are shown in Fig. 5.3.6.

PREFABRICATED WEB JOISTS FOR SUSPENDED FLOOR CONSTRUCTION

A double floor can be used for modest spans using a steel joist or large cross-section timber beam as intermediate floor joist support. For greater spans double flooring is uneconomic, as intermediate beams would need to be excessively large to carry the floor structure while still retaining minimal deflection (max 0.003 × span).

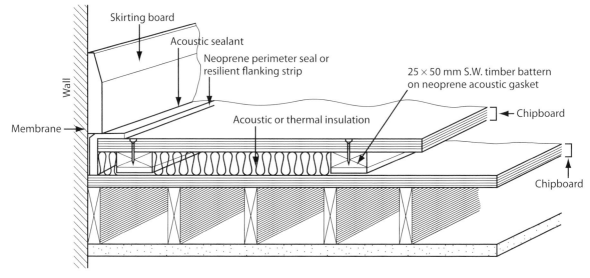

Figure 5.3.5 A typical arrangement for sound-proofing a timber floor.

With the increasing use of prefabricated elements there has been the development of engineered web-joist systems for upper floors as these are more economic than double floors. These can span up to about 8,000 mm at normal joist spacing of 400–600 mm.

Advantages of engineered web-joist systems

- Manufactured off-site to individual site/design dimensions; hence less site waste.
- High strength-to-weight ratio.
- Minimal shrinkage and movement, therefore limited use of strutting.
- Generally deeper than conventional joists, therefore less deflection.
- Wide section (typically ex. 75 mm flanges) provides large bearing area for floor deck and ceiling boards.
- Also suitable for other applications: for example, roofing construction.

Open-web joists have the additional benefit of providing virtually uninterrupted space for services, and they do not require end bearing; that is, joists can be top-flange suspended.

These web-joist systems are categorised as follows:

- open-web or steel web system joists;
- solid or boarded web joists.

OPEN-WEB JOISTS

These are a type of lattice frame, as shown in Fig. 5.3.7. They consist of a pair of parallel stress-graded timber flanges, spaced apart with V-shaped galvanised steel web members. Open-web joists are prefabricated off-site to specific spans in a factory quality-controlled environment. This system is particularly good for incorporating services such as pipes and small ducts within the floor void.

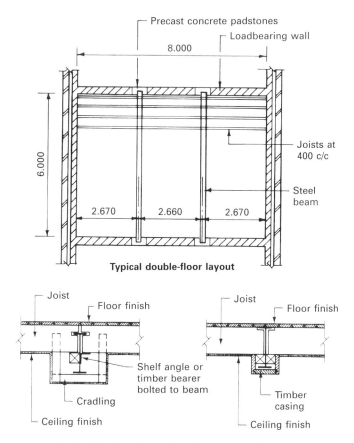

Typical double-floor layout

Typical details using timber binder or steel beam

Figure 5.3.6   Double floor details.

SOLID OR BOARDED WEB JOISTS

These are an alternative high-strength, relatively lightweight factory-produced joist system. A typical manufactured boarded web joist is shown in Fig. 5.3.8. This comprises a pair of stress-graded timber flanges separated by a central solid web of plywood or oriented strand board (OSB). At comparable joist depths, span potential is greater than with open web joists, with up to 10,000 mm possible. Solid web joists are less convenient for accommodating pipes and cables running at right angles to the joists. As for traditional timber-joisted floors, care and consideration must be exercised when cutting holes. The central area or neutral axis is the preferred location, away from the shear stress areas near supports. Flanges should never be notched.

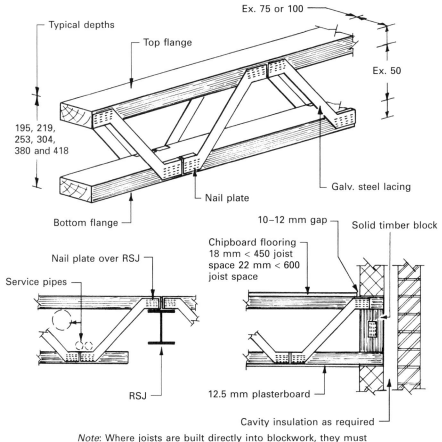

Note: Where joists are built directly into blockwork, they must be mastic sealed against air leakage.

Figure 5.3.7   Open-web joist.

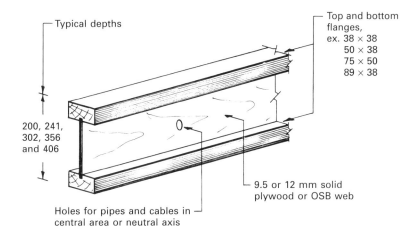

Typical depths

200, 241, 302, 356 and 406

Top and bottom flanges, ex. 38 × 38
50 × 38
75 × 50
89 × 38

9.5 or 12 mm solid plywood or OSB web

Holes for pipes and cables in central area or neutral axis

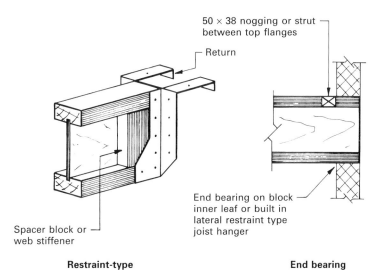

50 × 38 nogging or strut between top flanges

Return

Spacer block or web stiffener

End bearing on block inner leaf or built in lateral restraint type joist hanger

Restraint-type joist hanger

End bearing

Figure 5.3.8 Solid or boarded web joist.

Raised access floors 5.4

A raised floor is a system of elevated platforms suspended over the structural floor of in-situ or precast concrete. Used internally, they evolved primarily in response to the information technology revolution that occurred in the latter part of the twentieth century. Although the huge volume of power, telephone and data cabling in modern offices has been largely responsible for developments in raised platforms, location for service pipes and ventilation ducts has also been influential. The dominance of services issues has promoted a re-think for designers, with the need for greater floor-to-floor heights to incorporate these floor voids (and suspended ceilings).

PARTIAL ACCESS

This is where the floor finish is secured, and access is limited – possibly into simple ducts or trunking, as shown in Fig. 5.4.1. It is most suited for access to mains cabling for power and lighting, with void depths limited to about 100 mm. Accesses should be positioned to avoid furniture and should be strategically located over junction boxes. This provides flexibility for changes in cable distribution where workstations and work functions may change.

FULL ACCESS

This comprises standard-size (600 mm × 600 mm) interchangeable floor panels elevated on adjustable pedestals to provide multi-directional void space. Heights usually vary between 100 and 600 mm, but extremes of as little as 50 mm and as much as 2 m are possible. Adjustable-height pedestals are manufactured from steel or polypropylene, with support plates containing four lugs or projections to locate loose-fit decking panels. Threaded flat support plates may be specified where panels are to be screw-fixed in place (see Fig. 5.4.2).

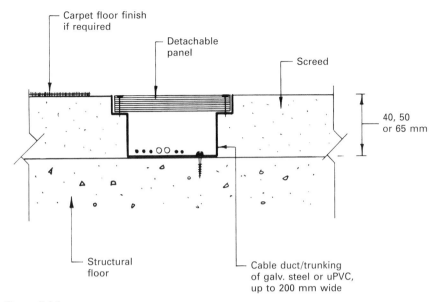

Figure 5.4.1   Floor duct.

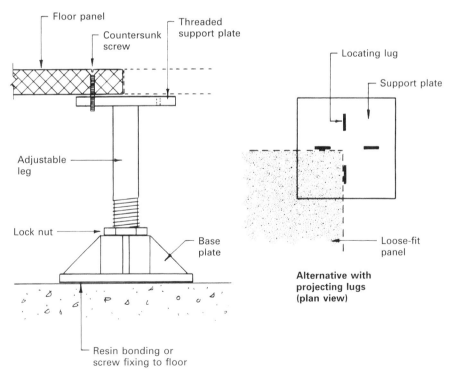

Alternative with
projecting lugs
(plan view)

Figure 5.4.2   Adjustable pedestal.

Floor panels or decking materials can vary considerably to suit application. Timber joists or battens may be used to support softwood boards or plywood/chipboard sheets, particularly if matching a traditional suspended floor. Office specification usually requires a fully interchangeable system with standard 600 mm × 600 mm panels. These are available in various gradings including light, medium and heavy, with corresponding maximum loadings of 10, 20 and 50 kN/m^2. Panel construction and application are shown in Fig. 5.4.3.

The combination of fully bonded chipboard within a steel casing provides structural soundness and integrity in the event of fire. The chipboard acts as a fire, acoustic and thermal insulator, and the steel resists tensile loading, also protecting the core against spread of flame. The presence of extraneous metal requires contact protection by electrical earth continuity through the complete floor system.

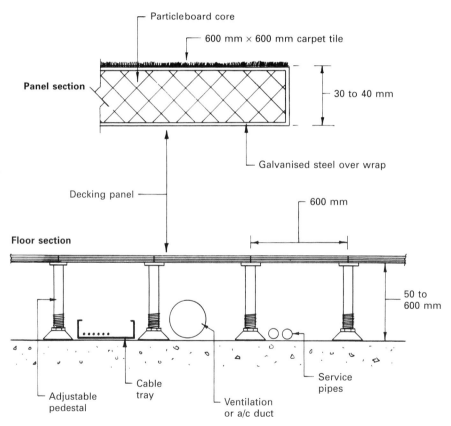

Figure 5.4.3   Raised access floor.

Precast concrete upper floors 5.5

The function of any floor is to provide a level surface that is capable of supporting all the imposed and dead loads. Reinforced concrete, with its flexibility in design, good fire resistance and sound-insulating properties, is widely used for the construction of suspended floors for all types of building. The disadvantages of in-situ concrete are:

- the need for formwork and supporting falsework;
- the provision for safe access and egress for fixing of the steelwork, and the placing, compacting and finishing of the wet concrete;
- the time taken for the concrete to cure before the formwork can be released for reuse and the floor made available as a working area;
- the very small contribution by a large proportion of the concrete to the strength of the floor.

Floors composed of reinforced precast concrete units have been developed over the years to overcome some or all of the disadvantages of in-situ reinforced concrete slab. To realise the full economy of any one particular precast flooring system, the design of the floors should be within the span, width, loading and layout limitations of the units under consideration, coupled with the advantages of repetition and the use of standard-sized units.

CHOICE OF SYSTEM

Before any system of precast concrete flooring can be considered in detail, the following factors must be taken into account:

- maximum span;
- nature of support;

- weight of units;

- thickness of units;

- thermal insulation properties;

- sound insulation properties;

- fire resistance of units;

- speed of construction;

- amount of temporary support required;

- having a safe lifting plan in place.

The systems available can be considered as either precast hollow floors or composite floors; further subdivision is possible by taking into account the amount of temporary support required during the construction period.

In each of these cases it is important that control measures are put in place to ensure that workers are not put at risk when working at height installing these types of floor. A number of systems can be used, ranging from fully scaffolded working platforms to safety nets slung below the working area or inflated air bags placed below that can break a worker's fall.

Precast hollow floors

Precast hollow floor units are available in a variety of sections such as box planks or beams, tee sections, I beam sections and channel sections (see Fig. 5.5.1). The economies that can reasonably be expected over the in-situ floor are:

- 50 per cent reduction in the volume of concrete;

- 25 per cent reduction in the weight of reinforcement;

- 10 per cent reduction in size of foundations.

The units are cast in precision moulds, around inflatable formers or foamed plastic cores. The units are laid side by side, with the edge joints being grouted together; a structural topping is not required, but the upper surface of the units is usually screeded to provide the correct surface for the applied finishes (see Fig. 5.5.1).

Little or no propping is required during the construction period, but usually some safe means of mechanical lifting is required to offload and position the units. Hollow units are normally the cheapest form of precast concrete suspended floor for simple straight spans, with beam or wall supports up to a maximum span of 20.000 m. They are not considered suitable where heavy point loads are encountered unless a structural topping is used to spread the load over a suitable area.

The hollow beams or planks give a flat soffit, which can be left in its natural state or be given a skim coat of plaster; the voids in the units can be used to house

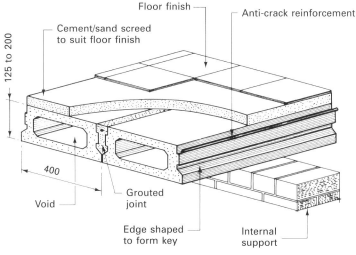

Floor finish

Anti-crack reinforcement

Cement/sand screed
to suit floor finish

125 to 200

400

Void

Grouted
joint

Edge shaped
to form key

Internal
support

Spans up to 13.000

Typical hollow floor unit details

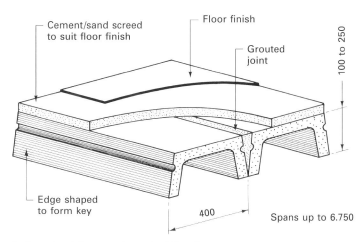

Cement/sand screed
to suit floor finish

Floor finish

Grouted
joint

100 to 250

Edge shaped
to form key

400

Spans up to 6.750

Anti-crack reinforcement required if units
are continuous over internal supports

Typical channel section floor unit details

Figure 5.5.1   Precast concrete hollow floors.

the services that are normally incorporated in the depth of the floor, or can be incorporated into the heating or cooling air ducting around the structure. The ribbed soffit of the channel and tee units can be masked by a suspended ceiling; again, the voids created can be utilised to house the services. Special units are available with fixing inserts for suspended ceilings, service outlets and edges to openings.

Composite floors

These floors are a combination of precast units and in-situ concrete. The precast units, which are usually pre-stressed or reinforced with high-yield steel bars, are used to provide the strength of the floor with the smallest depth practicable and at the same time act as permanent formwork to the in-situ topping, which provides the compressive strength required. It is essential that an adequate bond is achieved between the two components. In most cases this is provided by the upper surface texture of the precast units; alternatively a mild steel fabric can be fixed over the units before the in-situ topping is laid.

Composite floors generally take one of two forms:

- thin pre-stressed planks with a side key and covered with an in-situ topping (see Fig. 5.5.2);

- reinforced or pre-stressed narrow beams, which are placed at 600 mm centres and are bridged by concrete filler blocks known as **pots**. The whole combination is covered with in-situ structural concrete topping. Most of the beams used in this method have a shear reinforcing cage projecting from the precast beam section (see Fig. 5.5.3).

In both forms, temporary support should be given to the precast units by props at 1.800–2.400 m centres until the in-situ topping has cured.

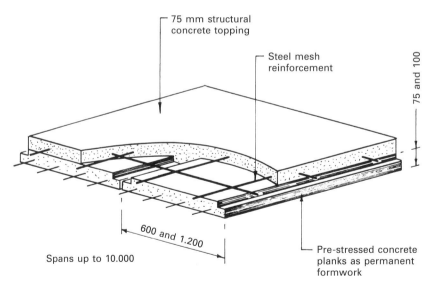

Figure 5.5.2   Composite floors – pre-stressed plank.

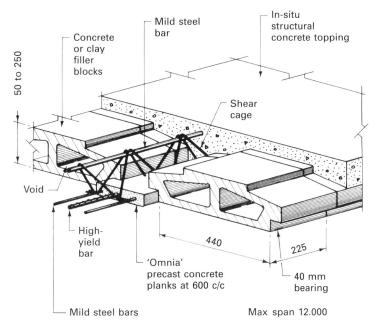

Typical composite floor using PCC planks

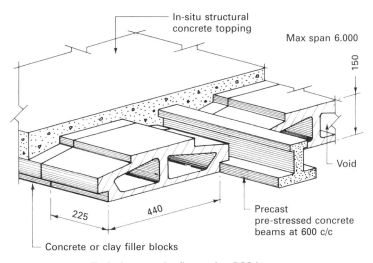

Typical composite floor using PCC beams

Figure 5.5.3   Composite floors – beam or pot.

COMPARISON OF SYSTEMS

Precast hollow floors are generally cheaper than composite; in-situ concrete is not required, and therefore the need for mixing plant and storage of materials is eliminated. The units are self-centring, and temporary support is not required; the construction period is considerably shorter; and generally the overall weight is less.

Composite floors will act in the same manner as an in-situ floor and can therefore be designed for more complex loadings. The formation of cantilevers is easier with this system, and support beams can be designed within the depth of the floor, giving a flat soffit. Services can be housed within the structural in-situ topping, or within the voids of the filler blocks. Like the precast hollow floor, composite floors are generally cheaper than a comparable in-situ floor, within the limitations of the system employed.

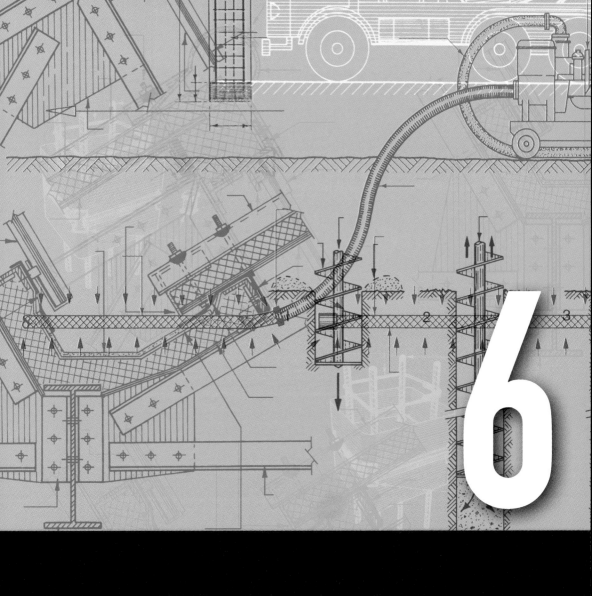

6

Roofs

Roofs: Timber, flat and pitched 6.1

The structural element that keeps water from entering our homes is, of course, the roof. A roof completes the building envelope and directs water into the guttering and down pipes to discharge into a drain. Roof eaves and overhangs protect the external walls from driving rain and wind, and provide an aesthetic look to a modern or traditional building. Traditional roofs tended to be pitched as the material used was a natural product, such as thatch, which relies on a pitch to direct rain down the roof covering. Modern roofing materials have now allowed the use of a flat roof design to be utilised in domestic extensions and commercial applications.

FUNCTIONS

The functions of any roof are to:

- keep out rain, wind, snow and dust;
- prevent excessive heat loss in winter;
- keep the interior of the building cool in summer;
- accommodate all stresses encountered;
- accept movement due to changes in temperature and moisture content;
- provide lateral restraint and stability to adjacent walls;
- resist penetration of fire and spread of flame from external sources.

There is a variety of roof types that will provide the above functions but the main domestic applications are pitched and flat.

TIMBER FLAT ROOF CONSTRUCTION

A flat roof is essentially a low-pitched roof, and is defined in BS 6100: *Glossary of building and civil engineering terms* as a pitch of 10° or less to the horizontal. Generally the angle of pitch is governed by the type of finish that is to be applied to the roof.

This form of roof is suitable for spans up to 4 m. With spans over 4 m, the depth of the timber joists would be excessive to counter the amount of bending that would occur. Larger spans like this are usually covered with a reinforced concrete slab or a patent form of decking.

The advantages of timber flat roofs are that they:

▨ are an economical solution for extensions to a main roof;

▨ can be designed under the 2.5 m height to meet the planning restriction rule;

▨ can be used for roof terrace applications.

The disadvantages of flat roofs are that they:

▨ increase the depth of construction due to the increases in U-values with over-joist insulation;

▨ may contrast in style with other buildings in the vicinity and, if an extension, the building to which it is attached;

▨ can encourage the collection of pools of water on the surface causing local variations in temperature, unless they are properly designed and constructed; this results in deterioration of the covering and, consequently, high maintenance costs;

▨ have little or no space to accommodate services.

CONSTRUCTION

The construction of a timber flat roof follows the same methods as those employed for the construction of timber upper floors (see page 223). Suitable joist sizes can be obtained by structural design or by reference to tables in publications recommended in the TRADA design tables for load span timber sizing (see page 251). The spacing of roof joists is controlled by the width of decking material to be used and/or the width of ceiling board on the underside. Timber flat roofs are usually constructed to fall in one direction towards a gutter or outlet. This is achieved by the application of **firrings** (pre-cut sloping timbers) set to a constant fall (see Fig. 6.1.1). The materials used in timber flat roof construction are generally poor thermal insulators, and therefore insulation has to be laid over the roof boards before the felting or behind the plasterboard ceiling or between joists to achieve the requirements of Part L of the Building Regulations.

Decking materials

Decking can be in the form of softwood boarding, chipboard, oriented strand board or plywood. Exterior grade water- and boil-proof (WBP) plywood is

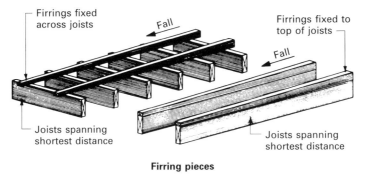

Firrings fixed
across joists

Fall

Firrings fixed to
top of joists

Fall

Joists spanning
shortest distance

Joists spanning
shortest distance

Firring pieces

Figure 6.1.1   Timber flat roof slope details.

available in sheet form, which requires fixing on all four edges. This means noggins will be required between joists to provide the bearing for end fixings. Chipboard, which is fixed in a similar manner to plywood, can be susceptible to moisture movement, so for roofing it should be specified moisture resistant to BS EN 312 2010, board type P3, P5, P7.

Flat roofs must have the void between ceiling and decking ventilated to prevent condensation occurring or the inclusion of vapour checks to prevent the passage of moisture. It is also advisable to use structural timbers that have been treated against fungal and insect attack. In certain areas, treatment to prevent softwood infestation by the house longhorn beetle is a requirement under the Building Regulations. Details can be found in the Approved Document to support Regulation 7.

Wood wool slabs
These are 610 mm-wide slabs of various lengths, which can span up to 1,200 mm. Thickness varies; for roof decking 51 mm is normally specified. The slabs are made of shredded wood fibres that have been chemically treated and are bound together with cement.

Insulating materials

There are many types of insulating material available, usually in the form of boards or quilts. Insulation boards laid over the decking create a **warm deck roof**, whereas quilted materials draped over the joists or placed between them make a **cold deck**. Great care needs to be taken with the design in order to achieve the U-value required and to prevent the build-up of any interstitial condensation within the roof void.

The development of cut-to-falls insulation has removed the need for firrings as the insulation has a fall manufactured within it. The insulation is bonded to the roof deck and the felt is laid and bonded over it to complete the roof. A vapour-check layer needs to be incorporated to prevent the build-up of condensation within the roof void.

Decking

Decking was originally made from lightly compressed vegetable fibres, bonded with natural glues or resins. Compressed straw and wood wool slabs have been particularly successful as decking materials with the benefit of in-built insulation.

For insulation-only purposes, the most popular materials for roof decking boards are high-density mineral wool with a tissue membrane bonded to one surface, expanded polystyrene, or polyurethane slabs. A warm deck roof has insulation board placed over the decking, but below the waterproof membrane. Conversely, an inverted warm deck has the insulation board above the waterproof membrane. The inverted warm deck insulation board must be unaffected by water and must be capable of receiving a surface treatment of stone granules or ceramic pavings. With this technique, the waterproof membrane is protected from the stresses caused by exposure to weather extremes. Examples of the various forms of construction are shown in Fig. 6.1.2.

Quilts

These are made from mineral or glass wool that is loosely rolled, with the option of a kraft-paper facing. The paper facing is useful as the wool is in fine shreds, which give rise to irritating scratches if handled. Quilts rely on the loose way in which the core is packed for their effectiveness, and therefore the best results are obtained when they are laid between joists.

A variety of loose fills is also available for placing between the joists and over the ceiling to act as thermal insulators. Existing flat roofs can have their resistance to heat loss improved by applying thermal insulation laminated plasterboard to the ceiling.

Weatherproof finishes

Suitable materials are asphalt, glass fibre, green roof construction, EPDM (ethylene propylene diene Monomer) synthetic rubber, zinc and built-up roofing felt; only the latter will be considered at this stage as this is the most common and economical method to use.

Built-up roofing felt

Most roofing felts consist essentially of a base sheet of glass fibre or polyester reinforcement, impregnated with hot bitumen during manufacture. This is coated on both sides with a stabilised weatherproof bitumen compound. The outer coating is dusted with sand while still hot and tacky. The underlayer may receive a thin layer of polythene to prevent the sheet from sticking to itself when rolled. After cooling the felt is cut to form rolls 1 m wide and 10 or 20 m long before being wrapped for dispatch.

BS EN 13707: *Reinforced bitumen sheets for roof waterproofing* covers the characteristics and specification of this roof covering.

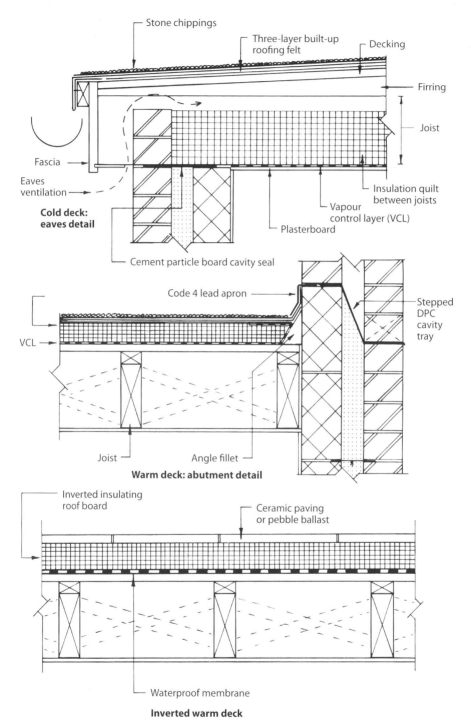

Cold deck: eaves detail

- Stone chippings
- Three-layer built-up roofing felt
- Decking
- Firring
- Joist
- Fascia
- Eaves ventilation
- Insulation quilt between joists
- Vapour control layer (VCL)
- Plasterboard
- Cement particle board cavity seal

Warm deck: abutment detail

- Cement particle board cavity seal
- Code 4 lead apron
- Stepped DPC cavity tray
- VCL
- Joist
- Angle fillet

Inverted warm deck

- Inverted insulating roof board
- Ceramic paving or pebble ballast
- Waterproof membrane

Figure 6.1.2 Timber flat roofs.

For flat roofs, three layers of felt should be used, the first being laid at right angles to the fall, commencing at the eaves. If the decking is timber, the first layer is secured with large flat-head felt nails, and subsequent layers are bonded to it with a hot bitumen compound by a roll-and-pour method. This involves pouring molten bitumen on the decking or underlayer and unrolling the sheet over it. Torch-on is for specially made sheets that are heated on the underside to produce a wave of molten bitumen while the sheet is unrolled. If the decking is of a material other than timber, all three layers are bonded with hot bitumen compound. It is usually recommended that a vented first layer be used in case moisture is trapped during construction; this recommendation does not normally apply to roofs with a timber deck, as timber has the ability to 'breathe'. The minimum fall recommended for built-up roofing felt is 17 mm in 1,000 mm or 1°.

Flat roofs, are often finished with chippings that protect the underlying felt, provide additional fire resistance and give increased solar reflection, which reduces the effects of heating on the felt. A typical application would be 12.5 mm stone chippings at approximately 50 kg to each 2.5 m² of roof area. Chippings of limestone, granite and light-coloured gravel would be suitable.

Vapour-control layer/membrane

The problem of condensation should always be considered when constructing a flat roof. The position of the vapour-control layer will be determined by the cold or warm roof construction. When the insulation is below the built-up roofing felt, it will not prevent condensation occurring: as the insulation is a permeable material, water and vapour will pass upwards through it and condense on the underside of the roofing felt. The drops of moisture so formed will soak into the insulating material, lowering its insulation value and possibly causing staining on the underside. To prevent this occurring, a vapour-control layer should be placed on the underside of the insulating material. Purpose-made plasterboard with a metallised polyester backing can be specified in this situation, to combine water-vapour resistance with the ceiling lining. The inverted warm deck roof will not require a separate vapour-control layer, as the waterproof membrane will provide this function. For typical timber flat roof details, see Figs. 6.1.1 and 6.1.2.

FLAT ROOF JOIST SIZING

Joist sizes can be determined by calculation or by reference to design tables. The Timber Research and Development Association (TRADA) publish *Span tables for solid timber members in floors, ceilings and roofs (excluding trussed rafter roofs) for dwellings*. These tables give a timber size for a defined span and timber centres. Table 6.1.1 on page 252 gives some guidance from one of the tables for flat roof joists illustrating loading and the spans permissible with the cross-section sizes.

Table 6.1.1   Guide to span and loading potential for GS grade softwood timber flat roof joists (for timber grading categories, see Table 4.8.2, page 200).

Permissible clear spans for joists for flat roofs with access for maintenance and repair only

Imposed load 0.75 kN/m²
Strength class C16   Service class 1 or 2

Size of joist		Dead loads gk (kNm²) excluding self-weight of joist								
		Not more than 0.50			More than 0.50 but not more than 0.75			More than 0.75 but not more than 1.00		
		Spacing of joints (mm)								
Breadth (mm)	Depth (mm)	400	450	600	400	450	600	400	450	600
		Maximum clear span (m)								
38	97	1.74	1.72	1.67	1.67	1.65	1.58	1.61	1.58	1.51
38	120	2.33	2.30	2.21	2.21	2.17	2.08	2.12	2.08	1.97
38	145	2.98	2.94	2.82	2.82	2.76	2.63	2.68	2.63	2.48
38	170	3.66	3.60	3.38	3.43	3.36	3.18	3.26	3.18	2.99
38	195	4.34	4.26	3.88	4.06	3.97	3.65	3.83	3.74	3.46
38	220	4.99	4.80	4.37	4.68	4.52	4.11	4.42	4.30	3.90
47	97	1.93	1.90	1.84	1.84	1.81	1.74	1.77	1.74	1.66
47	120	2.56	2.52	2.43	2.43	2.39	2.27	2.32	2.27	2.16
47	145	3.27	3.22	3.08	3.08	3.02	2.87	2.93	2.87	2.70
47	170	4.00	3.93	3.63	3.75	3.67	3.42	3.55	3.47	3.25
47	195	4.73	4.57	4.16	4.41	4.31	3.92	4.17	4.07	3.72
47	220	5.34	5.14	4.68	5.04	4.85	4.41	4.79	4.61	4.19
63	97	2.20	2.17	2.10	2.10	2.06	1.98	2.01	1.98	1.88
63	120	2.91	2.87	2.75	2.75	2.70	2.57	2.63	2.57	2.43
63	145	3.70	3.64	3.42	3.48	3.41	3.22	3.30	3.23	3.04
63	170	4.50	4.39	4.00	4.21	4.12	3.77	3.98	3.89	3.58
63	195	5.21	5.02	4.58	4.92	4.74	4.31	4.66	4.51	4.10
63	220	5.85	5.64	5.15	5.53	5.33	4.86	5.27	5.07	4.62
75	120	3.13	3.06	2.96	2.96	2.90	2.76	2.82	2.76	2.61
75	145	3.97	3.90	3.62	3.72	3.65	3.41	3.53	3.45	3.24
75	170	4.81	4.64	4.23	4.50	4.38	3.99	4.25	4.15	3.79
75	195	5.50	5.30	4.84	5.19	5.01	4.57	4.95	4.77	4.34
75	220	6.17	5.96	5.45	5.84	5.63	5.14	5.57	5.36	4.89
American/Canadian lumber sizes										
38	140	2.85	2.82	2.69	2.69	2.64	2.51	2.57	2.51	2.38
38	184	4.04	3.97	3.66	3.78	3.70	3.44	3.58	3.49	3.27
38	235	5.32	5.12	4.66	5.02	4.83	4.38	4.76	4.59	4.16

PITCHED ROOFS

The term **pitched roof** includes any roof whose angle of slope to the horizontal lies between 10° and 70°; below this range it would be called a flat roof and above 70° it would be classified as a wall.

The pitch is generally determined by the roof covering material that is specified to be used. The terminology used in timber roof work and the basic members for various spans are shown in Figs. 6.1.3 and 6.1.4.

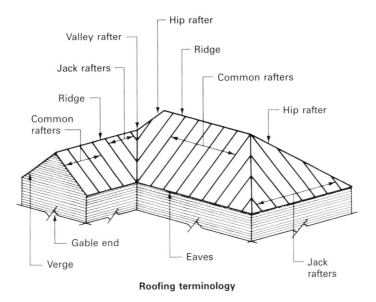

Roofing terminology

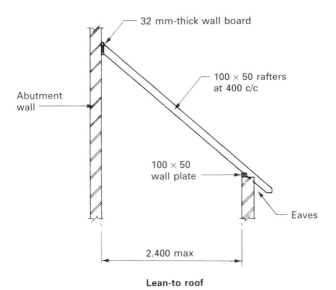

Lean-to roof

Figure 6.1.3   Roofing terminology and lean-to roof.

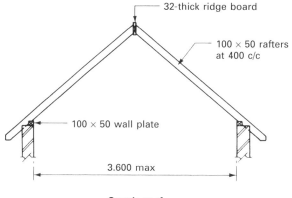

Couple roof

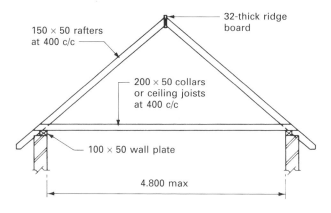

Closed couple roof

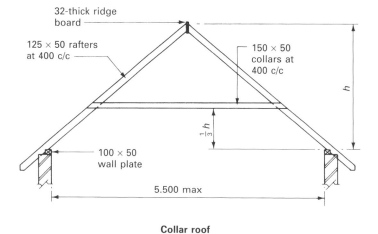

Collar roof

Figure 6.1.4 Pitched roofs for small spans.

The advantages of pitched roofs are that they:

- create a roof space that can be used;
- have a longer lifespan than flat roofs;
- are more aesthetically pleasing than flat roofs;
- are more effective than flat roofs in draining rainwater.

The disadvantages of pitched roofs are that:

- the cost of constructing a pitched roof is greater than for a flat roof;
- large spans will require designed roof trusses and calculations.

The timber used in roof work is structural softwood, the members being joined together with nails. The sloping components or rafters are used to transfer the covering, wind, rain and snow loads to the loadbearing walls on which they rest. The rafters are sometimes assisted in this function by struts and purlins in what is called a **purlin roof** or **double roof** (see Fig. 6.1.5 on page 258). As with other forms of roof, the spacing of the rafters and consequently the ceiling joists is determined by the module size of the ceiling boards that are to be fixed on the underside of the joists.

ROOF MEMBERS

- **Ridge** This is the spine of a roof and is essentially a board to ensure that the rafters are lined up and nailed together. The depth of ridge board is governed by the pitch of the roof: the steeper the pitch, the deeper will be the vertical or plumb cuts on the rafters abutting the ridge.
- **Common rafters** These are the main loadbearing members of a roof; they span between a wall plate at eaves level and the ridge. Rafters have a tendency to thrust out the walls on which they rest, and this must be resisted by the walls and the ceiling joists. Rafters are notched over and nailed to a wall plate situated on top of a loadbearing wall; the depth of the notch should not exceed one-third of the depth of the rafter.
- **Jack rafters** These fulfil the same function as common rafters but span from ridge to valley rafter or from hip rafter to wall plate, and are therefore shorter than common rafters.
- **Hip rafters** These are similar to a ridge but form the spine of an external angle and are similar to a rafter spanning from ridge to wall plate.
- **Valley rafters** These are similar to hip rafters but form an internal angle.
- **Wall plates** These provide the bearing and fixing medium for the various roof members, and distribute the loads evenly over the supporting walls; they are bedded in cement mortar on top of the loadbearing walls and strapped down using galvanised straps and stainless steel screws.

- **Dragon ties** These ties are placed across the corners and over the wall plates to help provide resistance to the thrust of a hip rafter.
- **Ceiling joists** These fulfil the dual function of acting as ties to the feet of pairs of rafters and providing support for the ceiling boards on the underside and any cisterns housed within the roof void.
- **Purlins** These act as beams, reducing the span of the rafters and enabling an economic section to be used. If the roof has a gable end they can be supported on a corbel or built in, but in a hipped roof they are mitred at the corners and act as a ring beam.
- **Struts** These are compression members that transfer the load of a purlin to a suitable loadbearing support within the span of the roof.
- **Collars** These are extra ties to give additional strength, and are placed at purlin level.
- **Binders** These are beams used to give support to ceiling joists and counteract excessive deflections, and are used if the span of the ceiling joist exceeds 2,400 mm.
- **Hangers** These vertical members are used to give support to the binders and allow an economic section to be used; they are included in the design if the span of the binder exceeds 3,600 mm.

Note: The arrangement of struts, collars and hangers occurs only on every fourth or fifth pair of rafters.

PITCHED ROOF COMPONENT SIZING

The size of the principal timber members constituting a roof structure can be determined by calculation or from design tables as referred to under flat roof joist sizing on page 252. The TRADA tables again provide details of the various roof components, grades of timber, different roof pitches and loadings. As an example, Table 6.1.2 is provided for guidance on selection of rafters for pitches between 22.5° and 30° with an imposed loading of 1.00 kN/m².

EAVES

The eaves of a roof is the lowest edge that overhangs the wall, thus giving the wall a degree of protection; it also provides the fixing medium for the rainwater gutter. The amount of projection from the wall of the eaves is a matter of choice but is generally in the region of 300–450 mm.

Table 6.1.2 Guide to span and loading potential for GS grade softwood timber rafters.

Permissible clear spans for common or jack rafters

Slope of roof 22.5° or more but less than 30.0°
Strength class C16   Service class 1 or 2

Size of joist		Dead loads gk (kNm²) excluding self-weight of joist								
		Not more than 0.5			More than 0.5 but not more than 0.75			More than 0.75 but not more than 1.00		
		Spacing of rafters (mm)								
Breadth (mm)	Depth (mm)	400	450	600	400	450	600	400	450	600
		Maximum clear span (m)								
38	100	2.19	2.14	1.97	2.02	1.96	1.83	1.88	1.83	1.69
38	125	2.81	2.70	2.45	2.67	2.56	2.31	2.54	2.44	2.15
38	150	3.37	3.24	2.94	3.19	3.07	2.74	3.05	2.93	2.56
47	100	2.42	2.33	2.11	2.29	2.21	2.00	2.19	2.10	1.91
47	125	3.02	2.90	2.64	2.86	2.75	2.50	2.73	2.63	2.38
47	150	3.61	3.48	3.16	3.43	3.29	2.99	3.27	3.15	2.85
ALS/CLS										
38	89	1.83	1.80	1.70	1.70	1.66	1.55	1.60	1.55	1.44
38	140	3.15	3.03	2.75	2.98	2.87	2.57	2.85	2.73	2.40

Note: Clear span is measured between wall plate and purlin and between purlin and ridge board.

There are two basic types of eaves finish: **open eaves** and **closed eaves**. The former is a less expensive and not very common method currently as the ends of the rafter are exposed. With both, the space between the rafters and the roof covering receives insulation over the top of the wall. A 50 mm air space must remain above the insulation for free air circulation through the roof in order to prevent condensation from occurring. This is easily achieved with a proprietary eaves ventilator secured between the rafters. A continuous triangular tilting fillet is fixed over the backs of the rafters to provide support for the bottom course of slates or tiles. A closed eaves is one in which the feet of the rafters are boxed in using a vertical fascia board, with the space between the fascia and the wall containing a ventilated soffit board or over the fascia board ventilator. In a cheaper variant, the rafters are cut marginally beyond the wall face to leave space for ventilation and only a fascia board fixed to the rafter ends. This is called a **flush eaves**. Figs. 6.1.5 and 6.1.6 indicate various rafter finishes, with provision for adequate through-ventilation.

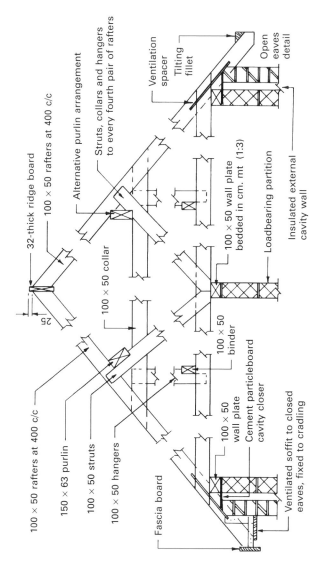

100 × 50 rafters at 400 c/c

32-thick ridge board

100 × 50 rafters at 400 c/c

Alternative purlin arrangement

Struts, collars and hangers
to every fourth pair of rafters

25

100 × 50 collar

150 × 63 purlin

100 × 50 struts

100 × 50 hangers

Fascia board

100 × 50
binder

100 × 50
wall plate

Cement particleboard
cavity closer

Ventilated soffit to closed
eaves, fixed to cradling

100 × 50 wall plate
bedded in cm. mt (1:3)

Loadbearing partition

Insulated external
cavity wall

Ventilation
spacer

Tilting
fillet

Open
eaves
detail

Note: Mineral wool insulation (or equivalent) placed between and over ceiling joists or between rafters.
Thickness in accordance with Building Regulations Approved Document L.

Figure 6.1.5   **Typical double or purlin roof details for spans up to 7.200 mm.**

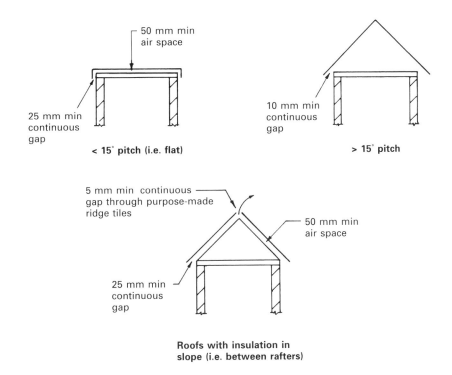

Note: Gaps need not be continuous, if staggered voids provide the equivalent area.

Figure 6.1.6 Roof ventilation requirements.

TRUSSED RAFTERS

This is the modern approach to the construction of a domestic timber roof, giving a clear span by using a triangulated frame. The timber members are jointed using galvanised truss plates, which are hydraulically fixed to produce a secure joint. All members in a trussed rafter are machined on all faces so that they are identical and of uniform thickness, ensuring a strong connection on both faces. The trussed rafters are placed at 600 mm centres and tied together over their backs with 38 mm × 25 mm tiling battens. Stability is achieved from the tile battens and 100 mm × 25 mm diagonal wind braces from the bottom corner to a top corner. Also, two 100 mm × 25 mm longitudinal ties should run horizontally along the ceiling ties. (See Figs. 6.1.7 and 6.1.8).

Truss manufacturers will design and prefabricate to the client's specification. This should include span, loading (type of tile), degree of exposure, pitch, spacing (if not 600 mm) and details of any special loadings such as water cistern location.

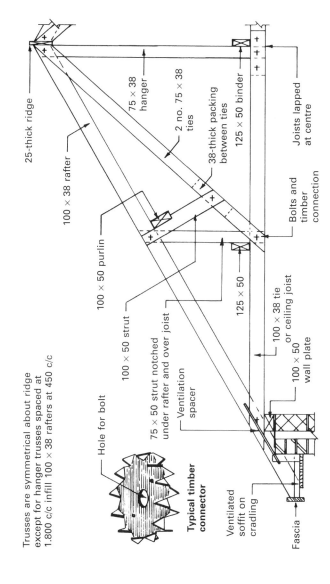

Trusses are symmetrical about ridge
except for hanger trusses spaced at
1.800 c/c infill 100 × 38 rafters at 450 c/c

25-thick ridge

100 × 38 rafter

75 × 38 hanger

2 no. 75 × 38 ties

38-thick packing between ties

125 × 50 binder

Joists lapped at centre

Bolts and timber connection

100 × 50 purlin

Hole for bolt

100 × 50 strut

75 × 50 strut notched under rafter and over joist

Ventilation spacer

125 × 50

100 × 38 tie or ceiling joist

100 × 50 wall plate

Typical timber connector

Ventilated soffit on cradling

Fascia

Figure 6.1.7   Typical truss details for spans up to 8.000 mm.

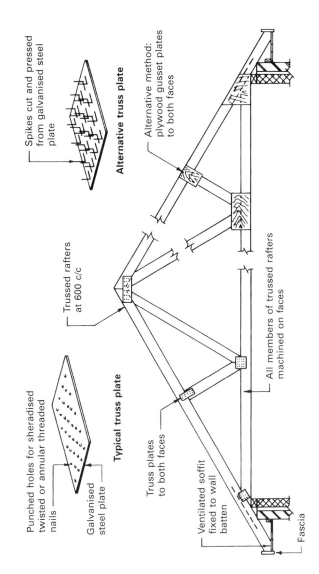

Spikes cut and pressed from galvanised steel plate

Alternative truss plate

Alternative method: plywood gusset plates to both faces

Trussed rafters at 600 c/c

All members of trussed rafters machined on faces

Punched holes for sheradised twisted or annular threaded nails

Galvanised steel plate

Typical truss plate

Truss plates to both faces

Ventilated soffit fixed to wall batten

Fascia

Figure 6.1.8   Typical trussed rafter details for spans up to 11.000 mm.

LATERAL BRACING

The roof trusses are secured to the structure at the gables using restraint straps and gable ladders, which also stabilises the external wall. Fig. 6.1.9 shows the application with straps at 2 m spacing. Straps may need to be more frequent, depending on building height and degree of exposure.

ROOFING AND BUILDING REGULATIONS

Approved Document A

This gives guidance on the sizing of roof timbers that should be specified from the TRADA publication *Span tables for solid timber members in floors, ceilings and roofs (excluding trussed rafter roofs) for dwellings.*

Table 1 lists geographical areas where the softwood timber used for roof construction should be adequately treated with a suitable preservative to prevent infestation by the house longhorn beetle.

Approved Document B1

This requires the roof to offer adequate resistance to the spread of fire over the roof. Table 5 gives limitations on roof coverings for dwelling houses by designations and distance from the boundary.

Approved Document C

This requires that the roof of a building shall adequately resist the passage of moisture to the inside of the building.

Approved Document L1

This states that reasonable provision shall be made for the conservation of fuel and power in buildings. To satisfy this requirement, Approved Document L requires that the roof construction is enhanced with suitable insulation material.

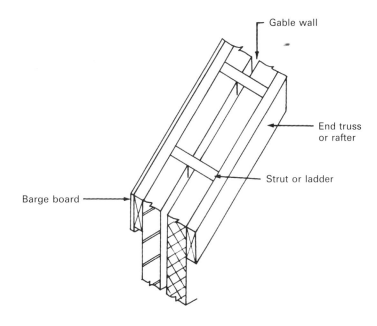

Gable wall

End truss
or rafter

Strut or ladder

Barge board

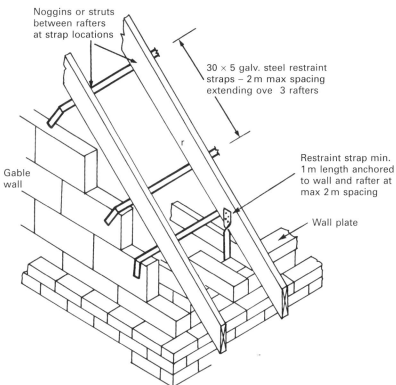

Noggins or struts
between rafters
at strap locations

30 × 5 galv. steel restraint
straps – 2 m max spacing
extending ove 3 rafters

r

Restraint strap min.
1 m length anchored
to wall and rafter at
max 2 m spacing

Gable
wall

Wall plate

Figure 6.1.9   Lateral restraint to gable walls.

Roof tiling and slating 6.2

The roof structures detailed in Part 6.1 have to be covered with a roof finish to provide a structure that protects the building from the effects of rain, snow, wind, sun and general atmosphere. The roof covering must therefore have good durability to meet these often extreme conditions. Roofs are subjected to the pressure exerted by the wind, both positive and negative. The latter can cause uplift and suction, which can be overcome by firmly nailing lightweight coverings to the structure or by relying upon the deadweight of the covering material. Domestic roofs cannot be easily accessed and so the roof covering material needs to be very durable and easily replaced when damaged. The total dead load of the covering will affect the type of support structure required and, ultimately, the total load on the foundations, so careful consideration must be given to the medium selected for the roof covering in terms of weight, maintenance and cost.

TILING

Tiles are manufactured from clay and concrete to a wide range of designs and colours suitable for pitches from 15° to 45°, and work upon the principle of either double or single lap. The vital factor for the efficient performance of any tile or slate is the pitch, and it should be noted that the pitch of a tile is always less than the pitch of the rafters owing to the overlapping technique with which they are fixed.

Tiles are laid in overlapping courses and rely upon the water being shed off the surface of one tile onto the exposed surface of the tile in the next course. The problem of water entering by capillary action between the tiles is overcome by the camber of the tile, by the method of laying, by the interlocking fixing or by overlapping side joints. In all methods of tiling a wide range of fittings are produced to enable all roof shapes to be adequately protected.

PLAIN TILING

This is a common method in the UK and in other countries and works on the double lap principle. The first row of tiles is laid across a roof and the second row starts halfway along the first row. This forms a lap between rows and prevents water egress into the roof structure.

Modern plain tiles are manufactured through a pressed clay process then fired. Modern clay roof tiles are now available in a variety of colours and shapes to provide all the finishing details required for a roof. Machine-pressed tiles are harder, more durable, denser and more uniform in shape than handmade varieties and can be laid to a minimum pitch of 35°.

As an alternative to clay, concrete is commonly used for the manufacture of roof tiles. These are produced in a range of colours to the same size specifications as the clay tiles and with the same range of fittings. The main advantage of concrete tiles is their lower costs; the main disadvantage is the extra weight.

Plain clay tile details are specified in BS EN 1304 and BS EN 538. Concrete tiles have separate designation in BS EN 490 and 491.

Plain tiling, in common with other forms of tiling, provides an effective barrier to rain and snow penetration, but wind is able to penetrate into the building through the gaps between the tiling units. Bitumen-based underfelt is normally laid below the tile battens to prevent wind and rain penetrating through to the roof structure. The use of a breather membrane layer that allows moisture through one way to the outside prevents wind and water from entering the roof structure.

The rule for plain tiling is that there must always be at least two thicknesses of tiles covering any part of the roof, and bonded so that no 'vertical' joint is immediately over a 'vertical' joint in the course below. This obviously increases the weight on the roof structure. To enable this rule to be maintained, shorter-length special tiles are required at the eaves and the ridge; each alternate course is commenced with a wider tile of one-and-a-half tile widths. The apex or ridge is capped with a special tile bedded in cement mortar over the general tile surface.

The hips can be covered with a ridge tile, in which case the plain tiling is laid underneath and cut to the line of the ridge. A cement mortar joint between the ridge tile and the main roof prevents water ingress and bonds the hip tiles in position. A hip iron must be placed on the end of the hip rafter to hold the first tile in position; alternatively a special bonnet tile can be used where the plain tiles bond with the edges of the bonnet tiles.

Valleys can be formed by using special tiles or mitred plain tiles, or by forming an open gutter with a durable material such as lead, which is dressed under the adjacent tiles each side of the valley.

The verge at the gable end can be formed by bedding plain tiles face down on the gable wall as an undercloak, and bedding the plain tiles in cement mortar on the upper surface of the undercloak. The verge tiling should overhang its support by at least 50 mm to allow water to drip off without staining the wall below.

Abutments are made watertight by dressing a flashing over the upper surface of tiling, between which is sandwiched a soaker. The soaker in effect forms a gutter.

The support or fixing battens are of treated softwood extended over and fixed to at least three rafters, the spacing or gauge being determined by the lap given to the tiles thus:

$$\text{Gauge} = \frac{\text{length of tile} - \text{lap}}{2}$$

for example, $\text{Gauge} = \dfrac{265 - 65}{2} = 100 \text{ mm}$

Plain tiles are fixed with two galvanised or copper nails to each tile in every fourth or fifth course.

Details of plain tiles, fittings and methods of laying are shown in Figs. 6.2.1 to 6.2.5.

SINGLE-LAP TILING

Single-lap tiles are laid with overlapping side joints to a minimum pitch of 35° and are not bonded like the butt-jointed single-lap plain tiles; this gives an overall reduction in weight as fewer tiles are used. A common form of single-lap tile is the pantile, which has opposite corners mitred to overcome the problem of four tile thicknesses at the corners (see Fig. 6.2.6). The pantile is a larger unit than the plain tile and is best employed on large roofs with gabled ends, because the formation of hips and valleys is difficult and expensive. Other forms of single-lap tiling are Roman tiling, Spanish tiling and interlocking tiling. The latter types are produced in both concrete and clay and have one or two grooves in the overlapping edge to give greater resistance to wind penetration and allow captured water to drain out; they can generally be laid as low as 15° pitch (see Fig. 6.2.6).

SLATING

Slate is a naturally dense material, which has been used for hundreds of years; it can be split into thin sheets and is used to provide a long-lasting and durable covering to a pitched roof. Slates are laid to the same basic principles as double-lap roofing tiles except that every slate should be twice nailed. Slates are sourced mainly from Wales, Cornwall and the Lake District, and are cut to a wide variety of sizes – the Westmorland slates are harder to cut and are usually supplied in random sizes. Slates can be laid to a minimum pitch of 25° and are fixed by head nailing or centre nailing. Centre nailing is used to overcome the problem of vibration caused by the wind and tending to snap the slate at the fixing if nailed at the head; it is used mainly on the long slates and pitches below 35°.

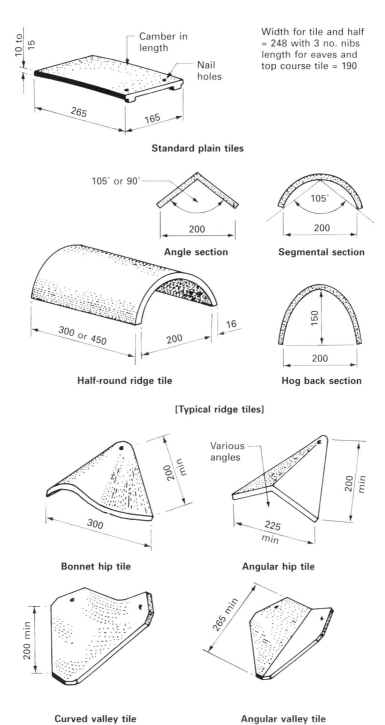

Width for tile and half
= 248 with 3 no. nibs
length for eaves and
top course tile = 190

Standard plain tiles

Angle section

Segmental section

Half-round ridge tile

Hog back section

[Typical ridge tiles]

Bonnet hip tile

Angular hip tile

Curved valley tile

Angular valley tile

Figure 6.2.1   Standard plain tiles and fittings.

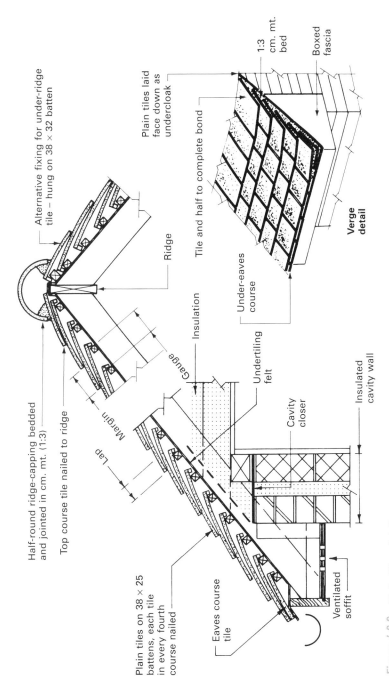

Alternative fixing for under-ridge
tile – hung on 38 × 32 batten

Plain tiles laid
face down as
undercloak

1:3
cm. mt.
bed

Boxed
fascia

Tile and half to complete bond

**Verge
detail**

Half-round ridge-capping bedded
and jointed in cm. mt. (1:3)

Top course tile nailed to ridge

Ridge

Under-eaves
course

Insulation

Undertiling
felt

Insulated
cavity wall

Cavity
closer

Gauge

Margin

Lap

Plain tiles on 38 × 25
battens, each tile
in every fourth
course nailed

Eaves course
tile

Ventilated
soffit

Figure 6.2.2 Plain tiling details.

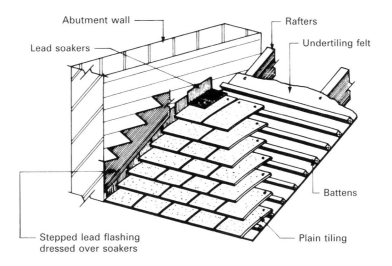

Typical abutment detail

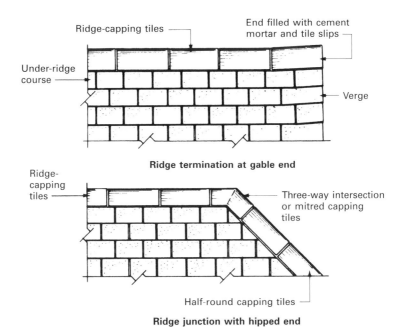

Ridge termination at gable end

Ridge junction with hipped end

Figure 6.2.3 Abutment and ridge details.

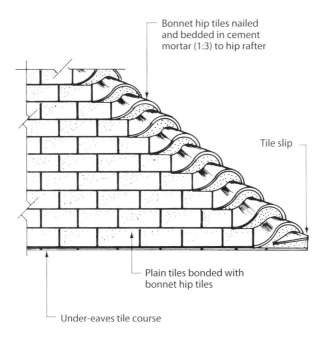

Bonnet hip tiles nailed and bedded in cement mortar (1:3) to hip rafter

Tile slip

Plain tiles bonded with bonnet hip tiles

Under-eaves tile course

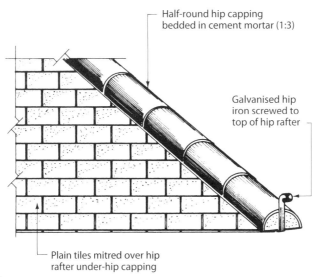

Half-round hip capping bedded in cement mortar (1:3)

Galvanised hip iron screwed to top of hip rafter

Plain tiles mitred over hip rafter under-hip capping

Figure 6.2.4   Hip treatments.

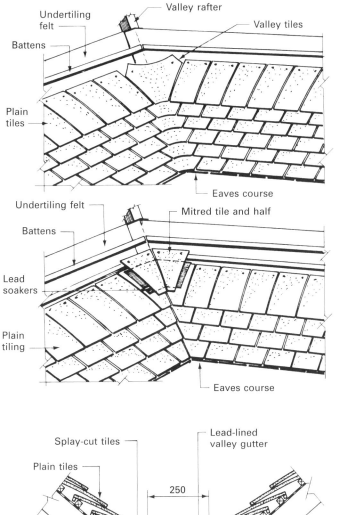

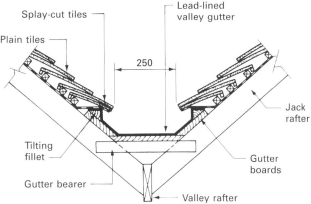

Figure 6.2.5 **Valley treatments**.

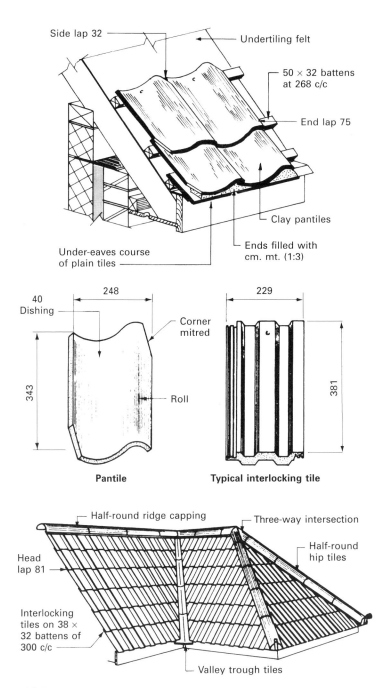

Side lap 32

Undertiling felt

50 × 32 battens
at 268 c/c

End lap 75

Clay pantiles

Ends filled with
cm. mt. (1:3)

Under-eaves course
of plain tiles

40
Dishing

248

Corner
mitred

343

Roll

Pantile

229

381

Typical interlocking tile

Half-round ridge capping

Three-way intersection

Half-round
hip tiles

Head
lap 81

Interlocking
tiles on 38 ×
32 battens of
300 c/c

Valley trough tiles

Figure 6.2.6   Examples of single-lap tiling.

The gauge of the battens is calculated thus for a 400 mm-long slate:

$$\text{Head-nailed gauge} = \frac{\text{length of slate} - (\text{lap} + 25 \text{ mm})}{2}$$

$$= \frac{400 - (75 + 25)}{2} = 150 \text{ mm}$$

$$\text{Centre-nailed gauge} = \frac{\text{length of slate} - \text{lap}}{2}$$

$$= \frac{400 - 76}{2} = 162 \text{ mm}$$

Roofing slates are covered by BS EN 12326-1, which gives details of standard sizes, thicknesses and quality. Typical details of slating are shown in Fig. 6.2.7.

ROOFING UNDERLAYS

The roof tile finish has an underlay layer below the tile battens. This layer is used to keep out the dust and the wind. It also provides a secondary waterproof layer for limited protection against rain if tiles are damaged. Underlays can be manufactured from two materials in use today: bituminous-based felts and breather membranes.

Bituminous felts

These consist basically of a bituminous-impregnated matted fibre sheet, which can be reinforced with a layer of jute hessian embedded in the coating on one side to overcome the tendency of felts to tear readily. Undertiling felts are supplied in rolls 1 m wide and 10 or 20 m long depending upon type. They should be laid over the rafters and parallel to the eaves with 150 mm laps and temporarily fixed with large-head felt nails until finally secured by the battens. Traditional first-generation undertiling felts are also known as sarking felts and should conform to the requirements of BS 747. They should also be permeable to water vapour, to relieve any possibility of condensation in the roof space.

Subsequently there has been a second generation of lightweight tiling underlays produced from reinforced plastics. As for traditional bituminous felt, there is a risk that condensation can form on the underside. Third-generation underlays are of triple-ply construction, comprising a waterproof and vapour-permeable core between layers of non-woven, spun-bonded polypropylene. In effect, they permit internal moisture to pass through but remain watertight to external conditions.

Breather membranes

A breather membrane relies on its structure to allow water vapour to pass through from one side and to resist the passage of water back the other way.

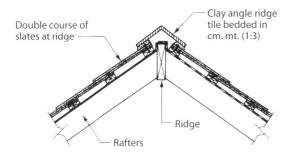

Double course of slates at ridge

Clay angle ridge tile bedded in cm. mt. (1:3)

Ridge

Rafters

Typical ridge detail

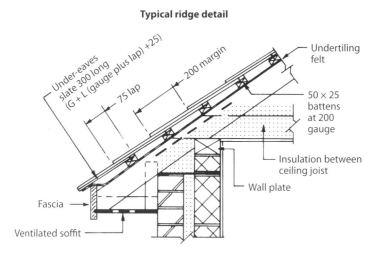

Under-eaves slate 300 long (G + L (gauge plus lap) +25)

75 lap

200 margin

Undertiling felt

50 × 25 battens at 200 gauge

Insulation between ceiling joist

Wall plate

Fascia

Ventilated soffit

Head-nailed slating using 500 × 250 slates

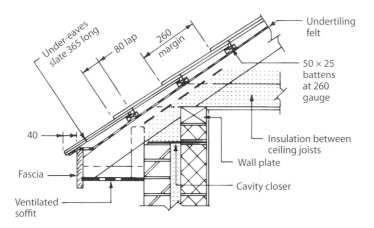

Under-eaves slate 365 long

80 lap

260 margin

Undertiling felt

50 × 25 battens at 260 gauge

Insulation between ceiling joists

Wall plate

40

Fascia

Cavity closer

Ventilated soffit

Centre-nailed slating using 600 × 300 slates

Figure 6.2.7   Typical slating details.

They act in a similar way to a felt in resisting the wind and dust that might blow through between the roof tiles. They are manufactured from a woven material containing three layers bonded together. A breather membrane should always be fixed in accordance with the manufacturer's instructions. This modern approach has the effect of reducing the amount of ventilation that is now required within the roof void, as the water vapour can escape through the membrane.

COUNTER BATTENS

Use of counter battening is common where pitched roofs are of warm construction; the insulation is above the roof slope to create a habitable room in the roof void. This is often done in order to achieve the required U-values without restricting the inner headroom (see Fig. 6.2.8). Cold roof construction with insulation at ceiling joist level is shown in Figs. 6.2.2 and 6.2.7.

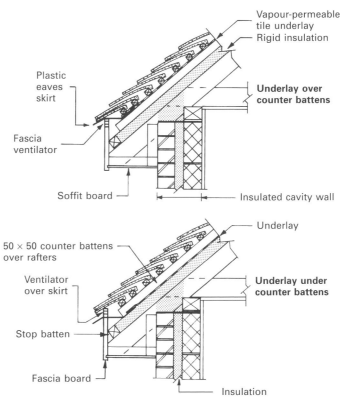

Figure 6.2.8   Plain tiled warm roof.

Lead-covered flat roofs 6.3

Lead as a building material has been used extensively for thousands of years as it is a material that is durable and flexible enough to be formed and bent around roof structures. Lead is mainly produced in flat sheets as rolls of varying widths and thicknesses.

Lead is a very dense material (11,340 kg/m^3) of low strength, but is extremely malleable and can be worked without heating into complicated shapes, without splitting or fracturing. In common with other non-ferrous metals, lead oxidises on exposure to the atmosphere and forms a thin, protective film or coating over its surface. When in contact with other metals there is seldom any corrosion by electrolysis, and therefore fixing is usually carried out using durable copper nails.

With domestic applications lead is used as a flashing material where one roof abuts another, as a waterproof lining for rainwater capture and discharge, and as a roof-covering material for dormer cheeks and bay window applications.

For application to flat roofs, milled lead sheet should comply with the recommendations of BS EN 12588. The sheet is supplied in rolls of widths varying between 150 mm and 2,400 mm, in 3 m and 6 m standard lengths. For easy identification, lead sheet carries a colour guide for each code number (see Table 6.3.1).

Table 6.3.1 **Lead sheet colour guide.**

BS Code No.	Thickness (mm)	Weight (kg/m²)	Colour
3	1.32	14.97	green
4	1.80	20.41	blue
5	2.24	25.40	red
6	2.65	30.10	black
7	3.15	35.72	white
8	3.55	40.26	orange

The code number is derived from the former imperial notation of 5 lb/ft² =
No. 5 lead.

The thickness or code number of lead sheet for any particular situation will
depend upon the protection required against mechanical damage and the
shape required. The extent to which the application is exposed to the weather
and thermal heating and cooling should also be carefully considered, along
with the cost of access in its eventual replacement. The following thicknesses
can, therefore, be considered as a general guide for flat roofs:

- small areas without foot traffic, No. 4 or 5;
- small areas with foot traffic, No. 5, 6 or 7;
- large areas with or without foot traffic, No. 5, 6 or 7;
- flashings, No. 4 or 5;
- aprons, No. 4 or 5.

Milled lead sheet may be used as a covering over timber or similar deckings
and over smooth, screeded surfaces. In all cases an underlay of felt or stout
building paper should be used to reduce frictional resistances, decrease
surface irregularities and, in the case of a screeded surface, isolate the lead
from any free lime present that might cause corrosion. This layer in effect
allows the movement of the lead through thermal forces independently of the
substrate. Provision must also be made for the expansion and contraction of
the metal covering. This can be achieved by limiting the area and/or length of
the sheets being used. Table 6.3.2 indicates the maximum recommended area
and length for any one piece of lead.

Table 6.3.2 **Maximum recommended area and length for any one piece of lead.**

BS Code No.	Max length between drips (m)	Max area (m²)
3	Only suitable for soakers	–
4	1.50	1.13
5	2.00	1.60
6	2.25	1.91
7	2.50	2.25
8	3.00	3.00

Joints that can accommodate the anticipated thermal movements are in the
form of rolls running parallel to the fall and drips at right angles to the fall,
positioned so that they can be cut economically from a standard sheet: for
layout and construction details, see Figs. 6.3.1 and 6.3.2.

Note: Current legislation will require these details to incorporate cavity
insulation in the walls with an insulative concrete block inner leaf. Insulation
must also be included within the roof construction, as shown in Part 6.1.

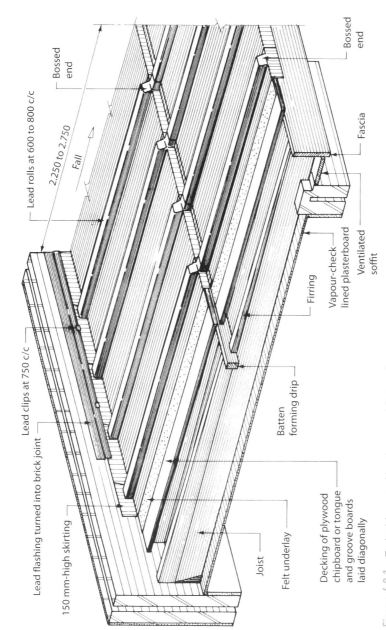

Bossed end

Bossed end

Lead rolls at 600 to 800 c/c

2.250 to 2.750

Fall

Fascia

Lead clips at 750 c/c

Vapour-check lined plasterboard

Ventilated soffit

Firring

Lead flashing turned into brick joint

150 mm-high skirting

Batten forming drip

Joist

Felt underlay

Decking of plywood chipboard or tongue and groove boards laid diagonally

Figure 6.3.1 Typical layout of lead-covered flat roof.

Advantages of lead roof coverings

- durable and long-lasting product;
- water-resistant;
- flexible and can be formed into three dimensions;
- can easily be repaired by adding in sections;
- can be recycled.

Disadvantages of lead roof coverings

- a high-value product that is subject to theft;
- expensive in large areas.

FINISHING

Leadwork should always be cleaned down and then have an application of paternoster oil applied over all the exposed surfaces. This allows the lead to weather uniformly and prevents discolouration with age. Lead flashings should be wedged and pinned using strips of lead formed into wedges and finally pointed in a sand and cement joint, which should be of sufficient depth to allow the mortar to hold in position.

The Lead Association is a source of information regarding the use and application of lead as a roofing and flashing material.

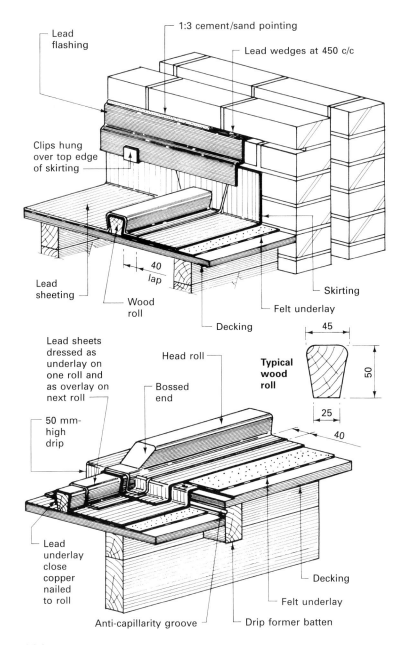

Lead flashing

1:3 cement/sand pointing

Lead wedges at 450 c/c

Clips hung over top edge of skirting

40 lap

Lead sheeting

Wood roll

Skirting

Felt underlay

Decking

Lead sheets dressed as underlay on one roll and as overlay on next roll

Head roll

Typical wood roll

Bossed end

45

50

25

40

50 mm-high drip

Lead underlay close copper nailed to roll

Anti-capillarity groove

Drip former batten

Felt underlay

Decking

Figure 6.3.2 **Lead-covered flat roof details.**

Rooflights in pitched roofs 6.4

Rooflights can be included in the design of flat and pitched roofs to provide daylight and ventilation to rooms within the roof space, or to supplement the daylight from windows in the walls of medium- and large-span single-storey buildings that may have natural light issues, because of a high density of surrounding buildings.

In domestic work, the provision of natural lighting generally takes one of two forms: the **dormer window** or the **skylight**.

A **dormer window** has a vertical sash, and therefore projects from the main roof; the cheeks or sides can be clad with a sheet material such as lead or tile hanging, and the roof can be pitched or flat of traditional construction (see Fig. 6.4.1). Dormer windows provide an aesthetic feature to a house and help create a living space when included with loft conversions.

A **skylight** can take several forms within domestic construction:

■ a fixed dead light;

■ an opening sash;

■ a light tube;

■ a lantern.

A skylight is fixed within a trimmed opening that fully supports the roof structurally, and follows the pitch of the roof. It can be constructed as an opening or dead light (see Fig. 6.4.2). Commercial off-the-shelf skylights are available, many of which now integrate balcony features to enhance the application. In common with all rooflights in pitched roofs, making the junctions watertight and weathertight presents the greatest problems, especially on a flat roof. Careful attention must be given to the detail and workmanship involved in the construction of dormer windows and rooflights and their abutment flashings, in order to prevent water ingress problems in the future.

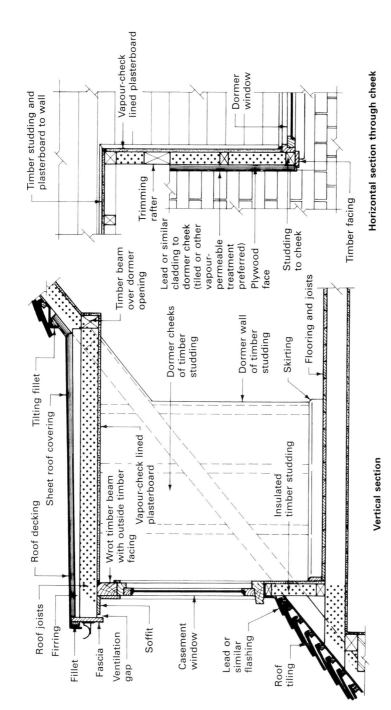

Horizontal section through cheek

Timber studding and plasterboard to wall

Vapour-check lined plasterboard

Dormer window

Trimming rafter

Lead or similar cladding to dormer cheek (tiled or other vapour-permeable treatment preferred)

Plywood face

Studding to cheek

Timber facing

Timber beam over dormer opening

Dormer cheeks of timber studding

Dormer wall of timber studding

Skirting

Flooring and joists

Roof joists

Firring

Fascia

Ventilation gap

Soffit

Casement window

Lead or similar flashing

Roof tiling

Fillet

Tilting fillet

Roof decking

Sheet roof covering

Wrot timber beam with outside timber facing

Vapour-check lined plasterboard

Insulated timber studding

Vertical section

Figure 6.4.1 Typical flat roof dormer window details.

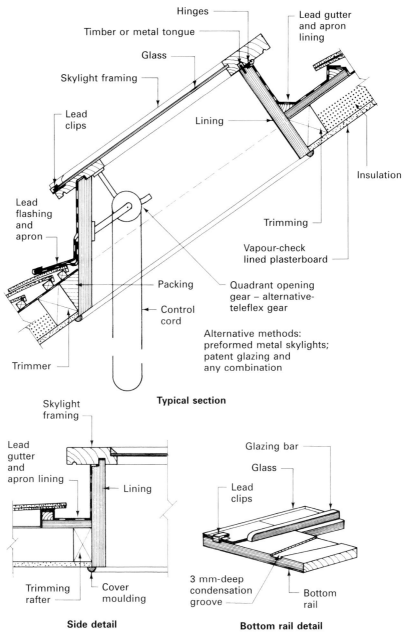

Hinges

Lead gutter and apron lining

Timber or metal tongue

Glass

Skylight framing

Lead clips

Lining

Lead flashing and apron

Insulation

Trimming

Vapour-check lined plasterboard

Packing

Control cord

Quadrant opening gear – alternative-teleflex gear

Trimmer

Alternative methods: preformed metal skylights; patent glazing and any combination

Typical section

Skylight framing

Lead gutter and apron lining

Lining

Glazing bar

Glass

Lead clips

Trimming rafter

Cover moulding

3 mm-deep condensation groove

Bottom rail

Side detail

Bottom rail detail

Figure 6.4.2   Typical timber opening skylight.

A different treatment is required for commercial roofs of the type used on medium-span industrial buildings with coverings such as corrugated fibre cement sheeting or metal profiled cladding supported by purlins and steel roof trusses or rafters. The amount of useful daylight entering the interior of such a building from windows fixed in the external walls will depend upon:

■ the size of the window;

■ the angle of the window;

■ the height of the window above the floor level;

■ the span of the building;

■ the building's orientation.

Generally, the maximum distance that useful daylight will penetrate is approximately 10.000 m; over this distance, artificial lights or rooflights will be required during the daylight period, so a safer environment can be provided within the building.

Special rooflight units of fibre cement consisting of an upstand kerb surmounted by either a fixed or opening glazed sash can be fixed instead of a standard profiled sheet. Commercial rooflight units are available for many applications and to fit a variety of standard roof sheet profiles. These units are useful where the design calls for a series of small, isolated glazed rooflights to supplement the natural daylight.

A more economical alternative is to use translucent profiled sheets that are of the same size and profile as the main roof covering. These are manufactured from a clear glass-reinforced plastic, but do suffer from long-term maintenance issues and safety concerns when they become discoloured. In selecting the type of sheet to be used, the requirements of Part B of the Building Regulations must be considered. Approved Document B4 (Section 14) deals specifically with the fire risks of roof coverings, and refers to the designations defined in BS 476-3. These designations consist of two letters: the first letter represents the time of penetration when subjected to external fire, and the second letter is the distance of spread of flame along the external surface. Each group of designations has four letters, A–D, and in both cases the letter A has the highest resistance. Specimens used in the BS 476 test are tested for use on either a flat surface or a sloping surface, and therefore the material designation is preceded by either EXT. F or EXT. S.

Most of the translucent profiled sheets have a high light transmission. When they are installed as new, they are light in weight, and can be fixed in the same manner as the general roof covering. It is advisable to weather-seal all lapping edges of profiled rooflights with silicon or mastic sealant to accommodate the variations in thickness and expansion rate of the adjacent materials. Here are some typical examples.

- **Polyester glass fibre sheets** Made from polyester resins reinforced with glass fibre and nylon to the recommendations of BS 4154. These sheets can be of natural colour or tinted, and are made to suit most corrugated fibre cement and metal profiles. Typical designations are EXT. S.AA for self-extinguishing sheets and EXT. S.AB for general-purpose sheets.

- **Wire-reinforced PVC sheets** Made from unplasticised PVC reinforced with a fine wire mesh to give a high resistance to shattering by impact. Designation is EXT. S.AA, and they can therefore be used for all roofing applications. Profiles are generally limited to Categories A and B defined in BS EN 494.

- **PVC sheets** Made from heavy-gauge clear unplasticised rigid PVC to the recommendations of BS 4203, they are classified as self-extinguishing when tested in accordance with method 508A of BS 2782: Part O, Annex C, and may be used on the roof of a building provided that part of the roof is at least 6.000 m from any boundary. If that part of the roof is less than 6.000 m from any boundary and covers a garage, conservatory or outhouse with a floor area of less than 40 m^2, or is on an open balcony, carport or the like, or a detached swimming pool, PVC sheets can be used without restriction. Table 6.3 of Approved Document B defines the use of these sheets for the roof covering of a canopy over a balcony, veranda, open carport, covered way or detached swimming pool.

With any rooflight comes the issue of health and safety. Access to an old roof with discoloured rooflights presents a risk of injury. A physical barrier must be placed to outline the extent of the roof light so any personnel are aware of their vulnerability. This is normally achieved by the use of small vertical rods in each corner with a line attached from each to identify the rooflight area. Personnel are also further protected by the use of safety harnesses and a permit to work system for working at height.

As an alternative to profiled rooflights in isolated areas, continuous rooflights can be incorporated into a corrugated or similar roof covering by using a Georgian flat-wired glass and patent glazing bars system. The bars are fixed to the purlins and spaced at 600 mm centres to carry either single or double glazing. Modern glazing systems use an aluminium alloy extrusion (see Fig. 6.4.3), with built-in seals and water drainage channels. There are of course a variety of different sections produced by manufacturers, but all have the same basic principles. The bar is generally an inverted 'T' section, the flange providing the bearing for the glass and the depth of the 'T' providing the spanning properties. Other standard components are fixing shoes, glass weathering springs or clips, and glass stops at the bottom end of the bar (see Fig. 6.4.3). This type of system is more expensive than the integrated translucent profiled sheets, but provides the opportunity to colour-coat the glazing bars and introduce a third dimension into the skylight feature by using angled or upstand panels.

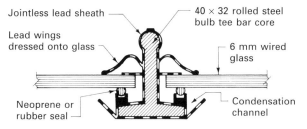

Crittall-Hope lead-clothed steel bar

Jointless lead sheath
Lead wings dressed onto glass
40 × 32 rolled steel bulb tee bar core
6 mm wired glass
Neoprene or rubber seal
Condensation channel

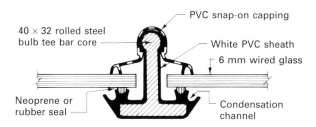

Crittall-Hope polyclad bar

40 × 32 rolled steel bulb tee bar core
PVC snap-on capping
White PVC sheath
6 mm wired glass
Neoprene or rubber seal
Condensation channel

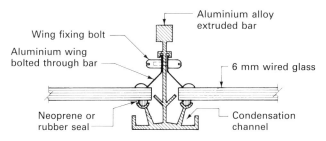

British Challenge aluminium bar

Wing fixing bolt
Aluminium wing bolted through bar
Aluminium alloy extruded bar
6 mm wired glass
Neoprene or rubber seal
Condensation channel

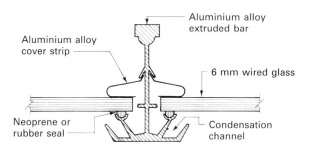

Heywood Williams 'Aluminex' bar

Aluminium alloy cover strip
Aluminium alloy extruded bar
6 mm wired glass
Neoprene or rubber seal
Condensation channel

Figure 6.4.3   Typical patent glazing bar sections.

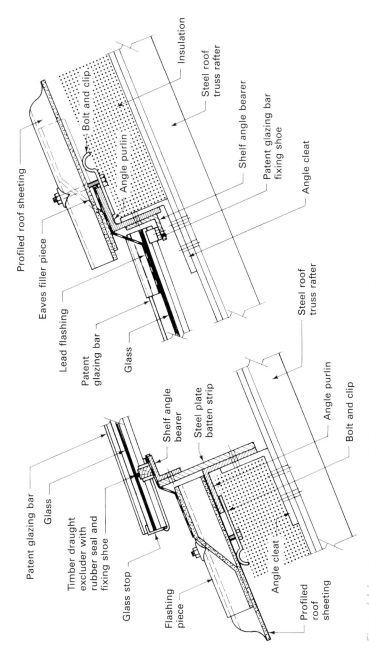

Insulation

Steel roof truss rafter

Bolt and clip

Angle purlin

Shelf angle bearer

Patent glazing bar fixing shoe

Angle cleat

Profiled roof sheeting

Eaves filler piece

Lead flashing

Patent glazing bar

Glass

Steel roof truss rafter

Patent glazing bar

Glass

Timber draught excluder with rubber seal and fixing shoe

Glass stop

Flashing piece

Shelf angle bearer

Steel plate batten strip

Angle purlin

Bolt and clip

Angle cleat

Profiled roof sheeting

Figure 6.4.4 Patent glazing and profiled roof covering connection details.

As the glass and the glazing bar are straight, they cannot simply replace a standard profiled sheet; they must be fixed below the general covering at the upper end and above the covering at the lower end to enable the rainwater to discharge onto the general roof surface. Care must, therefore, be taken with the side detailing in providing adequate weather resistance. Great care must be taken with the quality of workmanship on site if a durable and satisfactory junction is to be made. Typical details were shown in Fig. 6.4.4.

The amount of natural light required is a matter for the designer with regard to the type of operations that will be undertaken within the building. The general rule of thumb is to design for 10 per cent rooflight provision within the roof structure. The glass specified is usually a Georgian wired glass of suitable thickness for the area of pane being used. Wired glass is selected so that it will give the best protection should an outbreak of fire occur; the splinters caused by the heat cracking the glass will adhere to the wire mesh and not shatter onto the floor below. A successful alternative is laminated glass composed of two or more sheets of float glass with a plastic interlayer. Polyvinylbutyral (PVB) is the most common interlayer, which is bonded to the glass laminates by heat and pressure. If broken, the glass remains stuck to the PVB to provide the benefit of safety and security. It also remains intact following impact from bullets or explosive blasts.

As with timber skylights, provision should be made to collect the condensation that can occur on the underside of the any single glazing to prevent the annoyance of droplets of water falling to the floor below and causing any damage to internal finishes. Most patent glazing bars for single glazing have condensation channels attached to the edges of the flange, which directs the collected condensation to the upper surface of the roof below the glazing line by the force of gravity (see Figs. 6.4.3 and 6.4.4).

Contemporary frames can be produced from uPVC or extruded aluminium. The latter is often coated in polyester powder for colour enhancement and protection. A double-glazed example is shown in Fig. 6.4.5. This applies the

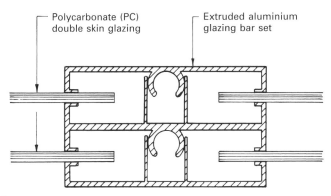

Figure 6.4.5 'Glidevale' patent double-glazing unit.

Polycarbonate (PC) double skin glazing

Extruded aluminium glazing bar set

principle of tension between the glazing frame and glass for security of fit, with the ability to respond to thermal movement without loss of seal.

Polycarbonate (PC) can be an economic option in preference to glass, particularly for rooflight applications. It has the benefit of a double skin if required and, although produced in flat sheet format, can be moulded to domed profiles. Colour variants are possible, in addition to transparent, translucent and opaque composition. Dimensions, types and characteristics are defined in BS EN ISO 11963.

LIGHT TUBES

This modern invention uses the ability of a mirrored surface to reflect light. It uses polished steel tubes whose internal surfaces have been mirrored such that a dome lens skylight can be placed on the roof end and an internal diffuser on the ceiling end of the tube. This allows natural light to be let into a habitable room using an economical method, where a normal rooflight would be a distance away from the ceiling level. Essentially it provides free light into a darkened room at low cost.

Green roofs 6.5

With the drive to reduce carbon emissions has come the development of the design and installation of the 'green roof', so called because it has a 'living' structure built into its outer layer. Vegetation is encouraged to grow on the roof in order to provide several environmental benefits. This method of roof finish has been used by Scandinavian countries as a traditional method of completing a roof that suits the environment surrounding the buildings. Before the industrial revolution, we used thatch and turf to cover many of our domestic roofs.

A green roof essentially consists of:

- structural loadbearing element;
- decking;
- waterproof layer;
- root-protection layer;
- moisture-retention layer;
- drainage layer;
- growing layer.

Advantages

- Low embodied energy.
- Reduction in sound transmission.
- Reduction in rainwater run-off.
- Increase in thermal insulation.
- Air quality improvements.

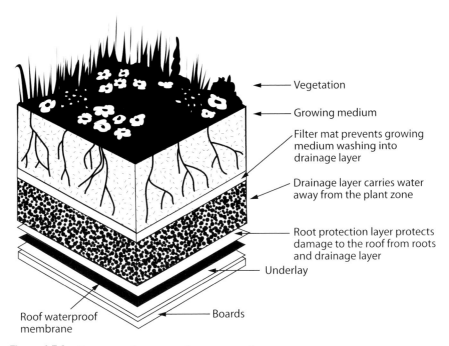

Vegetation

Growing medium

Filter mat prevents growing medium washing into drainage layer

Drainage layer carries water away from the plant zone

Root protection layer protects damage to the roof from roots and drainage layer

Underlay

Boards

Roof waterproof membrane

Figure 6.5.1 **Structure of an extensive green roof.**
Source: www.delston.co.uk

- Aesthetic and attractive method.
- Wildlife and biodiversity encouraged.
- Fire-resistance increased.

Disadvantages

- Requires maintenance to planted areas.
- Increased weight on structure may require calculations.
- Repair of waterproof barrier may prove difficult.

ROOF SUPPORT STRUCTURE

The supporting structure below a green roof is similar to that of a traditional tiled roof. Prefabricated roof trusses with an increased section size to carry the additional loading may be required for a pitched roof. For a flat roof application a solid concrete structure or beam and block structure can be effectively used. Steel beams with purlin rails and sheeting also provide an adequate structure to cover with a green roof.

The key with all the roof structures is to provide a waterproof layer that will stand up to the rigours of any freeze and thaw action, thermal movements and age. It needs to be flexible and durable enough to last the lifetime of the green roof without maintenance.

There are two basic industry-recognised methods of providing a green roof: intensive and extensive.

INTENSIVE

An intensive green roof has a much greater depth than an extensive one. This increased depth gives it the ability to sustain plants, small trees, grass and shrubs, and therefore provides a wide variety of biodiversity as it matures. With the increased depth comes the greater weight that the substructure has to bear, as the depth of soil will also absorb moisture, adding to the total loading. An intensive roof also acts as a space that occupants can use for various social activities. Roofs like these are often placed on the tops of office blocks and as domestic roof gardens where inner city space is at a premium cost.

The structure of the roof consists of a growing layer of topsoil, which may be sculptured for contouring and can contain lightweight aggregates to increase the depth without adding extra loadings. A layer of filter material is then placed below the soil to prevent fines from washing out of the soil and blocking the drainage layer. Below this is a drainage layer which draws away excess water, relieving the wet loads on the roof. Under the drainage layer is the waterproof layer, which protects the roof structure from water ingress.

EXTENSIVE

An extensive roof is a self-supporting roof that is seeded with less soil than an intensive one, and therefore has less depth and is lighter. Alpine sedums are used, which are specially grown to provide a low-maintenance and slow-growing medium that looks attractive with its multiple colours. This type of roof is more applicable to the domestic roof market where its low maintenance is of benefit to the occupants. The only maintenance required is an annual weeding and the application of a slow-release fertiliser to encourage growth. The extensive roof has the same construction methods as the intensive roof, but the growing medium layer is considerably smaller and lighter, which is ideal for domestic applications.

TURF

Using a turf roll has several advantages.

- An instant roof is created.
- After watering, roots will develop, growing down into the substrate and holding the turf in place.

- A wildflower meadow-effect can be created with the introduction of flower seeds into the turf when it is sown.

With a turf roof, maintenance will be required to reduce the length of the grass created and to feed the nutrient bed of the turf to promote growth.

TREES AND SHRUBS

An intensive roof is suited to this type of planting, as the greater depth of soil allows for the root system of small trees and shrubs. Pruning will be required as this type of planting matures and develops.

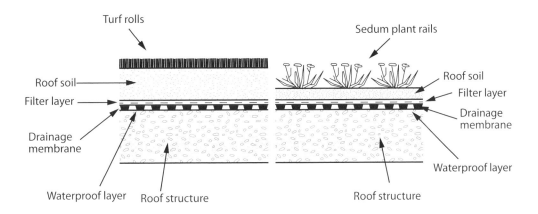

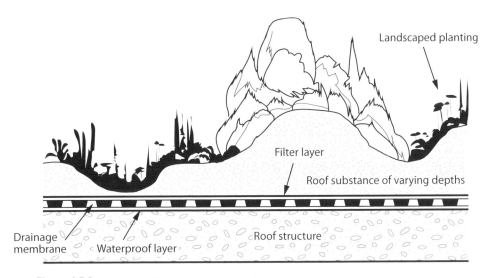

Figure 6.5.2 **Types of planting for green roofs.**

Trees and shrubs can be mixed with turf to provide a biodiverse environment that is ideal for attracting wildlife. This type of roof may require some form of irrigation system to provide water to the heavier, landscaped areas.

SEDUM

Sedum is a mixture of small growing plants, mainly alpine-based, that typically reach a height of 20 mm to 50 mm (up to 100 mm when in flower). Sedum grows relatively slowly, and therefore does not become overgrown. Sedum does not require regular cutting or pruning, which is ideal where accessibility to the roof is limited. The plants react well to periods of intense heat and drought, and can resist extreme climates, which makes them ideal for a roofing application.

The evergreen sedum mix provides a number of different colours and textures that remain all year round, providing an attractive roof to look at and one that blends in with a green environment.

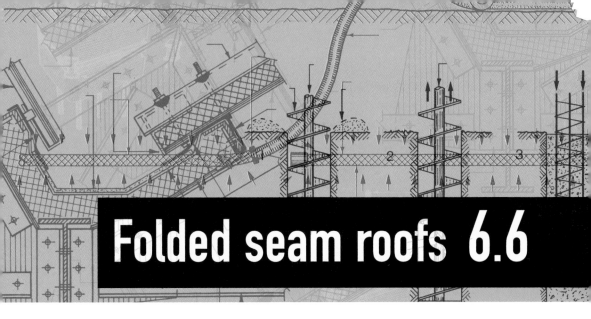

Folded seam roofs 6.6

Modern roof construction can now make full use of zinc, which can now be produced in sheets in sufficient width to provide a covering that is a sustainable product. Zinc is a naturally occurring product that has been used for many applications. Through innovative design, it can be used to cover pitched roofs and flat roofs at a 3° pitch.

STRUCTURE

A typical roof structure will depend on the application of a warm or cold roof application, the difference being the position of the insulation layer. (See Fig. 6.6.1)

WARM ROOF APPLICATION

- Zinc roof covering.
- Insulation.
- Vapour control layer.
- Structural support.

COLD ROOF APPLICATION

- Zinc roof covering.
- Vapour control layer.
- Structural support.
- Insulation between supports.

Standing seam roofing: warm roof on mineral
wool slabs supported by a steel deck

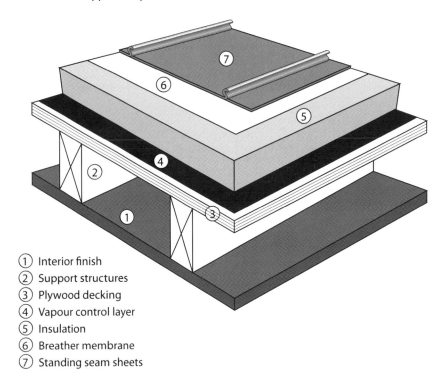

1. Interior finish
2. Support structures
3. Plywood decking
4. Vapour control layer
5. Insulation
6. Breather membrane
7. Standing seam sheets

Figure 6.6.1   Typical commercial application.

Advantages

- Zinc is a totally recyclable material that can be fully recovered.
- Reduced amount of energy required in the manufacturing process.
- Has a lifespan of 100 years.
- Can be easily bent and formed.
- Can be installed vertically as a facade system.
- Ventilated roof spaces can be accommodated.
- Compatible rainwater goods available.

Disadvantages

- Care has to be taken with contact against any other metals.
- Discolouration can be caused from exhaust gases on boiler flues.
- Great difficulty in replacing sheets once seamed.
- Expansion and contraction must be accounted for.
- Limited sheet width and the number of folds required to complete a roof.
- Limited range of roof finish colours available.
- Zinc is expensive.

CONSTRUCTION

The sheets of zinc have to be seamed to provide a waterproof seal against driven rain and the capillary action of water through the joints. The zinc sheets have a fold formed on each edge. Each sheet interlocks with the next sheet's edge fold, and the seam is then compressed, either by hand crimping or by the use of a seaming machine, to form a watertight joint.

Careful detailing of the sheets must be undertaken with abutments, verges and guttering. The crimping of the seams is undertaken using the sequence shown in Fig. 6.6.2.

HAND-CRIMPING

The sequence for hand-crimping is as follows.

- A sheet is butted together against the previous one and the leading edge is retained using clips.
- The joint formed between them has the initial fold made.
- The double fold is completed.
- The joint is finished with a mallet and hand-roofing seamer.

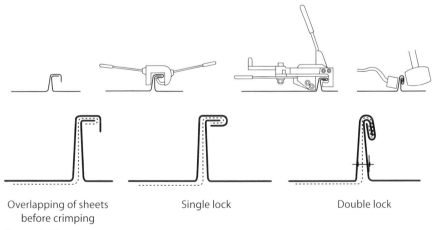

Overlapping of sheets before crimping   Single lock   Double lock

Figure 6.6.2   **Hand-crimping of seams on a zinc roof.**

The joint can be finished using an electric seamer, which automatically closes the joint to produce a uniform seal.

VENTILATION

The zinc roofing system can accommodate the provision of ventilation within the roof void by the use of proprietary ventilated flashings and infills to produce a passage for airflow through the roof void structure. Ventilation is often required to remove high levels of humidity and to prevent deterioration of the roof structure through rot and insect attack if the substructure is timber-based. The insulation would be laid below this ventilated void and would incorporate a vapour barrier to the internal finish.

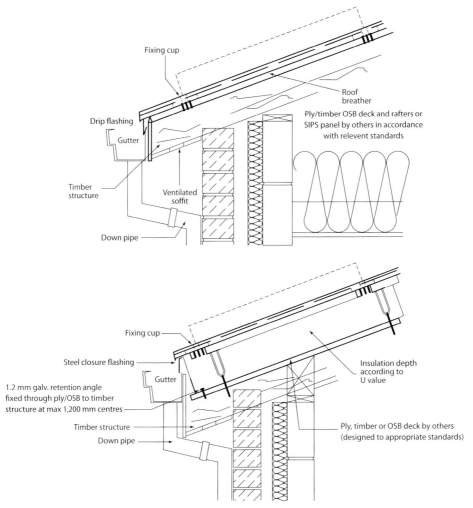

Figure 6.6.3 Typical details of ventilated zinc roof construction.

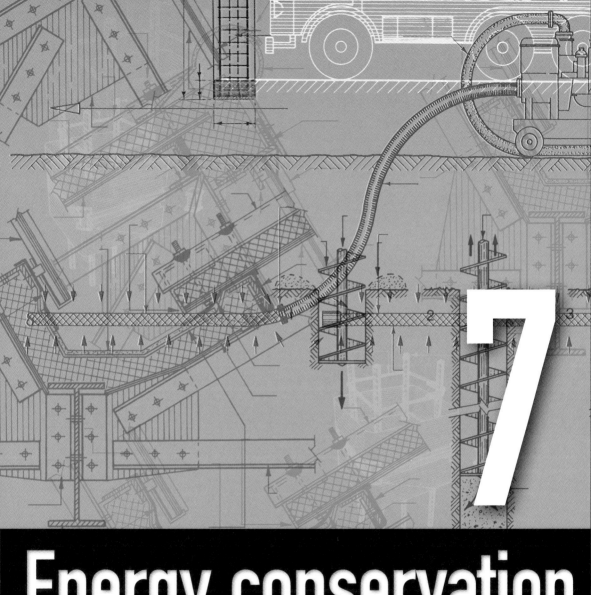

Energy conservation and sustainability in buildings

7

Construction and the environment 7.1

The construction industry has a direct impact on the natural and built environment through a variety of energy-intensive processes involved in the lifecycle of building project. These include:

- mineral extraction;
- materials processing and refining;
- manufacturing of components and their assembly;
- haulage, transport and distribution;
- erection, construction and finishing activities;
- use of the facility and maintenance works, and finally the decommissioning or refurbishment of the facility.

BASIC RESOURCES

The three basic resources that are affected by these built environment processes are:

- energy;
- water;
- materials.

The construction industry needs to consider the long-term sustainability of these resources, so that it meets the needs of the present population without compromising the ability of future generations to meet their own needs.

ENERGY

Recent research by various government bodies and private organisations has shown that in the UK the main sources of CO_2 emissions are approximately:

- 50 per cent due to heating, lighting and ventilating buildings;
- 25 per cent due to transportation of goods and people;
- 20 per cent due to industry and agriculture;
- 5 per cent due to building construction activities.

Although the fossil fuel energy supplied by the power stations cannot be directly controlled by the construction industry, it is important that those professionals involved in the design, construction and maintenance of new buildings consider the sustainable use of energy and the reduction of CO_2 emissions that are within their control. One of the key areas that needs to be addressed is the development and implementation of new materials and technologies to enable higher standards of energy efficiency within the 10 million or so pre-1944 buildings still in use today; and as such, carbon emissions from the domestic housing stock contribute to approximately 10 per cent of all UK CO_2 emissions. This is the main reason that the Government has set a target that, by 2016, all new homes will be 'zero carbon'. 'Zero carbon' essentially means having no net carbon emission; however, the standard requirements of what 'zero carbon' will entail have not yet been defined in current legislation.

To reduce the carbon emissions from building construction operations, the construction processes are being closely examined and a number of logistical and planning strategies are being implemented to reduce transport costs and energy. One such initiative is the use of Construction Consolidation Centres. These are small, local warehousing and distribution centres that serve one or several construction sites. These centres receive and hold deliveries from suppliers, then make a 'milk round' to several sites. This round can be co-ordinated so that the return journey to the Centre can pick up and bring back packaging waste or reusable packaging and pallets. Wherever possible, locally sourced materials of the appropriate quality should be used to reduce transportation pollution and to conserve energy.

WATER

Water is a basic requirement for life and is just as important as energy; water scarcity is becoming an increasing problem as global warming progresses. Unlike energy, water impacts directly on human health and food production, and despite the traditional water cycle it is often wasted with fresh potable water being used for inappropriate ends, such as to flush toilets. Flooding is also an issue with a combination of rising sea levels and the use of Victorian under-capacity public storm water systems. This is particularly true in urban areas with paved and hard landscapes where rainwater cannot drain away into the natural soil. These sewer systems reach their capacity quickly and frequently burst and cause localised flooding. In a measure to deal with these problems, localised solutions such as sustainable urban drainage systems (SUDS) are used. These collect, attenuate and allow harvesting or reuse of water. Other solutions to prevent flooding in new and refurbishment projects include the use of water butts, green roofs and permeable paving.

MATERIALS

The finite amount of raw materials that the Earth possesses needs to be carefully marshalled and managed if the construction industry is to help future generations to survive and prosper. A study in 2005 estimated that the UK construction industry uses six tonnes of construction materials per head of population per year, and that 107 million tonnes of inert construction and demolition waste were produced in that year alone. It is difficult to foresee that this type of material usage is sustainable in the long term, without changing the way construction materials are used and recycled from project to project.

WASTE HIERARCHY

There is a staged process for dealing with materials, which is known as the waste hierarchy (see Fig. 7.1.1). The ideal is to have an efficient design that avoids unnecessary use of materials, designing out the material or reducing how much material is needed as the first stage. The next stage is to see if any existing materials or components can be reused without having to re-manufacture them: for example, reusing second-hand roof tiles for a new roof. The third stage is specifying materials which, although they cannot be used for their original function, can be re-processed to serve an alternative use: for example, recycling old crushed float glass as a bedding material for paving slabs. Where a material has no recycling value, the alternative is to take that waste and extract the energy from it by incinerating it. The heat given out can then be recovered and transferred to generate power, as with burning timber off-cuts or tree chippings. The final stage is to take the inert waste material to infill worked-out quarry sites or similar locations. However, facilities in some regions of the UK, such as the south-east, are fast reaching their allowable capacity.

CHAIN OF CUSTODY

In a number of schemes, materials must be shown to originate from sustainable sources. This requires that there is a 'chain of custody' from the original supplier through the distribution network to the arrival and use on the construction site. The chain of custody requires independent auditing and rigorously certified documents for each stage. For example, for timber there must be documentation to show how it has got from the forest to the final point of delivery. The process tracks timber through each stage of the supply chain, from logging in the forest through to the sawmill, then the distributor and the timber merchant, and finally the contractor. This provides a level of transparency and traceability to guarantee compliance with demands for ethically sourced timber products. Similarly there must be a chain of custody for all waste products that leave a site to go to a waste transfer station, specialist waste contractor and landfill.

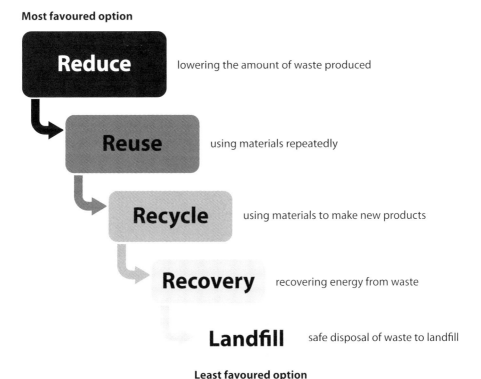

Most favoured option

Reduce — lowering the amount of waste produced

Reuse — using materials repeatedly

Recycle — using materials to make new products

Recovery — recovering energy from waste

Landfill — safe disposal of waste to landfill

Least favoured option

Figure 7.1.1   The waste hierarchy.

Considerations in energy conservation   7.2

Heat is a form of energy consisting of the ceaseless movement of tiny particles of matter called molecules; if these particles are moving fast, they collide frequently with one another and the substance becomes hot. Temperature is the measure of hotness and should not be confused with heat.

The transfer of heat can occur in three ways: conduction, convection and radiation.

- **Conduction** Vibrating molecules come into contact with adjoining molecules and set them vibrating faster, so they become hotter. This process is carried on throughout the substance without appreciable displacement of the particles.

- **Convection** This is the transmission of heat within a fluid (gas or liquid) caused by the movement of particles. The fluid becomes less dense when heated and rises, setting up a current or circulation.

- **Radiation** This is the transmission of thermal energy by electromagnetic waves from one body to another. The rate at which heat is emitted and absorbed will depend on the respective surface temperatures.

In a building all three methods of heat transfer can take place, because the heat will be conducted through the fabric of the building and dissipated on the external surface by convection and/or radiation.

U-VALUES

Thermal insulation may be defined as a barrier to the natural flow of heat from an area of high temperature to an area of low temperature. In buildings this flow is generally from the interior to the exterior. This conductivity of heat in buildings is quantified by determining the U-value through the building elements that make up its external envelope. The U-value for the element, such as a cavity wall, is defined as the rate of heat transfer in watts (joules/second) through 1 m² of the structure for one unit of temperature difference between the air on the two sides of the element.

DETERMINING THE U-VALUE

To calculate a U-value, the complete constructional detail must be known, together with the following thermal properties of the materials and voids involved: thermal conductivity, thermal resistance and surface or standard resistances.

- **Thermal conductivity** Represented by the symbol λ (lambda), this is the measure of a material's ability to transmit heat. It is expressed as the energy flow in watts per square metre of surface area for a temperature gradient of one Kelvin per metre thickness: W/m² K.

- **Thermal resistance** Represented by the letter R, this is a property of a material that indicates its ability to conduct heat. It is calculated by dividing its thickness in metres by its thermal conductivity: $R = m/\lambda$ or m² K/W.

  The values for λ and R can be obtained by reference to tables published by the Chartered Institution of Building Services Engineers or from listings in product manufacturers' catalogues.

- **Surface or standard resistances** These are values for surface and air space (cavity) resistances. They vary with direction of energy flow, building elevation, surface emissivity and degree of exposure. They are expressed in the same units as thermal resistance.

In calculations, surface resistances are represented as shown in Table 7.2.1.

Table 7.2.1 **Surface resistances.**

Heat at flow direction	Element type	Surface resistance internal (R_{si}) m² K/W	Surface resistance external (R_{se}) m² K/W
Horizontal	Wall, window	0.13	0.04
Upwards	Roof	0.10	0.04
Downwards	Floor	0.17	0.04

In addition to the internal and external surface resistances, cavity air spaces also provide resistance. In a standard non-insulated cavity, the air resistance R_a is taken as 0.18 m² K/W.

However, the value of the surface resistance depends on the assumed direction of heat flow and whether the surfaces are high or low resistivity. For example, for foil-faced products, with the foil adjacent to an unventilated air space of width at least 25 mm, the thermal resistance of this air space may be taken as 0.44 m² K/W due to the low-emissivity surface, while ventilated air spaces, such as voids behind rainscreen cladding with a high-emissivity surface, may only offer a thermal resistance of 0.13 m² K/W.

The method for calculating the U-value of any combination of materials will depend on the construction. The procedure is set out in the Building Research Establishment's guide BR 443 *Conventions for U-value calculations*.

For example, a traditional rendered and plastered one-brick wall is consistent: therefore the U-value is expressed as the reciprocal of the summation of resistances: $U = 1/\Sigma R$. ΣR comprises internal surface resistance (R_{si}), thermal resistance of materials (R_1 = plaster, R_2 = brickwork including mortar, and R_3 = render) and external surface resistance (R_{se}). If an element of structure is inconsistent, such as lightweight concrete blockwork with dense mortar joints or timber framing with insulation infilling, cold bridging will occur through the denser parts. Inconsistencies can be incorporated into the calculation using the combined method from BS EN ISO 6946: *Building components and building elements. Thermal resistance and thermal transmittance. Calculation method.*

Sample calculation

A sample calculation for the U-value of a partially insulated cavity wall for a 'new build dwelling' is shown in Fig. 7.2.1.

1. Butterfly-pattern twisted-wire wall ties can be used where the gap between masonry leaves does not exceed 75 mm. With these ties no allowance for thermal conductivity or bridging is considered necessary.

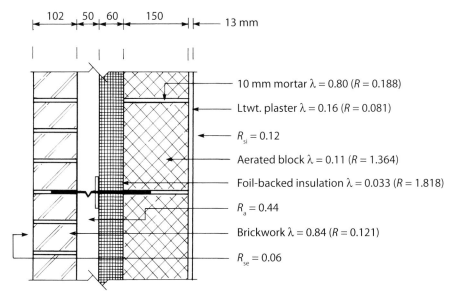

10 mm mortar $\lambda = 0.80$ ($R = 0.188$)

Ltwt. plaster $\lambda = 0.16$ ($R = 0.081$)

$R_{si} = 0.12$

Aerated block $\lambda = 0.11$ ($R = 1.364$)

Foil-backed insulation $\lambda = 0.033$ ($R = 1.818$)

$R_a = 0.44$

Brickwork $\lambda = 0.84$ ($R = 0.121$)

$R_{se} = 0.06$

Figure 7.2.1   Calculation of U-value through an external insulated cavity wall.

In Fig. 7.2.1 where the gap exceeds 75 mm, vertical twist-pattern ties are generally specified and an allowance of 0.02 W/m² K should be added to the U-value calculation.

2. As the thermal resistances of brick and mortar are similar (< 0.1 m² K/W), the outer leaf is considered consistent and unbridged.

3. Fractional area of bridged layer:

   Standard block nominal area = 450 × 225 = 101,250 mm²
   Standard block format area   = 440 × 215 = 94,600 mm²
   Difference or mortar area   =   =   6,650 mm²

   Block format = 94,600/101,250 = 0.934 or 93.4%
   Mortar area = 6,650/101,250 = 0.066 or 6.6%

4. The example shown has only one bridged layer, the mortar joints in the inner leaf of brickwork.

Material	Thermal resistance (R)
Internal surface	0.130
Plaster	0.081
Blockwork – mortar (6.6%)	0.188
Blockwork – blocks (93.4%)	1.364
Insulation	1.818
Cavity	0.440
Brickwork (inc. mortar)	0.121
External surface	0.040
Total	**4.182 m² K/W**

5. Upper resistance limit:

   Heat energy loss has two paths in the wall, through the blocks and through the mortar.

   (a) Resistance through blocks:
       4.182 − 0.188 = 3.994 m² K/W $[R_1]$ and
       Fractional area, 93.4% or 0.934 $[F_1]$

   (b) Resistance through mortar:
       4.182 − 1.364 = 2.818 m² K/W $[R_2]$ and
       Fractional area, 6.6% or 0.066 $[F_2]$

   $$\text{Upper limit of resistance} = \frac{1}{\dfrac{F_1}{R_1} + \dfrac{F_2}{R_2}} =$$

   $$\frac{1}{\dfrac{0.934}{3.994} + \dfrac{0.066}{2.818}} = \; = 3.886 \text{ m}^2 \text{ K/W}$$

6. Lower limit of resistance is a summation of resistance of all parts of the wall.

Internal surface	= 0.130
Plaster	= 0.081
Blockwork 1/[(0.934/1.364) + (0.066/0.188)]	= 0.965
Insulation	= 1.818
Cavity	= 0.440
Brickwork	= 0.121
External surface	= 0.040
Total	**3.595 m² K/W**

7. The total resistance of the wall is the average of upper and lower resistances:

(3.886 + 3.595)/2 = 3.741 m² K/W
U-value = $1/R$ = 1/3.741 = 0.267 W/m² K
To this value, add 0.020 (wall tie allowance) = 0.287 W/m² K

This value is very close to complying with the current building regulation as set out in the data shown in Fig. 7.3.1 where new external walls must have a U-value less than or equal to 0.28 W/m² K. To comply fully, the insulation thickness could be increased within the width of the cavity and the calculation repeated. Examples of current insulation standards are provided in Part 7.3 Insulating Materials (see following page).

Insulating materials

7.3

When selecting or specifying thermal insulation materials, the following must be taken into consideration:

- thermal resistance of the material;
- need for a vapour control layer, because insulating materials that become damp or wet, generally from condensation, rapidly lose their insulation properties. If condensation is likely to occur, a suitable vapour control layer should be included in the detail. Vapour control layers should always be located on the warm side of the construction;
- availability of material chosen;
- ease of fixing or including the material in the general construction;
- appearance if visible;
- cost in relation to the end result and ultimate savings on fuel and/or heating installation;
- fire risk – all wall and ceiling surfaces must comply with the requirements of Building Regulation B2: *Internal fire spread (linings)*.

Fig. 7.3.1 shows typical construction details for a dwelling's fabric to suit current thermal performance required by Part L Conservation of fuel and power. There are a variety of solutions depending on the type of insulation specified.

U-Values: walls

Wall type 1: full fill masonry

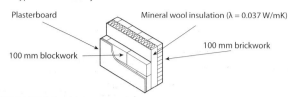

Plasterboard
Mineral wool insulation (λ = 0.037 W/mK)
100 mm brickwork
100 mm blockwork

U-value	Aircrete blockwork (λ = 0.15 W/mK)		Dense blockwork (λ = 1.33 W/mK)	
	Insulation thickness	Wall thickness	Insulation thickness	Wall thickness
0.25 W/ m²K	10 mm	340 mm	135 mm	365 mm
0.23 W/ m²K	25 mm	355 mm	150 mm	380 mm
0.20 W/ m²K	150 mm	380 mm	175 mm	405 mm
0.15 W/ m²K	220 mm	450 mm	240 mm	470 mm

Wall type 2: partial fill masonry

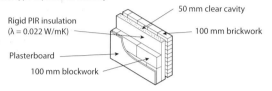

Rigid PIR insulation (λ = 0.022 W/mK)
50 mm clear cavity
100 mm brickwork
Plasterboard
100 mm blockwork

U-value	Aircrete blockwork (λ = 0.15 W/mK)		Dense blockwork (λ = 1.33 W/mK)	
	Insulation thickness	Wall thickness	Insulation thickness	Wall thickness
0.25 W/ m²K	55 mm	335 mm	70 mm	350 mm
0.23 W/ m²K	65 mm	345 mm	75 mm	355 mm
0.20 W/ m²K	80 mm	360 mm	90 mm	370 mm
0.15 W/ m²K	115 mm	395 mm	130 mm	410 mm

Wall type 3: timber frame mineral wool

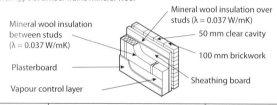

Mineral wool insulation over studs (λ = 0.037 W/mK)
50 mm clear cavity
100 mm brickwork
Mineral wool insulation between studs (λ = 0.037 W/mK)
Plasterboard
Sheathing board
Vapour control layer

U-value	Insulation thickness		Wall thickness
	In cavity	Between studs	
0.25 W/ m²K	15 mm	140 mm	360 mm
0.23 W/ m²K	25 mm	140 mm	370 mm
0.20 W/ m²K	50 mm	140 mm	395 mm
0.15 W/ m²K	110 mm	140 mm	455 mm

Wall type 4: timber frame rigid PIR

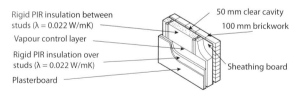

Rigid PIR insulation between studs (λ = 0.022 W/mK)
50 mm clear cavity
100 mm brickwork
Vapour control layer
Rigid PIR insulation over studs (λ = 0.022 W/mK)
Sheathing board
Plasterboard

U-value	Insulation thickness		Wall thickness
	Over studs	Between studs	
0.25 W/ m²K	0 mm	80 mm	340 mm
0.23 W/ m²K	0 mm	100 mm	340 mm
0.20 W/ m²K	0 mm	125 mm	340 mm
0.15 W/ m²K	30 mm	125 mm	370 mm

U-Values: floors

Floor type 1: suspended concrete beam with EPS insulation infill block

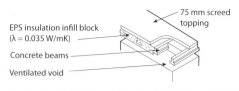

75 mm screed topping
EPS insulation infill block (λ = 0.035 W/mK)
Concrete beams
Ventilated void

U-value	Insulation thickness	
	Between beams	Below beam
0.18 W/ m²K	150 mm	40 mm
0.15 W/ m²K	180 mm	55 mm

Floor type 2: Solid ground floor U-value

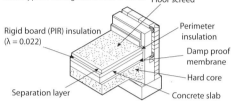

Floor screed
Rigid board (PIR) insulation (λ = 0.022)
Perimeter insulation
Damp proof membrane
Hard core
Separation layer
Concrete slab

U-value	Insulation thickness
0.25 W/m2K	60 mm
0.20 W/m2K	80 mm
0.12 W/m2K	140 mm

Figure 7.3.1 Construction fabric types to suit current thermal performance [based on Part L 2010 – where to start: an introduction for house builders and designers, NHBC Foundation www.nhbcfoundation.org]

U-Values: roofs

Roof type 1: warm roof with mineral wool

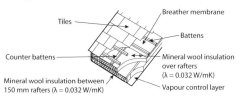

	Insulation thickness	
U-value	Between rafters	Over rafters
0.15 W/ m²K	200 mm	70 mm
0.13 W/ m²K	200 mm	100 mm
0.11 W/ m²K	250 mm	100 mm

Roof type 2: Warm roof with rigid PIR

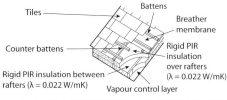

	Insulation thickness	
U-value	Between rafters	Over rafters
0.15 W/ m²K	150 mm	30 mm
0.13 W/ m²K	150 mm	50 mm
0.11 W/ m²K	175 mm	70 mm

Roof type 3: cold roof

Mineral wool insulation between joists (λ = 0.044 W/mK)

Mineral wool insulation over joists (λ = 0.044 W/mK)

Plasterboard

	Insulation thickness	
U-value	Between rafters	Over rafters
0.15 W/ m²K	150 mm	30 mm
0.13 W/ m²K	150 mm	50 mm
0.11 W/ m²K	175 mm	70 mm

Roof Type 4: Cold roof with safe access.

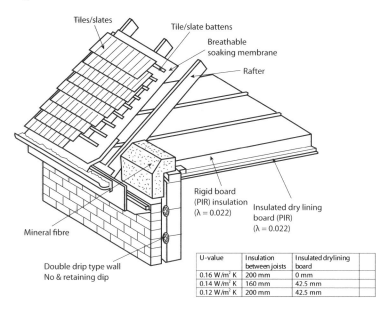

U-value	Insulation between joists	Insulated drylining board
0.16 W/m² K	200 mm	0 mm
0.14 W/m² K	160 mm	42.5 mm
0.12 W/m² K	200 mm	42.5 mm

Figure 7.3.1 Construction fabric types to suit current thermal performance. Construction fabric types to suit current thermal performance [based on Part L 2010 – where to start: an introduction for house builders and designers, NHBC Foundation www.nhbcfoundation.org]

INSULATING CONCRETE

Insulating concrete is a low-density concrete containing a large number of voids. This can be achieved using lightweight aggregates such as clinker, foamed slag, expanded clay, sintered pulverised fuel ash, exfoliated vermiculite and expanded perlite, or by using an aerated concrete made by the introduction of air or gas into the mix. For loadbearing walls, no-fines concrete made using lightweight or gravel aggregates between 20 and 10 mm size and omitting the fine aggregate is suitable. Generally, lightweight insulating concrete is used in the form of an in-situ screed to a structural roof or as lightweight concrete blocks for walls.

LOOSE FILLS

These are materials that can be easily poured from a bag and levelled off between the joists with a shaped template. Materials include exfoliated vermiculite, fine glass-fibre wool, mineral wool, expanded polystyrene beads and cork granules. The depth required to satisfy current legislation is usually well above the top level of ceiling joists, so this type of insulation is now better used as a vertical fill between stud framing to dormer windows and walls in loft conversions. Most loose fills are rot- and vermin-proof as well as being classed as non-combustible.

BOARDS

Boards are used mainly as drylinings to walls and ceilings, either for self-finish or for direct decoration. Types include metallised polyester-lined plasterboard, woodwool slabs, and thermal-backed expanded polystyrene (EPS)/extruded polystyrene (XPS) plasterboard. In use, most require additional vapour barriers to be included. However, there has been considerable development in the use of closed-cell insulation products using phenolic (PL) and polyisocyanurate (PIR) materials. These resist water and moisture ingress as well as defying air leakage. They are often used as insulated rigid drylining boards as a cost-effective replacement for plasterboarded ceilings and walls.

QUILTS

Quilts are made from glass fibre or mineral wool bonded or stitched between outer paper coverings for easy handling. The quilts are supplied in rolls from 6.000 to 13.000 m long and are cut to suit standard joist spacings, then laid between and over the ceiling joists. Quilts are available in two thicknesses: 100 mm for general use and 150 mm for use in roof spaces. They can be placed in two layers, the lower layer between the joists with another superimposed at right angles to the joists.

INSULATING PLASTERS

Insulating plasters are factory-produced, pre-mixed plasters that have lightweight perlite and vermiculite expanded minerals as aggregates, and require only the addition of clean water before application. They are only one-third the weight of sanded plasters, have three times the thermal insulation value and are highly resistant to fire. However, they can only be considered as a supplement to other insulation products as they have insufficient thickness to provide full insulation.

FOAMED CAVITY FILL

This is a method for improving the thermal insulation properties of an external cavity wall by filling the cavity wall with urea–formaldehyde resin foam on site. The foam is formed using special apparatus by combining urea–formaldehyde resin, a hardener, a foaming agent and warm water. Careful control with the mixing and application is of paramount importance if a successful result is to be achieved; specialist contractors are normally employed. The foam can be introduced into the cavity by means of 25 mm boreholes spaced 1.000 m apart in all directions, or by direct introduction into the open end of the cavity. The foam is a white, cellular material containing approximately 99 per cent by volume of air with open cells. The foam is considered to be impermeable, so unless fissures or cracks have occurred during the application, it will not constitute a bridge of the cavity in the practical sense. Non-combustible water-repellent glass- or rock-wool fibres are alternative cavity fill materials. Manufacturers' approved installers use compressors to blow the fibres into new or existing cavity walls, where both inner and outer leaves are constructed of masonry. The application of cavity fill must comply with the requirements of Building Regulation D1.

NATURAL INSULATION MATERIALS

There is a growing demand for the use of natural insulation products as opposed to the more conventional petro-chemical and mineral fibre based insulation materials. Natural insulation materials have low embodied energy, are often a recycled product, and have good thermal and acoustic properties. The main commercially available products are sheep's wool, organic hemp, woodwool and blown cellulose.

Generally these materials have the following advantages over mineral and oil-based products – they:

- use less energy and emit less CO_2 in manufacture than mineral fibre;
- are safer to handle than mineral fibre (e.g. glass fibres are irritants);
- have the ability to absorb moisture without losing their efficiency;
- can store CO_2 during growth (i.e. hemp);
- can be composted or incinerated for energy use at the end of their life.

NATURAL INSULATION PRODUCTS

Other natural products that are used for insulation include flax, which is plant fibres used usually to weave linen cloth. Although it has been successfully used in some projects, it remains largely untested and has limited commercial success. A more successful product is waste wheat straw, which has a thermal conductivity of 0.063–0.045 W/m² K and hence makes an excellent insulator. However, it is more linked to the construct of 'bale form' mass walling in timber-framed buildings and not as an infilling insulation component. The straw bale wall must be supported on a moisture-resistant plinth, such as brick or stone, and rendered with a breathable lime render. The top of the wall is usually protected with deep over-hanging eaves to prevent rainwater penetration.

A broad outline of the most common commercially available natural insulation products is given in Table 7.3.1.

Table 7.3.1 **Natural insulation products.**

Insulation material	Use	Treatment	Thermal conductivity (W/m² K)	Density (kg/m³)
Cellulose – recycled newspaper	Between rafters and joists and timber wall construction. Available in a loose format for pouring and dry or damp spraying as well as in slab format for fitting within metal or timber frames	Borax and boric acid added to provide fire resistance as well as to repel insects and fungi	0.038–0.040	32
Sheep's wool – in rolls and slabs	Between rafters, joists and timber studs in timber 'breathing' wall construction. Hygroscopic – provides a degree of humidity control by absorbing moisture	Polyester binder to bind fibres and treated for fire and insect resistance, beware possible use of pesticides in imported wool	0.039	25
Wood wool – sawmill residue and forestry thinnings	Breathable wall construction, ventilated pitched roofs and in ceilings and floors	Fibres binded with polyester resin and fire retardant added	0.038	50
Hemp insulation slabs – often mixed with recycled cotton or wood fibres	Breathing wall construction, ventilated pitched roofs and in ceilings and floors	Fibres binded with polyester resin and fire retardant added	0.038–0.040	40

Source: 'GreenSpec' Directory available at www.greenspec.co.uk, sponsored by the Building Research Establishment (BRE) and the Department of Trade and Industry (DTI).

Building Regulations Part L: fuel and power conservation

7.4

The Building Regulations Approved Document L: *The conservation of fuel and power* has a major effect on design and construction practices with regard to CO_2 emissions and energy use. The Approved Document has four main parts:

- L1A: Conservation of fuel and power in new dwellings;
- L1B: Conservation of fuel and power in existing dwellings;
- L2A: Conservation of fuel and power in new buildings other than dwellings;
- L2B: Conservation of fuel and power in existing buildings other than dwellings.

Here we will deal with the first two parts, dealing with domestic dwellings, in L1. L2 will be covered in *Advanced Construction Technology* as they apply to commercial and industrial buildings.

To comply with Part L, five basic criteria must be satisfied.

1. The new dwelling must have CO_2 emissions less than or equal to its target emission rate (TER), based on a standard energy loss calculation.

2. There are limits on design flexibility. In order to ensure the fabric and energy generation improvements are not bypassed by creative trade-offs with renewable sources, the thermal performance of building elements and the building services must not fall below minimum efficiency values.

3. The effect of solar gains in summer is limited, so that the building does not suffer from excessive summer solar heat gains.

4. The designed dwelling emission rate (DER) performance is achieved in the actual built dwelling; that is, the performance of the building is verified through specialist testing once the dwelling has been built.

5. Provisions are made for energy-efficient operation of the dwelling, where the owner is supplied with all relevant information to operate their new home efficiently.

NEW DWELLINGS TO APPROVED DOCUMENT L1A

Traditionally, construction elements were specified in accordance to their U-value (the thermal resistance to the passage of heat energy through the external fabric of the dwelling). The Building Regulations then specified maximum U-values for these elements such as walls, floors and roofs. However, the new legislation not only sets limiting U-values that have to be met for the element to ensure compliance (see Fig. 7.4.1), but also sets performance targets for the whole building, based on its CO_2 emissions as measured by the mass of carbon released per metre squared of floor area per year ($kg/m^2/year$).

The procedure involves doing detailed calculations for the designed **dwelling emission rate** (DER) and comparing it against a calculated **target emissions rate** (TER) for the same dwelling. Methods and formulae for these calculations are published in the Government's Standard Assessment Procedure for Energy Rating of Dwellings, known as SAP (or also Dwelling Energy Assessment Procedure, DEAP). SAP incorporates a number of considerations relating to energy consumption in addition to U-values and type of insulation. These include the effect of incidental solar gains, ventilation characteristics, boiler type, hot water and heating systems, energy controls and accessories, amount and orientation of glazing, fuel prices and chosen fuel. These calculations are very detailed and, although they can be carried out manually, they are usually undertaken using specialist software packages, available from a number of approved suppliers. Details can be found on the Building Research Establishment (BRE) website: www.bre.co.uk

In calculating the TER for the proposed dwellings, standard values for the above factors are used and applied to a dwelling of the same size and shape of the proposal. The current regulations require the new dwelling to have a minimum overall improvement of 25 per cent relative to 2006 Part L standards for this notional dwelling. The standard reference factors used for this notional dwelling are given in Tables 7.4.1 and 7.4.2.

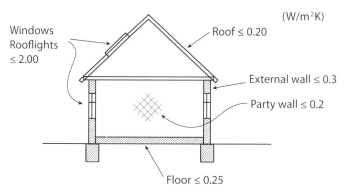

Windows
Rooflights
≤ 2.00

Roof ≤ 0.20

(W/m^2K)

External wall ≤ 0.3

Party wall ≤ 0.2

Floor ≤ 0.25

Also max air leakage $10m^3$/ hour. metre^2 at 50 Pa. and windows & doors should not exceed 25% of the floor area of the dwelling

Figure 7.4.1   The limiting values to reduce thermal conductivity (U-value) in new dwellings – Building Regulation Approved Documents L1A. (Units of U-value $W/m^2 K$)

In simple terms, the TER is found from the following relationship:

$$TER = (CH \times FF + CL) \times (1 - \text{improvement factors})$$

where:

- CH is the CO_2 emission due to the hot water and heating systems given by SAP 2009;

- CL is the CO_2 emission due to the internal fixed lighting given by SAP 2009;

- FF is the fuel factor taken from Table 1 of Part L1A to differentiate between different types of fossil fuels and their CO_2 footprint.

Table 7.4.1 Standard reference criteria for the determination of Target CO_2 Emission Rate (TER), SAP 2009.

Element or system	Value
Size and shape	Same as actual dwelling
Opening areas (windows and doors)	25% of total floor area (or, if total exposed façade area is less than 25% of the total floor area, the total exposed façade area) The above includes one opaque door of area 1.85 m², any other doors are fully glazed All glazing treated as windows (i.e. no roof windows)
External walls	U = 0.35 W/m² K
Party walls	U = 0
Floor	U = 0.25 W/m² K
Roofs	U = 0.16 W/m² K
Opaque door	U = 2.0 W/m² K
Windows and glazed doors	U = 2.0 W/m² K Double glazed, low-E hard coat Frame factor 0.7 Solar energy transmittance 0.72 Light transmittance 0.80
Thermal mass	Medium (TMP (thermal mass parameter) = 250 kJ/m² K)
Living area	Same as actual dwelling
Shading and orientation	All glazing orientated E/W/ average overshading
Number of sheltered sides	2
Allowance for thermal bridging	0.11 × total exposed surface area (W/K)
Ventilation system	Natural ventilation with intermittent extract fans
Air permeability	10 m³ m²/h at 50 Pa
Chimneys	None
Open flues	None
Extract fans	Three for dwelling with floor area greater than 80 m², two for smaller dwellings
Main heating fuel (space and water)	Mains gas
Heating system	Boiler and radiators Water pump in heated space

Table 7.4.2 Appendix R: The Government's Standard Assessment Procedure for Energy Rating of Dwellings 2009.

Boiler	SEDBUK (2009) 78% Room-sealed Fanned flue On/off burner control
Heating system controls	Programmer, room thermostat and TRVs (thermostatic radiator valves) boiler interlock
Hot water cylinder	150-litre cylinder insulated with 35 mm of factory-applied foam
Primary water heating losses	Primary pipework not insulated Cylinder temperature controlled by thermostat
Water use limited to 125 litres per person per day	No
Secondary space heating	10% electric (panel heaters)
Low energy light fittings	30% of fixed outlets

The as-built DER is found using the same calculation process but using site-based data from the constructed dwelling. This calculation would include any changes to the materials or construction specifications from the original design, and actual values of the air permeability of the completed dwelling measured by a certified air pressure testing company. The building designer should aim for an overall improvement of performance of the DER compared to the TER of 25 per cent, with the as-built performance matching or bettering the design performance.

An additional requirement for the 2010 Part L Regulations is that the developer must supply the building control authority with a set of SAP 2009 calculations for the DER along with a list of the material specifications used in the calculations before work starts. In this way the building control authority can ask for further measures to reduce CO_2 emissions to try to achieve the 25 per cent improvement if practical or possible.

Furthermore, where separate dwellings are combined into a single building, as in apartment blocks or terraced houses, the TER (or DER) may be calculated in a similar manner as follows:

$$\frac{(TER_1 \times Floor\ Area_1) + (TER_2 \times Floor\ Area_2) + etc.}{Area_1 + Area_2 + etc}$$

OTHER FACTORS

House builders and designers now have far more to consider when specifying the construction of new dwellings. No longer can energy efficiency be determined by insulation alone. There are now many other factors to consider, creating many benefits in terms of greater flexibility of design. CO_2 reduction is the key, and this may be achieved by including renewable energy technologies into the construction as a trade-off against limited glazing area or extensive insulation, for example. Photovoltaic solar panels and heat pumps rate highly in SAP terms, but even if these save energy in the long term, there is an inevitable initial increase in capital costs of construction. Designing new houses to be energy-efficient is a complex and specialised exercise.

AIR PERMEABILITY

Draught-proofing is required to limit heat losses through breaks in the continuity of construction and to prevent the infiltration of cold external air through leakage in the building envelope. For all new dwellings the worst acceptable air permeability allowed by AD Part L is 10 $m^3/h/m^2$ of floor area at 50 pascals (Pa) pressure. The objective is to be less than $7m^3/h/m^2$, with current best practice in construction achieving around $3m^3/h/m^2$.

Areas most vulnerable to air leakage are the construction interface between wall and floor, wall and roof, wall and window, and so on. For guidance, Accredited Construction Details (ACDs) are published to supplement the Building Regulation Approved Documents. ACDs include criteria such as density and thickness of materials, and application of sealants at junctions and openings (see the section on thermal bridging on pages 322–325).

Particular attention must be paid to the points where service pipes and ducts penetrate the structure. Gaps must be made good with appropriate filling and a flexible sealant applied between pipe or duct and adjacent filling. In drylining, continuous plaster seals are required at abutments with walls, floors, ceilings and openings for services, windows and doors. Some areas and methods of treatment are shown in Fig. 7.4.2.

Other areas vulnerable to air permeability occur at junctions between different elements of construction, particularly where materials with differing movement characteristics meet, such as timber and masonry. A specific area for attention is where floor joists build into the inner leaf of external cavity walls (see Fig. 7.4.3).

The most exposed areas occur at the junction of different components in the same element, such as door and window abutments with walls, access hatches into the roof space or eaves cupboards. The fit between window sashes and frames, and between doors and frames, is also very vulnerable to air leakage. The closeness of fit should be the subject of quality control at the point of manufacture, but this alone is insufficient to satisfy Part L of the Building Regulations. All opening units should have purpose-made seals or draught excluders.

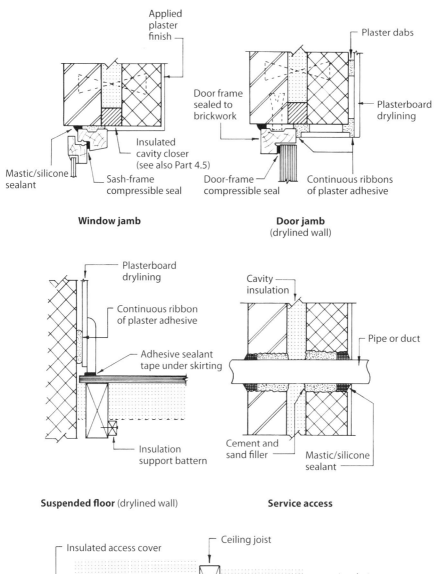

Window jamb

Door jamb
(drylined wall)

Suspended floor (drylined wall)

Service access

Loft hatch

Figure 7.4.2   Prevention of air infiltration.

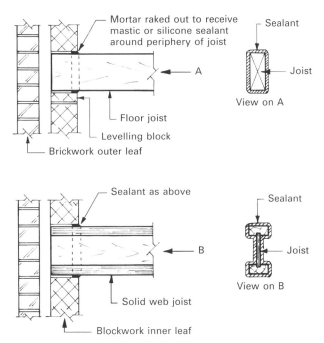

Figure 7.4.3 Air permeability seal at junction of floor joists and external wall.

The quality of workmanship involved in correctly installing thermal insulation is crucial. Common problems encountered on site include poorly positioned partial cavity insulation or mortar grout and other debris left on brick ties, creating a thermal bridge. Insulation within wall cavities may stop too short of the cavity closer or be discontinuous between the wall and ceiling at the roof eaves. Where services are to be fitted, care must be taken not to cut too large a hole that cannot be easily sealed around the pipe; this also applies to holes cut in drylined walls to receive electrical sockets. Where this is the case, extra dabs of adhesive should be placed around the socket to seal the air gap.

HEATING AND HOT WATER EQUIPMENT

The efficiency requirements for space heating and for hot-water heating to meet the requirement of Schedule 1: Part L are given in the Government's second-tier document *Domestic Services Compliance Guide*, which is available from www.planningportal.gov.uk. This document gives the minimum efficiencies for all types of domestic heating generation and distribution systems: oil-fired, electric, solid fuel, community, under-floor heating, mechanical ventilation, heat pumps, comfort cooling, solar water heating, and micro-combined heat and power.

All new installations must have high-efficiency boilers, graded on the SEDBUK scale at A or B energy rating (SEDBUK is the acronym for Seasonal Efficiency of a Domestic Boiler in the United Kingdom). All manufacturers' boilers are listed on a database developed by the Government's Energy Efficiency Best Practice programme, and can be found at: www.boilers.org.uk, which provides an impartial comparison of different manufacturers' models' websites.

All new boilers fitted today are high-efficiency boilers rated at A or B; most are **condensing boilers**, which operate on the principle of a double heating effect. After the initial heat exchange from burning the fuel, the flue gases are circulated by fan around the outside of the heat exchanger for a secondary effect before discharging to the atmosphere. Boilers rated at levels A or B are over 88 per cent efficient to SEDBUK 2009, so not more than 12 per cent of the fuel energy potential is lost in converting the fuel to heat energy in water.

FIXED LIGHTING

In most-used locations, such as the living room, light fittings and luminaires should also be fitted with lamps of minimum efficacy of 55 lumens per circuit watt. Unlike previous legislation, current legislation does not set out any specific requirements for the minimum number of these energy-efficient lights. This is because low-energy lighting, when used in calculating the DER, is tradable with the use of control factors based on use and occupancy of the building. The *Domestic Services Compliance Guide* contains data for the various control factors that can be used.

SOLAR OVERHEATING

Solar gains are beneficial in the winter as a means of offsetting heating demand but can contribute to overheating in the summer months. This must be avoided in the interests of comfort. Energy-consuming comfort cooling or mechanical ventilation may be an option, but this in theory contradicts the principles of conserving energy, so passive methods of control are usually the first solutions to be considered. Passive control can be achieved by limiting window areas, installing awnings and shading systems, using the thermal mass of the structure and limiting glazing exposed to the south.

THERMAL BRIDGES

The building fabric should be constructed so that there are no reasonably avoidable thermal bridges in the insulation layers. These are usually caused by gaps within the various components at the joints between components and at the edges of components, such as around the windows and doors.

The U-values of the building fabric could previously be calculated by assuming that an element was made up of a number of parallel layers, each with uniform thermal resistance; now, however, it is clear that features such as mortar joints or timber studs can cause thermal bridging of the insulation layer, and so contribute significantly to heat loss. Thermal bridges result in variations in thermal transmittance values, which should be avoided as much as possible. Where this does occur, the thermal or cold bridge will incur a cooler temperature than the remainder of the wall. As a result of this loss of heat energy, interstitial condensation may form, which could cause deterioration of the fabric. Also the growth of fungal mould may occur on the surface, causing problems to health and damage to the internal decorations. Examples of how thermal bridges are present in constructed dwellings are shown in Fig. 7.4.4.

Currently under SAP 2009 the heat conductivity at these thermal bridges is aggregated to give a Y-value, measured in $W/m^2 K$, for certain types of construction (such as single dwelling, terraced house or apartment block) and is included in the CO_2 emission calculations (TER/DER). To achieve zero carbon emissions in the future, it is the reduction of heat losses through thermal bridging that will be critically important. Revisions to Part L are likely to require all heat losses through thermal bridging to be identified and calculated separately.

To avoid these problems and to comply with the thermal bridging requirements of Part L, the junctions where construction elements and insulation meet must either:

- be specified as an Accredited Construction Detail (ACD) which have been developed by the Government in conjunction with the Energy Saving Trust and are available on the Planning Portal website: www.planningportal.gov.uk;
  or
- be developed and checked by a competent person using design software and calculation methods as specified in BR 497 *Conventions for calculating linear thermal transmittance and temperature factors*.

In both cases, the contractor must demonstrate that an appropriate system of site inspection is in place to confirm that the work has been carried out to the required standard.

Alternatively unassessed construction details can be used, but the U-values for inputting into the SAP assessment for the DER are highly conservative and will require large trade-offs to the other thermal capacity and energy generation components.

The details shown in Fig. 7.4.5 provide examples of some of the ACDs to prevent thermal bridging.

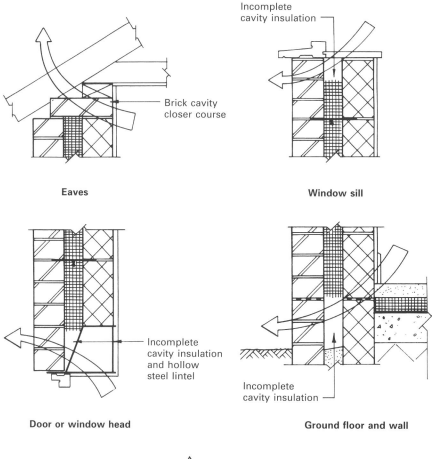

Eaves

Brick cavity closer course

Incomplete cavity insulation

Window sill

Incomplete cavity insulation and hollow steel lintel

Door or window head

Incomplete cavity insulation

Ground floor and wall

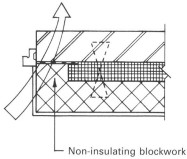

Non-insulating blockwork

Door or window jamb

Figure 7.4.4   Thermal bridging – areas for concern.

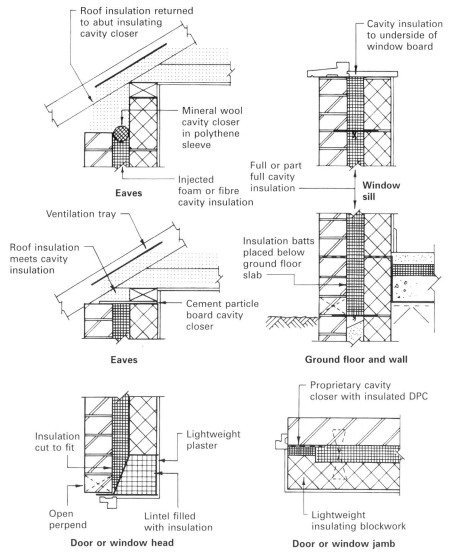

Roof insulation returned to abut insulating cavity closer

Mineral wool cavity closer in polythene sleeve

Injected foam or fibre cavity insulation

Eaves

Cavity insulation to underside of window board

Full or part full cavity insulation

Window sill

Ventilation tray

Roof insulation meets cavity insulation

Cement particle board cavity closer

Eaves

Insulation batts placed below ground floor slab

Ground floor and wall

Insulation cut to fit

Lightweight plaster

Open perpend

Lintel filled with insulation

Door or window head

Proprietary cavity closer with insulated DPC

Lightweight insulating blockwork

Door or window jamb

Figure 7.4.5 Thermal bridging – treatment at openings and junctions.

WORK ON EXISTING DWELLINGS TO APPROVED DOCUMENT L1B

This part covers energy conservation in work done on existing dwellings that are undergoing improvements in the following areas:

- enlarging a dwelling's size, as in the addition of an extension or 'room in the roof';

- renovation of the fabric of a dwelling due to maintenance or improving the thermal performance of a building, as in the installation of double glazing or increasing the energy capacity of the building services;
- a material change of use, as in the conversion of an integral garage into a bedroom.

When adding an extension to an existing building, the new elements must have thermal performance as set out in AD Part L1B: Table 2 and illustrated in Fig. 7.4.6. The usual way to show compliance with this is to determine that the area-weighted U-value for the proposed new extension does not exceed the area-weighted U-value for a similar extension with the thermal properties as set out in Fig. 7.4.6. However, if greater design flexibility is required (for example, to have large, glazed areas of wall in the new extension), then a SAP 2009 calculation can be undertaken on the whole dwelling, which will then allow upgrade improvements to the original property to be offset against any design requirements for the new extension.

In certain situations, the current regulations compel owners who wish to extend their dwelling to upgrade and improve the thermal properties of their original property. Regulation 17D requires that, if any existing dwelling is being extended and has a floor area greater than 1,000 m², additional work must be undertaken to improve the thermal performance of the whole dwelling. This additional requirement, known as 'consequential work', shall be of a value of not less that '10 per cent of the value of the principal work'. This could entail the local building control authority requiring an owner to install new windows, a new boiler or cavity wall insulation to their original dwelling, beyond that already proposed in the principal works.

Where an existing building is upgrading its current fabric, the thermal performance of the new situation is set out in Table 7.4.3. If the existing element has a U-value greater than the threshold value, it should be upgraded

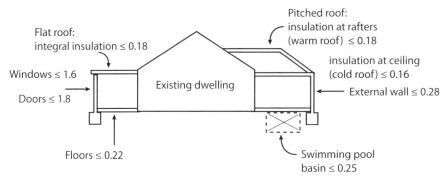

Figure 7.4.6 Limiting fabric parameters.

to achieve the specified improved U-value as noted. If the achievement of this improved value is not practically feasible nor would achieve a simple payback of 15 years higher U-values can be agreed, in consultation with the building control authority, to achieve the best standard that is technically possible.

Table 7.4.3 **Upgrading retained thermal elements.**

Element	Threshold U-value W/m² K	Improved U-value W/m² K
Wall – cavity insulation	0.70	0.55
Wall – internal or external insulation	0.70	0.30
Floor	0.70	0.25
Pitched roof – insulation at ceiling (cold roof)	0.35	0.16
Pitched roof – insulation at rafters (warm roof)	0.35	0.18
Flat roof	0.35	0.18

Adapted from Table 3 AD Part L

Where replacement or renewal work is being undertaken on existing elements of the building structure, AD Part L provides target values for improved fabric performance (see Table 7.4.4). If the achievement of these values is not practically feasible, or would not achieve a simple payback of 15 years, higher U-values can be targeted such that the best standard that is technically feasible is obtained in consultation with the building control authority.

Table 7.4.4 **Cost-effective U-value target when undertaking renovation (renewal) works to thermal elements.**

Proposed works	Target U-value W/m² K
Pitched roof: renewal of roof covering Existing insulation < 100 mm	0.16
Pitched roof: renewal of ceiling to cold loft	0.16
Dormer windows: cladding to side walls	0.30
Flat roof: existing insulation < 100 mm	0.18
Solid external wall: renewal of internal or external finish	0.30
Ground floor: solid or suspended	0.25

Adapted from Table A1 AD Part L

Where a material change of use is proposed, such as converting a barn or a garage into a habitable space, requirements of Tables 7.4.3 and 7.4.4 and the major clauses of AD Part L1B apply.

EXEMPTIONS DUE TO HISTORIC FABRIC

Although there is a great drive to improve the thermal performance of our new and existing buildings and to reduce our CO_2 emissions, there are certain buildings that have exemption from complying with Part L. Most of these buildings are historical and form the architectural character of our villages, towns and cities, and as such are set aside to protect the precious nature of their historic fabric. The categories of building that have this protection are:

■ listed buildings, classified as Grade I, Grade II or Grade II*;

■ buildings within a conservation area, such as scheduled buildings and ancient monuments;

■ other buildings noted for their historical or architectural value by the local authority or built with traditional materials and techniques that must 'breathe', such as medieval timber-frame buildings.

However, where any of the buildings are extended, AD Part L1B still applies to the new works. Advice should be sought from the local conservation officer or English Heritage on the most practical solution to reasonable upgrades in thermal performance at the interface of historic and new work.

AIR PRESSURE TESTING

Testing of air permeability is required for all new dwellings under Part L and is carried out in accordance with the procedure as given in the Air Tightness Testing and Measurement Association's (ATTMA) publication *Measuring air permeability of building envelopes*. On each development site, the test needs to be carried out on three units of each dwelling type or half of the number of dwelling, whichever is the less. In preparation for the test of the air tightness of the dwelling's fabric, all external windows and doors are closed; trickle vents are also closed, but not sealed. All chimney and boiler flues are sealed, and all internal doors are wedged open to allow equalisation of pressure. Finally all service ducts, pipes and overflows are sealed, and toilet and sink traps are filled with water.

A fan unit is sealed into the external door frame and is brought up to a positive pressure of 50 Pa, sufficient to force air through any gaps in the structure. The pressure input is monitored and any pressure drop recorded. This provides direct data to calculate the amount of air lost through the building's fabric. To pass the current AD Part L 2010 test requirements, the fabric must have no air leakage greater than $10m^3/m^2/hr$.

Air leakage in dwellings noted for best performance is around $3m^3/m^2/hr$. This is undoubtedly going to be reduced in later revisions as we approach the 2016 target of zero carbon emissions. Therefore, it is in the best interest of all developers to be designing for the worst case, with the least air permeability possible. However, designers also need to be mindful to design controlled ventilation systems with heat exchange mechanisms to ensure a healthy flow to the occupants, thus reducing condensation and the build-up of harmful stale air.

ENERGY PERFORMANCE CERTIFICATES

The Building Regulations for Part L clause 17E require that any new dwellings or refurbishment projects must have a certificate of energy performance that quantifies the energy-efficiency of the dwelling. The requirements for these certificates are enacted through separate legislation in the Energy Performance of Buildings (Certificates and Inspections) Regulations 2007. The assessments are based on the same calculations as undertaken for the SAP by a qualified energy assessor, who assesses the building on a sliding scale of 1 to 100 in terms of its Energy Efficiency Rating and its Environmental Impact (CO_2) Rating. The final score is published as a letter value ranging from A (best) to G (worst). The assessments also include guidance on how each of these ratings can be improved in terms of additional insulation or the use of improved energy generation methods. An example of an Energy Performance Certificate rating is provided in Fig. 7.4.7.

This home's performance is rated in terms of the energy use per square metre of floor area, energy efficiency based on fuel costs and environmental impact based on carbon dioxide (CO_2) emissions.

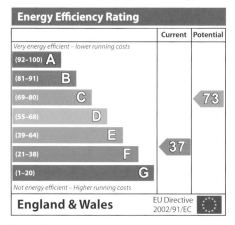
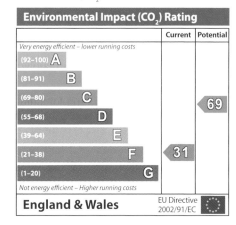

Figure 7.4.7 Example of an Energy Performance Certificate rating.

Code for Sustainable Homes 7.5

The Code for Sustainable Homes was launched in May 2008 by the Government as part of a package of measures to drive down CO_2 emissions in new dwellings. The scheme has been developed by BRE Global under contract to the Department of Communities and Local Government. It replaces the earlier scheme known as the EcoHomes scheme.

The Code gives new homebuyers better information about the environmental impact of their new home and its potential running costs. It also offers house-builders a tool with which to differentiate themselves in sustainability terms within the housing market.

The Code uses a 1 to 6 star rating system to communicate the overall sustainability performance of a new home, setting minimum standards for energy and water use at each level within England, and rating the dwelling as a complete package. Each level includes mandatory requirements for energy performance and water usage, together with tradable requirements for other aspects of sustainability to be assessed at the design and post-construction stages.

In September 2010, the Code became mandatory for all new-build dwellings, with a minimum code level 3 specified, which is currently enshrined in AD part L (2010).

Table 7.5.1 sets out the code levels for energy and water use, which account for nearly half of the final award.

In addition to meeting mandatory standards, achievement of the requirements in each design category scores a number of percentage points (see Table 7.5.1). This establishes the code level or rating for the dwelling. The code certificate illustrates the rating achieved with a row of stars, with a blue star

Table 7.5.1   Code for Sustainable Homes levels for energy and water use.

Code level	Minimum percentage improvement in Dwelling Emission Rate over Target Emission Rate (set by AD Part L: 2006)	Maximum water consumption per person per day (litres)
1 star ★	0% (compliance with Part L 2010 only is required)	120
2 star ★ ★	0% (compliance with Part L 2010 only is required)	120
3 star ★ ★ ★	0% (compliance with Part L 2010 only is required)	105
4 star ★ ★ ★ ★	44% improvement on Part L	105
5 star ★ ★ ★ ★ ★	100% improvement on Part L	80
6 star ★ ★ ★ ★ ★ ★	More than 100%, effectively 'zero carbon' where the dwelling generates more energy than it needs, and does not lose any	80

awarded for each level achieved. Table 7.5.2 summarises what is examined and the relative weightings of each issue.

Each issue is a source of environmental impact that can be assessed against a performance target and awarded one or more credits. Performance targets are more demanding than the minimum standard needed to satisfy Building Regulations. They represent best practice, are technically feasible and can be delivered by the construction industry today.

The code level is assessed on the credits awarded for each category (with weighting applied), which is then expressed as a percentage of the total available in each category. These are then summated and rounded to two decimal places. The final code assessment is then identified from the minimum threshold values shown in Table 7.5.3.

Fig. 7.5.1 (page 334) offers some examples of the build-up of credits under the Code for Sustainable Homes for a two-bed end-terrace house based on Part L 2010. Consultation on the next revisions for 2013 are currently underway, with the aim of achieving the zero carbon emission requirements for 2016. This will mean even tighter controls on emissions and air permeability, involving more detailed calculations and the use of evolving low-energy technologies.

PASSIVHAUS STANDARDS

One voluntary scheme that has become influential in driving forward the zero carbon agenda is the Passivhaus standards scheme for domestic housing. The Passivhaus methodology originated from joint research done by Scandinavian and German researchers. The first dwelling to be built to these early standards was in 1990 in Germany. In Europe's temperate climate, the Passivhaus has been very popular, due to its super-insulation, low air leakage

Table 7.5.2 Requirements for star rating under the Code for Sustainable Homes.

Categories	Issue (those in *italics* are mandatory)	Total credits available	Weighted value of each credit	Net weighting (% point contribution)
Energy and CO$_2$ emissions	***Dwelling emission rate – DER*** ***Fabric energy efficiency*** Energy display devices Drying space Energy-labelled white goods External lighting Low- and zero-carbon technologies Cycle storage Home office	31	1.17	36.4
Water	***Indoor water use*** External water use	6	1.5	9
Materials	***Environmental impact of materials*** Responsible sourcing of materials – basic building elements Responsible sourcing of materials – finishing elements	24	0.3	7.2
Surface water run-off	***Management of surface water run-off from developments*** Flood risk	4	0.55	2.2
Waste	***Storage of non-recyclable waste and recyclable household waste*** Construction site waste management Composting	8	0.8	6.4
Pollution	Global warming potential of insulants NOX (nitrogen oxides) emissions	4	0.7	2.8
Health and well-being	Daylighting Sound insulation Private space **Lifetime Homes** *(mandatory for level 6 only)*	12	1.17	14
Management	Home user guide Considerate Constructors Scheme Construction site impacts Security	9	1.11	10
Ecology	Ecological value of site Ecological enhancement Protection of ecological features Change in ecological value of site Building footprint	9	1.33	12

Table 7.5.3 Minimum scores to achieve code levels.

Code level	Min % score to achieve code level
1 star ★	36
2 star ★ ★	48
3 star ★ ★ ★	57
4 star ★ ★ ★ ★	68
5 star ★ ★ ★ ★ ★	84
6 star ★ ★ ★ ★ ★ ★	90

and use of mechanical ventilation heat recovery. The main factor is that the heating demand is very low – in the order of 15kWh per square metre per year – which reduces fuel costs to a negligible amount. It has been shown that a Passivhaus will not fall below 16 °C, even without heating during the coldest winter months.

Here are the main components of a Passivhaus.

- **Compact form and good insulation** All components of the exterior shell of the house are insulated to achieve a U-value that does not exceed 0.15 W/m² K and 0.8 W/m² K for external windows and doors. By comparison, the current Building Regulation requirements to Part L 2010 only require 0.3 W/m² K for fabric and 1.6-1.8 W/m² K for windows.

- **Passive use of solar energy** Careful design of southern orientation and shading methods maximise solar gain in winter but reduce overheating in summer.

- **Building envelope air-tightness** Air leakage through unsealed joints must be less than 0.6 times the house volume per hour. A typical small, detached house under current Building Regulations is allowed to leak 1,600 m³ of air, while a similar Passivhaus must only leak 200 m³ in the same time.

- **An efficient air-to-air heat exchanger** Here around 80 per cent of the heat in the exhaust air is transferred to the incoming fresh air.

- **Energy-saving household appliances** As well as using modern energy-saving household appliances, Passivhaus buildings use hot water supplied from solar collectors or heat pumps.

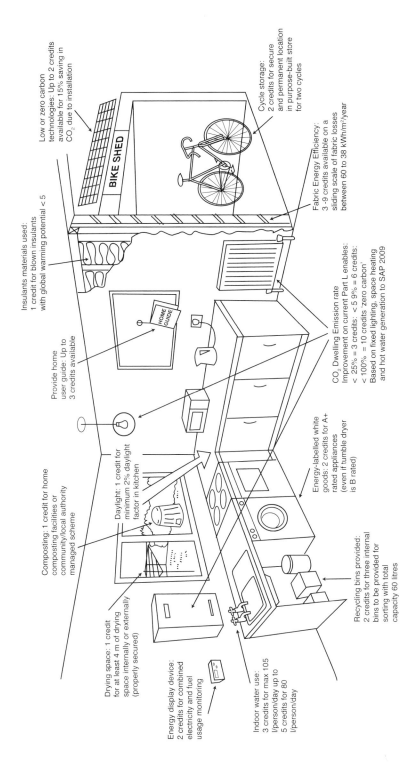

Low or zero carbon technologies: Up to 2 credits available for 15% saving in CO_2 due to installation

Cycle storage: 2 credits for secure and permanent location in purpose-built store for two cycles

BIKE SHED

Insulants materials used: 1 credit for blown insulants with global warming potential < 5

Fabric Energy Efficiency: 3 -9 credits available on a sliding scale of fabric losses between 60 to 38 kWh/m²/year

Provide home user guide: Up to 3 credits available

HOME GUIDE

CO_2 Dwelling Emission rate Improvement on current Part L enables: < 25% = 3 credits: < 5.9% = 6 credits: < 100% = 10 credits 'zero carbon' Based on fixed lighting, space heating and hot water generation to SAP 2009

Composting: 1 credit for home composting facilities or community/local authority managed scheme

Daylight: 1 credit for minimum 2% daylight factor in kitchen

Energy-labelled white goods: 2 credits for A+ rated appliances (even if tumble dryer is B rated)

Drying space: 1 credit for at least 4 m of drying space internally or externally (properly secured)

Recycling bins provided: 2 credits for three internal bins to be provided for sorting with total capacity 60 litres

Energy display device: 2 credits for combined electricity and fuel usage monitoring

Indoor water use: 3 credits for max 105 l/person/day up to 5 credits for 80 l/person/day

Figure 7.5.1   Credits available for a two-bedroom end-terrace house in the Code for Sustainable Homes.

Low- or zero-carbon energy sources 7.6

Most householders receive energy by being connected up to the national energy distribution grid. The Office for National Statistics reports that, in 2008, the energy supplied to households was in excess of 83M tonnes of oil-equivalent energy: 68M tonnes were supplied by gas, oil and coal-fired electricity power stations, and around an additional 15M tonnes were needed to cover generation and distribution losses. The total energy supplied for all sectors was 227M tonnes, of which only 5M tonnes were generated from renewable sources (around 2 per cent).

Low- or zero-carbon (LZC) technologies aim to cut down on our reliance on these fossil fuels with no, or limited, carbon by-products. Another advantage to LZC technologies is that they can involve small, local generation installations with reduced distribution losses and with the ability to meet local environmental conditions. The UK is aiming to achieve 20 per cent of its energy from renewable sources by 2020, so there is a considerable push to bring these technologies into both new and existing dwellings. To assist this process the Government has introduced the 'feed-in tariff' (FiT), under which all installations of LZC installed under the Microgeneration Certification Scheme (MCS) will be able to receive payment for any energy generation, whether it is used or sold back to the energy supply companies.

The most popular LZC technologies currently being applied to small-scale domestic dwellings are:

- solar thermal hot water;
- solar photovoltaic energy;
- heat pumps;
- wind turbines;
- greywater recycling.

This is one of the most popular LZC technologies used in both new-build and refit dwellings. In simple terms they collect thermal energy direct from the sun, which is then used to heat up the domestic hot water for washing and bathing. There are a number of systems on the market that can be open-loop or closed-loop systems. The open-loop is the cheapest as it warms up the water that is supplied to the end user. A main disadvantage is that the system may become quickly furred up, particularly in hard-water areas where limescale is deposited on the inside of the piping when the water is heated. Also the installation needs to be robust enough not to corrode and allow potable water to be contaminated by external agents. Closed-loop systems are preferred as the above problems can be avoided and the water in the closed loop only has to give up its contained limescale once, so furring is not a major issue. The components of a typical solar collector closed-loop system are shown in Fig. 7.6.1.

It is important to monitor the temperatures at the collector and at the tank as these can inform the central controller on the circulating pump operation. The heat-exchange liquid can be just water; however, in these systems, a drain-down tank is included in the circuit so that during freezing conditions, when the pump is switched off, water drains out of the exposed roof collectors to an insulated tank, thus avoiding expensive water leaks through freeze-thaw action within the exposed roof collector. If a drain-down tank is not used, the water needs a suitable concentration of glycol-based non-toxic antifreeze present. An expansion tank and pressure relief valve are also required, as in any sealed system, to prevent dangerous expansion and possible rupture of the piping. It is also necessary to include a secondary heating source to supplement the solar heat exchange, which can be by electric immersion heater or via secondary heat coil from a conventional gas-fired condensing boiler.

Two main types of solar panel are used in the UK: the flat plate collector and the vacuum tube collector.

- The **flat plate collector** is the traditional solar heat collector. It consists of a network of black painted pipes bedded in insulation within a black box with a glazed top, constructed in accordance with ENV 12977-1 and 2. A typical flat plate collector is shown in Fig. 7.6.2. These collectors heat the circulating fluid to a temperature considerably less than that of the boiling point of water. They are best suited to applications where the demand temperature is 30–70 °C, or for applications that require heat during the winter months, such as low-heat under-floor space heating.

- The **vacuum tube collector** is about 30–40 per cent more thermally efficient than a flat plate collector and achieves higher temperatures. They are also easier to maintain and repair without having to decommission the whole system. Like flat plates they are fitted at an angle, usually to the sloping roof of a dwelling. They operate by the solar energy heating up liquid contained in a sealed tube within the vacuum tube. A typical vacuum tube collector element is shown in Fig. 7.6.3.

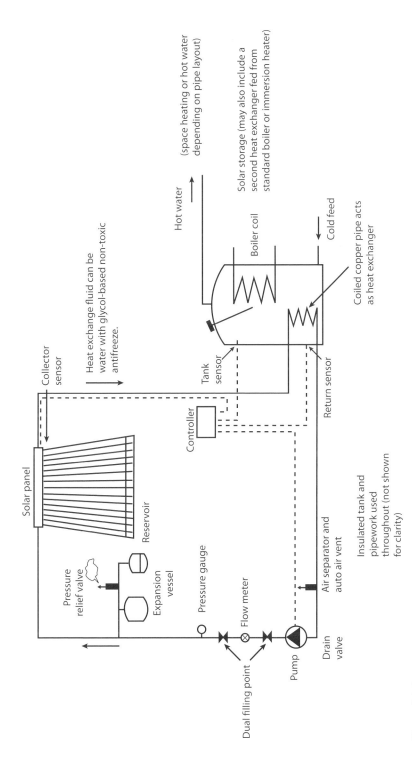

Figure 7.6.1 Typical solar heat collector system.

Solar panel

Collector sensor

Heat exchange fluid can be water with glycol-based non-toxic antifreeze.

Reservoir

Pressure relief valve

Expansion vessel

Pressure gauge

Dual filling point

Flow meter

Pump

Drain valve

Air separator and auto air vent

Insulated tank and pipework used throughout (not shown for clarity)

Controller

Tank sensor

Return sensor

Coiled copper pipe acts as heat exchanger

Cold feed

Boiler coil

Hot water

Solar storage (may also include a second heat exchanger fed from standard boiler or immersion heater)

(space heating or hot water depending on pipe layout)

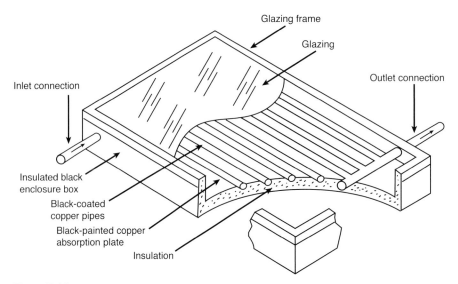

Glazing frame

Glazing

Outlet connection

Inlet connection

Insulated black
enclosure box

Black-coated
copper pipes

Black-painted copper
absorption plate

Insulation

Figure 7.6.2 **Flat plate collector.**

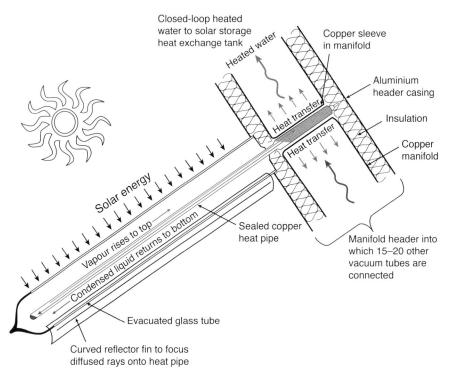

Closed-loop heated
water to solar storage
heat exchange tank

Copper sleeve
in manifold

Heated water

Aluminium
header casing

Heat transfer

Insulation

Heat transfer

Copper
manifold

Solar energy

Vapour rises to top

Condensed liquid returns to bottom

Sealed copper
heat pipe

Manifold header into
which 15–20 other
vacuum tubes are
connected

Evacuated glass tube

Curved reflector fin to focus
diffused rays onto heat pipe

Figure 7.6.3 **Vacuum tube collector.**

Photovoltaic (PV) cells convert sunlight directly into direct current (DC) electricity. They are not to be confused with thermal solar panels, which essentially use the sun's thermal energy to heat water. The PV cells comprise a thin piece of silicon semiconductor material that has been engineered into two different layers by a process known as doping, where impurities are added to the semiconductor to form a negative layer (n-layer) and a positive layer (p-type). When sunlight hits these two layers, a flow of electrons is released between them that causes an electric current to flow.

A standard PV cell will produce a voltage of around 0.5 V. To generate a higher voltage, a number of cells are connected in series to form typically a 150 x 150 mm module; these are then further connected together and embedded in glass, to protect and electrically insulate them. The final configuration of embedded modules is known as a photovoltaic array. To be useful in domestic situations, the DC power has to be converted to alternating current (AC) power using an in-line device known as an inverter. Most systems currently being installed are standalone systems that supplement public power-grid supplies.

Traditionally PV arrays have been installed on existing buildings as a retrofitted addition, usually fixed on top of an existing sloping roof. However, they are increasingly being used within the main fabric of new buildings, thanks to the development of new products such as integrated roof tiles, glass curtain walling, or 'in-roof' glass panels (which often serve a dual purpose as a solar thermal shade).

Advantages of using PV solar panels

- Few moving parts to undergo wear and tear.
- PV modules can be integrated into the building fabric, which can help to reduce costs.
- Can be grid connected and earn money by supplying electricity via a 'feed-in tariff'.
- Can produce around 100 kWh/m² power per year throughout the whole year.

Disadvantages of using PV solar panels

- Relatively high capital and installation cost and long payback time, where no grants available.
- Requires specialist fitting and careful positioning to obtain optimum performance.
- Must be sited away from shade created by adjacent buildings and trees, and will require regular cleaning.
- Solar panels only generate at full capacity 10–30 per cent of the time.

There are two main types of system that use highly efficient refrigerant technologies to maximise heat gain (or cooling):

- ground-source heat pumps;
- air-source heat pumps.

The schematic elements of a heat pump system are shown in Fig. 7.6.4. The heat pump contains an evaporator – a heat exchanger that causes a circulating refrigerant, such as propane, to turn to vapour as it absorbs the air or ground source heat energy. This warm propane vapour is compressed by the pump and its temperature is further boosted. It then enters the condenser and releases the heat to water via the second heat exchanger within the pump system. This heat is then distributed to space or hot water outlets. The vapour, which has now lost its heat, condenses back into liquid propane, which is then allowed to expand via the expansion valve, ready for the next cycle of heat transference.

The temperature of water for distribution released at the condenser typically ranges from 35 to 60 °C, depending on the season. It needs additional heating in winter, which can be via an integrated solar thermal or traditional gas- or electric-fired heating; even so, the contribution made by heat pumps alone can make very useful savings for the owner.

Systems are rated on their coefficient of performance (CoP), which means the ratio of heat delivered to the amount of electricity consumed (the amount of energy needed to drive the compressor and controls as well as any losses in the system). Table 7.6.1 gives typical CoP values for a water-to-water heat pump operation with various heat distribution systems.

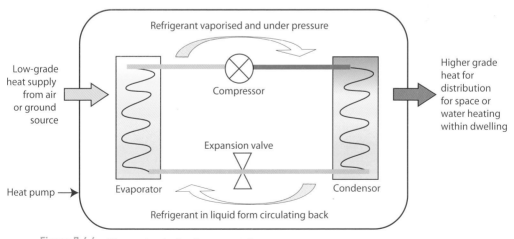

Figure 7.6.4 Elements of a heat pump system.

Table 7.6.1 Typical CoP values for a water-to-water heat pump.

Heat distribution system	Supply/Return temperature (°C)	CoP (heat sources at 5 °C)
Under-floor heating	30/35	4.0
Low temperature radiators	35/45	3.5
Conventional radiators	50/60	2.5

Table from Building Services Research and Information Association (BSRIA) *Illustrated Guide to Renewable Technologies*

GROUND SOURCE HEAT PUMPS

This technology uses the low heat energy that arises from both the deep-down heat contained within the Earth's crust and trapped solar energy within the top subsoil layers. The general principle is that the temperature below ground is on average around 10 °C over the annual seasonal temperatures. By using pipes buried under the ground, in either deep boreholes (> 100 m deep) or large, shallow excavation (1.5–3 m deep), heat can be used in summer to cool dwellings down and in winter to help heat them up. The shallow system is generally preferred because no specialist plant or skill is required and hence it is cheaper to construct. For a large domestic property, with an average heat demand for both space and hot water heating, an area of buried coiled piping of approximately 700 m in length is needed. This would typically fit into an area roughly equivalent to two tennis courts. Generally they are more efficient than air source heat pumps, but more expensive to install.

AIR SOURCE HEAT PUMPS

These units are usually mounted outside the dwelling adjacent to an external wall on rubber pads to reduce noise and vibration. Unlike ground source systems they require no additional groundworks and are quicker to install. In terms of efficiency they are very similar to ground source systems. They work well for underfloor heating systems and radiators, and can be docked to other heating systems to provide a back-up or just more flexibility in the choice of energy used.

OTHER HEAT PUMP SYSTEMS

There is a variety of other heat pump systems available on the market. Any medium that is carrying heat energy can be passed through a specialist heat pump to generate useful energy. These systems include heat pumps for exhaust air and heat recovery, and heat pumps for air-to-air heat recovery.

EXHAUST AIR AND HEAT RECOVERY

Exhaust air systems have air ventilation ducts situated in wet rooms, bathrooms and kitchens, but not cooker hoods. Heat energy is removed using an exhaust heat pump from the stale air and is recirculated into other rooms with exhaust air diffusers. The heat recovery (often known as mechanical ventilated heat recovery (MVHR)) systems are those that are specified in the Passivhaus as previously described. Basically a fresh supply of air is drawn in from the outside, usually by passive means (not pumped), using chimney cowls with a combined heat exchanger. This air is then heated by the exhaust stale air in the heat exchanger as it leaves the building. This heat is then redirected through passive pressure back into internal rooms, and is used to supply fresh, warm air to the habitable rooms. This ventilation method can recover around 80–90 per cent of the heat that would otherwise have been lost, and provides a healthy air supply to the occupants.

AIR-TO-AIR

These are the least efficient of the heat pumps that have been described here (although still better than traditional methods). They do not use a wet medium to transfer the heat but just pump out ambient air to boost the space heating in the dwelling. One good advantage to these supplementary systems is that they can incorporate air filters improving directly the air quality, which is particularly useful for asthma or allergy sufferers.

WIND ENERGY TURBINES

Wind energy turbines convert kinetic energy in the wind into mechanical energy, which is then converted into electricity. There are two basic types of turbine, each with a different type of turbine axis: horizontal (HAWT) or vertical (VAWT). Wind energy turbines can be either free-standing, mounted on a building or integrated into a building structure. Free-standing turbines are generally more efficient, as buildings that are close by can interfere with a consistent air flow. Best performance will be obtained in open, non-urban locations, away from other buildings and structures that can cause damaging turbulence and low functionality. For domestic purposes, the ones mounted on buildings have to be carefully designed to reduce the vibration that may be caused in the structure. Also the blades need to sweep out an area in excess of 1.5 m² to produce electricity in excess of that need to convert the direct current produced into useful AC power. HAWT systems have a yaw mechanism that directs the turbine into the path of the oncoming wind and, as such, are self-starting; VAWT systems need to be started by an electric motor. Wind turbines can provide electrical power directly to a load, via a battery system, or can be linked up to the national power grid to benefit from the 'Feed-in Tariff' scheme. The typical system for the installation of a domestic wind turbine is given in Fig. 7.6.5.

GREYWATER RECYCLING AND RAINWATER HARVESTING

Greywater is potable water originally used for bathing, washing or laundry, which would ordinarily go for waste water. With greywater recycling, the

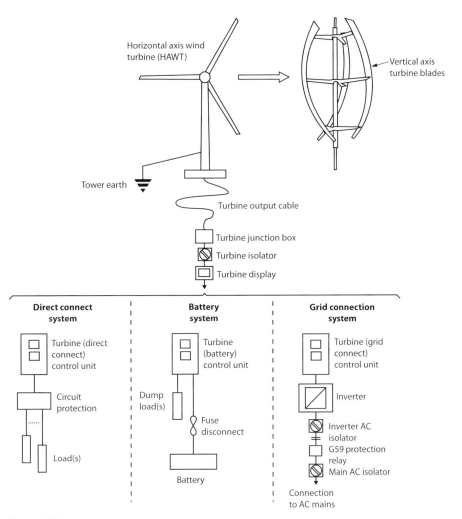

Figure 7.6.5 Typical domestic wind turbine system.

waste water is collected and treated, then used for toilet flushing or irrigation of non-edible plants. Rainwater can be harvested from roofs and hard-standing areas, then stored and treated for toilet flushing and irrigation of vegetables. In practice, greywater recycling can be specified for individual dwellings, multi-tenanted apartments or communal schemes.

Greywater must be treated before it is re-used. In single dwelling schemes, this may mean the introduction of a disinfecting agent, fed into the greywater storage tank. However, larger schemes may require a separate biological aerated filter and sand beds. The system must also have a back-up mains supply if there is insufficient greywater supply, and it must have an anti-syphon filter to prevent contamination of the mains. Fig. 7.6.6 shows the typical system for greywater recycling.

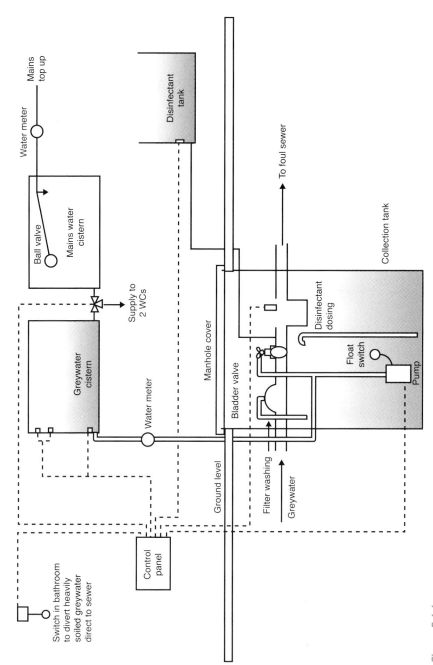

Figure 7.6.6 Typical greywater recycling system.

Internal fixtures and fittings

Doors, door frames and linings 8.1

DOORS

A door is a screen used to seal an opening into a building or between rooms within a building. It is used to provide privacy, sound reduction and fire resistance. It can be made of timber, glass, metal or plastic, or any combination of these materials. Doors can be designed to swing from one edge, slide, slide and fold, or roll to close an opening. The doors to be considered in this part of the book are those made of timber and those made of timber and glass that are hung so that they swing from one edge.

Doors may be classified by:

- their position in a building, such as security;

- their function, such as a fire door;

- their method of construction, such as panelled or flush.

EXTERNAL DOORS

These are used to close the access to the interior of a building and provide a measure of security. Because of their external exposed position they need to be weather-resistant, as in general they are exposed to water, heat, freezing and frost; this resistance is provided by the thickness, stability and durability of the construction and materials used, together with protective coatings of paint stain, or encapsulated finishes such as uPVC or powder-coated steel. The external walls of a building are designed to give the interior of a building a degree of thermal and sound insulation; doors in such walls should, therefore, be constructed, as far as is practicable, to maintain the insulation properties of the external enclosure, to avoid a cold spot. With uPVC external doors, this can be achieved by using foam insulation cores within the construction.

The standard sizes for external timber doors are 1,981 mm high x 762 or 838 mm wide x 45 mm thick, which is a metric conversion of the old imperial door sizes. Metric doors are produced so that, together with the frame, they fit into a modular co-ordinated opening size. Doors can be supplied as doorsets with the door already pre-hung within the frame ready for fixing.

INTERNAL DOORS

These are used to close the access through internal walls and partitions, and to the inside of cupboards. As with external doors, the aim of the design should be to maintain the properties of the wall in which they are housed. Internal doors are available in a variety of pre-finished sizes and finishes, such as pressed hardboard and veneered doors. Generally, internal doors are thinner than their external counterparts as weather protection is no longer a requirement, and they may be constructed using a hollow core, which makes them lighter to handle. Standard sizes are similar to those of external doors but with a wider range of widths to cater for narrow cupboard openings and for disabled adaptations, where 910 mm-wide doors are required for wheelchair access to rooms.

PURPOSE-MADE DOORS

The design and construction of these doors are usually the same as those of manufacturers' standard doors but to non-standard sizes, shapes or designs. Most door manufacturers produce a range of non-standard doors, which are often ornate and are used mainly for the front elevation doors of domestic buildings to provide an aesthetic entrance opening. Purpose-made doors are also used in buildings such as banks, civic buildings, shops, theatres and hotels, to blend with or emphasise the external façade design or internal decor (see Fig. 8.1.1).

METHODS OF CONSTRUCTION

BS EN 14221:2006 provides information on the specification of timber and wood-based materials in internal windows, internal door leaves and internal door frames.

Standard doors are used extensively because they are mass-produced to known requirements, are readily available from stock and are cheaper than purpose-made doors.

Panelled and glazed wood doors
These are constructed of timber, which should be in accordance with the recommendations of BS 1186 and BS EN 942, with plywood or glass panels. External doors with panels of plywood should be constructed using an external-quality plywood, which is more resistant to the elements (see Fig. 8.1.2).

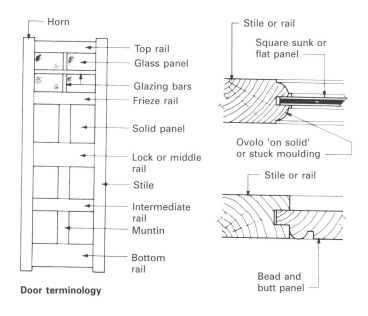

Horn

Top rail

Glass panel

Glazing bars

Frieze rail

Solid panel

Lock or middle rail

Stile

Intermediate rail

Muntin

Bottom rail

Door terminology

Stile or rail

Square sunk or flat panel

Ovolo 'on solid' or stuck moulding

Stile or rail

Bead and butt panel

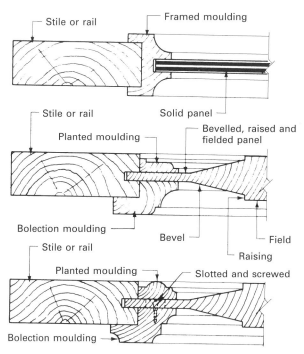

Stile or rail

Framed moulding

Solid panel

Stile or rail

Planted moulding

Bevelled, raised and fielded panel

Bolection moulding

Bevel

Field

Raising

Stile or rail

Planted moulding

Slotted and screwed

Bolection moulding

Figure 8.1.1 Purpose-made doors and mouldings.

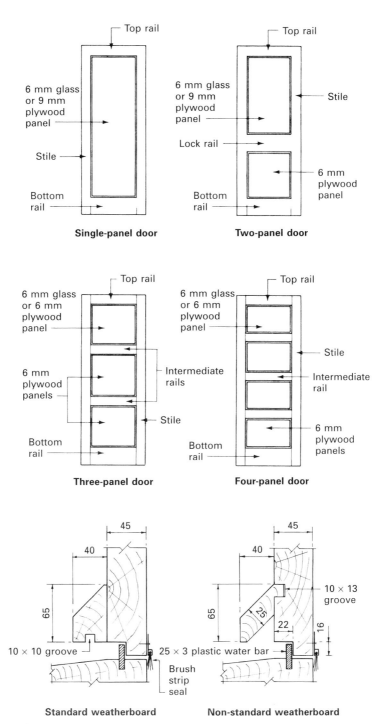

Figure 8.1.2 Standard panelled doors and weatherboards.

The joints used in framing the doors can be a dowelled joint or a mortise and tenon joint. The dowelled joint is considered superior to the mortise and tenon joint, and is cheaper when used in the mass production of standard doors. Bottom and lock rails have three dowels, top rails have two dowels, and intermediate rails have a single dowel connection (see Fig. 8.1.3). The plywood panels are framed into grooves with closely fitting sides, with a movement allowance within the depth of the groove of 2 mm. The mouldings at the rail intersections are scribed, whereas the loose glazing beads are mitred. Weatherboards for use on external doors can be supplied to fit onto the bottom rail of the door, which can also be rebated to close over a water bar (see Fig. 8.1.2).

Flush doors

This type of door has the advantage of a plain face that is easy to clean and decorate; it is also free of the mouldings that collect dust. Flush doors tend to be used as internal doors since they can be faced with a variety of finishes, such as hardboard, plywood or a plastic laminate; by using a thin sheet veneer of good-quality timber, the appearance of high-class joinery can be created.

The method of constructing flush doors is left to the manufacturer, which provides complete freedom for design. Therefore, the forms of flush door construction are many and varied, but basically they can be considered as either hollow-core doors or solid-core doors.

Hollow-core

A hollow-core door consists of an outer frame with small-section intermediate members over which the facing material is fixed. The facing has a tendency to deflect between the core members, and this can be very noticeable on the surface, especially if the facing is coated with gloss paint.

Solid-core

As they have no void within their construction, solid-core doors are stronger than hollow-core doors, and have improved sound resistance. They also provide a higher degree of fire resistance and additional security. Solid doors of suitably faced block or lamin board are available for internal and external use. Another method of construction is to infill the voids created by a hollow core with a lightweight material, such as foamed plastic, which will give support to the facings but will not add appreciably to the weight of the door.

The facings of flush doors are very vulnerable to damage at the edges, so a lipping of solid material should be fixed to at least the vertical edges (good-quality doors have lippings on all four edges).

Small glazed observation panels can be incorporated in flush doors when the glass panel is secured by loose fixing beads (see Figs. 8.1.4 and 8.1.5).

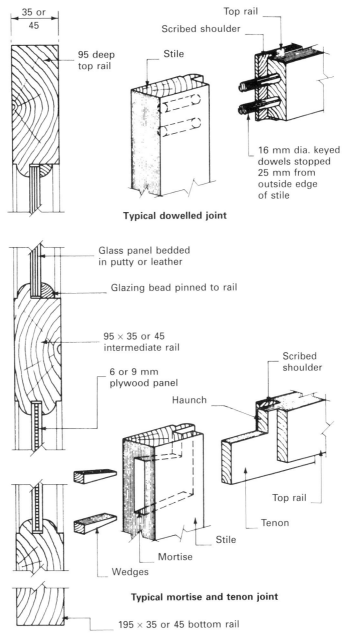

35 or 45

95 deep top rail

Top rail

Scribed shoulder

Stile

16 mm dia. keyed dowels stopped 25 mm from outside edge of stile

Typical dowelled joint

Glass panel bedded in putty or leather

Glazing bead pinned to rail

95 × 35 or 45 intermediate rail

6 or 9 mm plywood panel

Haunch

Scribed shoulder

Top rail

Tenon

Stile

Mortise

Wedges

Typical mortise and tenon joint

195 × 35 or 45 bottom rail

Figure 8.1.3   Panelled door details.

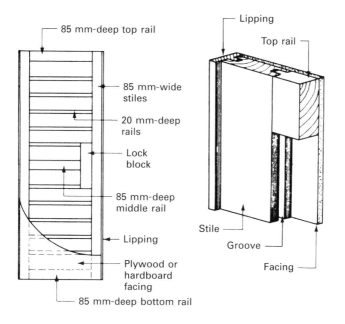

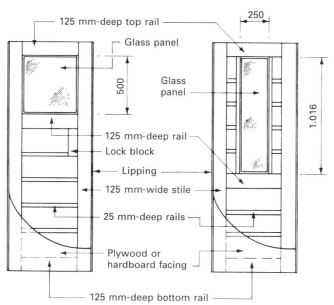

Figure 8.1.4   Hollow-core flush doors.

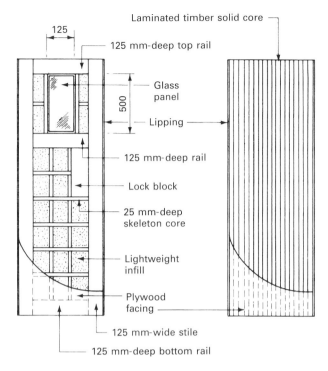

Laminated timber solid core

125

125 mm-deep top rail

Glass panel

500

Lipping

125 mm-deep rail

Lock block

25 mm-deep skeleton core

Lightweight infill

Plywood facing

125 mm-wide stile

125 mm-deep bottom rail

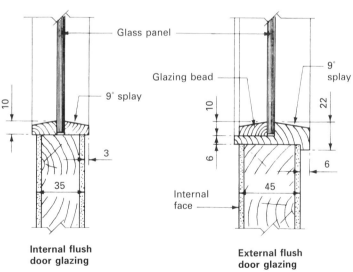

Glass panel

Glazing bead

9° splay

9° splay

10

10

6

22

3

35

6

45

Internal face

Internal flush door glazing

External flush door glazing

Figure 8.1.5   Solid-core doors.

Flush fire doors

With the application of fire regulations to commercial premises and multiple-occupation buildings with protected escape routes, doors have a requirement to be fire-rated. To achieve this, they must be used in conjunction with the correct frame, with the provision of intumescent fire and smoke strips, and a door closer.

BS 476-22: *Methods for determination of the fire resistance of non-loadbearing elements of construction* provides the details on the specification of fire integrity and insulation. Furthermore, BS 9999: *Code of practice for fire safety in the design, management and use of buildings* and the Building Regulations Approved Document B (see Appendix B, Table B1) designate fire doors by integrity performance only.

The performance of fire doors is simplified by expressing the category with the initials FD and the integrity in minutes numerically. For example, FD 60 indicates a fire door with 60-minute integrity, and if the coding is followed by the letter S – that is, FD 60S – this indicates the door or frame has a facility to resist the passage of smoke. This means that the frame must be fitted with smoke seals on all three sides. Standard ratings are 20, 30, 60 and 90, but other ratings are available to order. One-hour fire doors will generally require the thickness increasing from 44 mm to 54 mm.

Fire and smoke seals

Resistance to smoke penetration can be achieved by fitting a brush strip seal to the frame. These are primarily for draught proofing, but will be effective in providing a barrier to cold smoke. A more effective barrier to smoke and flame is an intumescent strip fitted to the door edge or integral with the frame. The latter is usually the most practical, being secured in a frame recess with a rubber-based or PVA (polyvinyl acetate) adhesive. At temperatures of about 150 °C the intumescent material within the seal expands and fills the gap to prevent the passage of fire and smoke for the designated period. The seal will not prevent door movement for access and escape of personnel. Fig. 8.1.6 shows fire doors and frames.

Notes

1. Doorsets/hardware and glass must be selected with compatibility to the door specification (see BS 476-31 for conformity of doorsets).

2. A colour-coded plastic plug is fitted into the hanging style of the door to indicate fire resistance. For example:

   ■ blue core/white background:

     ● FD 20 without intumescent strip in frame;

     ● FD 30 with intumescent strip in frame;

   ■ red core/blue background:

     ● FD 60 with two intumescent strips in frame.

   A full list of coloured core and background codes can be found in BS 8214, the *Code of practice for fire door assemblies* (see Table 1).

3. Half-hour fire-rated doors are hung using one pair of hinges, whereas one-hour doors require one and a half pairs of hinges.

4. All fire doors should have an automatic self-closing device, except cupboards and accesses to service ducts that are normally kept locked.

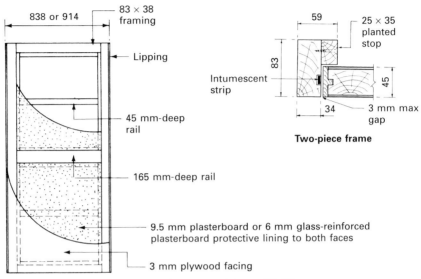

Half-hour fire door and frame (FD 30)

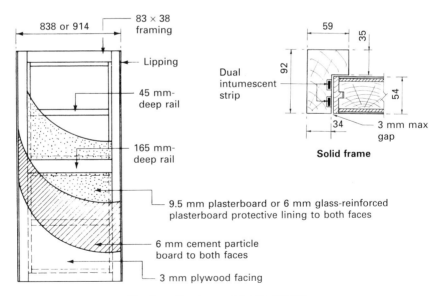

One-hour fire door and frame (FD 60)

Figure 8.1.6 Fire doors and frames.

Matchboarded doors

These doors are mainly used for external outbuilding applications and, as a standard door, take one of two forms: a ledged and braced door or a framed, ledged and braced door (a stronger and more attractive version).

The face of the door is made from tongued-and-grooved boarding that has edge chamfers to one or both faces; these form a V joint between consecutive boards. Three horizontal members called **ledges** clamp the boards together, and in this form a non-standard door, called a **ledged and battened door**, has been made. It is simple and cheap to construct but has the disadvantage of being able to drop at the closing edge, thus pulling the door out of square; the only resistance offered is that of the nails holding the boards to the ledges. In the standard door, braces are added to resist the tendency to drop out of square; the braces are fixed between the ledges so that they are parallel to one another and slope downwards towards the hanging edge (see Fig. 8.1.7).

In the second standard type a mortise and tenon frame surrounds the matchboarded panel, giving the door added strength and rigidity (see Fig. 8.1.7). If wide doors of this form are required, the angle of the braces becomes too low to be of value as an effective restraint, and the brace must therefore be framed as a diagonal between the top and bottom rails. Wide doors of this design are not covered by the British Standards but are often used in pairs as garage doors or as wide entrance doors to workshops and similar buildings (see Fig. 8.1.8).

The operation of fixing a door to its frame or lining is termed **hanging**, and entails the following sequence:

- removing the protective horns from the top and bottom of the stiles;
- planing the stiles to reduce the door to a suitable width;
- cutting and planing the top and bottom to the desired height;
- marking out and fitting the butts or hinges that attach the door to the frame;
- fitting any locks and door furniture that are required.

The hinges should be positioned 225 mm from the top and bottom of the door; where one and a half pairs of hinges are specified for heavy doors, the third hinge is positioned midway between the bottom and top hinges.

A door, irrespective of the soundness of its construction, will deteriorate if improperly treated during transportation and storage, and after hanging. Doors may require leaving stacked for 24 hours, in accordance with the manufacturer's instructions, in order to adjust to the environment that they are going to be hung within. The door should receive a wood-priming coat of paint before or immediately after delivery, and be stored in the dry and in a flat position so that it does not twist; it should also receive the finishing coats of paint as soon as practicable after hanging.

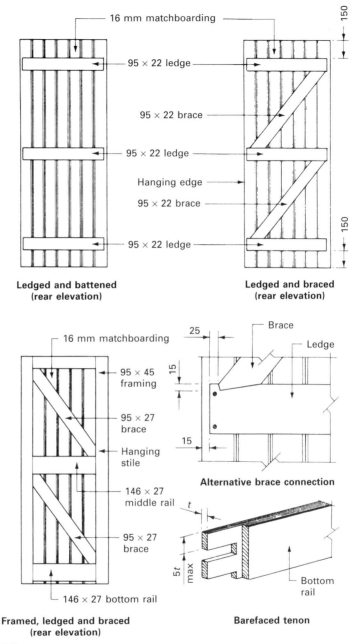

Figure 8.1.7 Matchboarded doors.

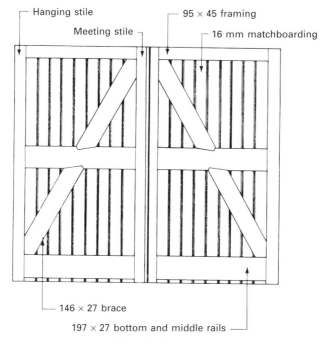

Hanging stile

Meeting stile

95 × 45 framing

16 mm matchboarding

146 × 27 brace

197 × 27 bottom and middle rails

Rear elevation

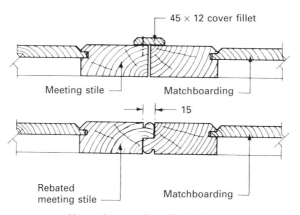

45 × 12 cover fillet

Meeting stile

Matchboarding

15

Rebated
meeting stile

Matchboarding

Alternative meeting stile treatments

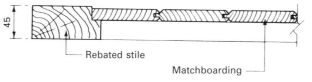

45

Rebated stile

Matchboarding

Alternative stile treatment

Figure 8.1.8   Matchboarded double doors.

A door frame or lining is attached to the opening in which a door is to be fitted; it provides a surround for the door and is the member to which a door is fixed or hung. Doorsets consisting of a storey-height frame with a solid or glazed panel over the door head are also available; these come supplied with the door ready hung on lift-off hinges (see Fig. 8.1.9). Frames should be securely fixed to the wall in which they sit, using frame fixings for masonry and screws for timber.

TIMBER DOOR FRAMES

These are made from rectangular-section timber in which a rebate is formed or to which a planted door stop is fixed to provide the housing for the door. Generally a door frame is approximately twice as wide as its thickness plus the stop. Frames are used for most external doors, heavy doors, doors situated in thin, non-loadbearing partitions and internal fire doors.

A timber door frame consists of three or four members: the head, two posts or jambs, and a sill or threshold. The members can be joined together by wedged mortise and tenon joints, combed joints, or mortise and tenon joints pinned with a metal star-shaped dowel or a round timber dowel (see Fig. 8.1.10). All joints should have a coating of adhesive. Modern frames should be fitted with draught seals to prevent the loss of valuable heat from a home or the ingress of cold air through any gaps between the frame and the door.

If the frame is in an exposed position, it is advisable to sit the feet of the jambs on a damp-proof pad, such as lead or bituminous felt, to prevent moisture soaking into the frame and creating the conditions for fungal attack. Alternatively, the timber can be preservation-treated with an anti-fungal chemical to prevent rot. The joint between the frame and the surround will require an external silicone seal to prevent water penetration behind the frame.

Door frames fitted with a sill are designed for one of two conditions:

- doors opening out;
- doors opening in.

In both cases the sill must be designed to prevent the entry of rain and wind under the bottom edge of the door. Doors opening out close onto a rebate in the sill, whereas doors opening in have a rebated bottom rail, and close over a water bar set into the sill (see Fig. 8.1.10). With both applications a suitable weather bar will be required at the base of the door to throw any rainwater away from the base of the door and onto the external sill.

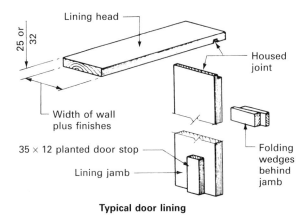

Lining head

25 or 32

Width of wall plus finishes

Housed joint

35 × 12 planted door stop

Lining jamb

Folding wedges behind jamb

Typical door lining

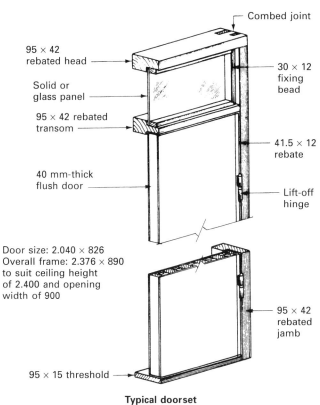

Combed joint

95 × 42 rebated head

Solid or glass panel

95 × 42 rebated transom

30 × 12 fixing bead

41.5 × 12 rebate

40 mm-thick flush door

Lift-off hinge

Door size: 2.040 × 826
Overall frame: 2.376 × 890
to suit ceiling height
of 2.400 and opening
width of 900

95 × 42 rebated jamb

95 × 15 threshold

Typical doorset

Figure 8.1.9   Door linings and doorsets.

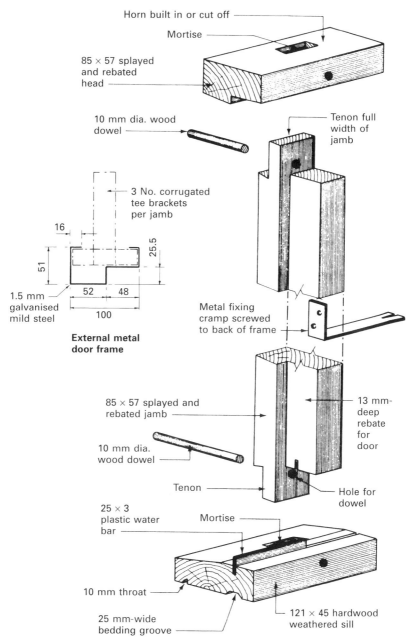

Horn built in or cut off

Mortise

85 × 57 splayed and rebated head

10 mm dia. wood dowel

Tenon full width of jamb

3 No. corrugated tee brackets per jamb

16

25.5

51

1.5 mm galvanised mild steel

52 48

100

External metal door frame

Metal fixing cramp screwed to back of frame

85 × 57 splayed and rebated jamb

13 mm-deep rebate for door

10 mm dia. wood dowel

Tenon

Hole for dowel

25 × 3 plastic water bar

Mortise

10 mm throat

25 mm-wide bedding groove

121 × 45 hardwood weathered sill

Figure 8.1.10   Door frames.

Timber door frames can be fixed to a wall using a range of methods. They can be:

- built into the brick or block wall as the work proceeds using L-shaped ties or cramps. The ties are made from galvanised mild steel with one end turned up 50 mm, with two holes for wood screws, the other end being 100–225 mm long and fishtailed for building into brick or block bed joints. The ties are fixed to the back of the frame for building in at 450 mm centres;

- fixed into a brick opening at a later stage in the contract to prevent damage to the frame during the construction period. This is a more expensive method and is usually employed only when high-class joinery using good-quality timber is involved. Frames are fixed using proprietary frame fixings inserted into the brickwork or blockwork, and the heads are sunk and filled before decoration.

Timber door frames of softwood are usually finished with several applications of paint, whereas frames of hardwood are either polished or oiled. Frames with a factory coating of plastic are also available.

DOOR LININGS

Linings are used for internal applications. They are made from timber 25 or 32 mm thick and as wide as the wall plus any wall finishes. Door linings are not built in but are fixed into an opening screwing directly into brick or block walls. Timber packing pieces or folding wedges are used to straighten and plumb up the sides or jambs of the lining (see Fig. 8.1.9).

Glass and glazing 8.2

GLASS

Glass is made mainly from soda, lime, silica and other minor ingredients such as magnesia and alumina, to produce a material suitable for general window glazing. The materials are heated in a furnace to a temperature range of 1,490–1,550 °C, where they fuse together in a molten state; they are then formed into sheets by a process of floating or rolling. BS 952-1: *Glass for glazing* classifies glass for use in buildings by composition, dimensions, mass and available sizes.

FLOAT GLASS

This is the main method of producing large sheets of near-perfect glass with undistorted, clear vision. The raw materials (including cullet, which is an element of recycled glass) are heated in a furnace to 1,500 °C until the mix is molten. The temperature is then reduced to 1,200 °C, where the glass is then formed by floating a continuous ribbon of molten glass over a bath of liquid metal at a controlled rate and temperature. As the glass moves down the bath, its temperature is reduced until it is picked up by rollers and processed into sheets. This method produces glass that has perfectly smooth surfaces on both sides. A general glazing quality and a selected quality are produced in thicknesses ranging from 3 mm to 25 mm.

ROLLED GLASS

'Rolled glass' is a term applied to a flat glass produced by a rolling process. Generally the glass produced in this manner is translucent, and transmits light with varying degrees of diffusion so that vision through it is obscured. This is because the rolling process introduces a pattern into one side of the glass that distorts the light. It is used for privacy for certain rooms, such as bathrooms and WCs, and to provide a decorative affect.

Georgian Wired Polished Plate (GWPP) is a glass that is produced with a wire cast into the centre of the sheet. This has the added bonus of providing reinforcement to the glass and holding it together should it break under load. It is available in a clear or obscured pattern. GWPP is therefore ideal in a fire-resistant application and is commonly used as vision panelling in fire doors.

SPECIAL GLASS PRODUCTS

Numerous special glasses are available. Some of the most significant include:

- toughened glass;
- laminated glass;
- fire-resisting glass.

TOUGHENED GLASS

Toughened glass is produced to meet the requirements of the Building Regulations, which need glass to withstand shock and pressure when used in windows below 800 mm above floor level: for example, patio doors which may have full-height panels of glass. The toughening process involves heating the glass, which is then cooled by jets of cold air. This creates compressive stresses at the surface with balancing tensile stresses in the centre. The result is a glass of a strength four to five times greater than that of annealed glass (annealing is the process of slowly cooling the glass after manufacture to relieve the stresses within the glass, making it more durable). The glass must be cut to size before the toughening treatment as it cannot be cut after this process. Toughened glass can, however, still fracture under extreme bending loads or severe impact from a sharp implement. When broken it shatters into small, blunt-edged fragments, so the risk of personal injury is reduced. Because of its great strength and potential for large glazed areas, it provides considerable scope for freedom of design.

LAMINATED GLASS

As the name suggests, laminated glass is built up in layers. It is composed of an outer and inner layer of annealed float glass with a heat- and pressure-sealed intermediate bonding layer of polyvinylbutyral (PVB). This plastic flexible layer retains the glass when it is broken. Performance can be varied by changing the number and thickness of glass layers to provide a wide range of applications and purposes. These include rooflights, internal doors and partitions, noise and solar controls, as well as security and safety situations, where it can be effective because of its inherent resistance to bullets and explosives.

FIRE-RESISTING GLASS

Fire resistance can be measured in terms of **integrity** and **insulation**, which is similar to the specification of fire doors. BS 476-20 defines integrity as 'the ability of a specimen of a separating element to contain a fire to specified criteria of collapse, freedom from holes, cracks and fissures and sustained flaming on the unexposed face'. Insulation is also defined in the same BS as 'the ability of a specimen of a separating element to restrict the temperature rise of the unexposed face to below specified levels'. 'Specified levels' means an average of no more than 140 °C and, in any specific position, 180 °C.

Wired glass is well established as an effective means of resisting fire. It is frequently used in doors for security reasons and because of its ability to remain integral in fire for up to two hours. The glass will fracture with heat but the wires retain its integrity in position. Clear fire-resisting glass is produced by Pilkington UK Ltd under the trade names of Pyrodur® and Pyrostop®. Pyrodur® is suitable for use internally or externally in doors and screens. Standard thicknesses are 10 and 13 mm, composed of three or four glass layers respectively. Interlayers are of intumescent material, with an ultraviolet bonding laminate of polyvinylbutyral to ensure impact-resistance performance. It is primarily intended to satisfy the Building Regulations' requirements for integrity, as shown in Table 8.2.1, although it has insulation properties for up to 22 minutes. Pyrostop® is produced in two grades corresponding to internal or external application. For internal use it is made up of four glass layers with three intumescent interlayers. External specification has an additional glass layer and an ultraviolet filter interlayer positioned to the outside to protect the intumescent material against degradation. It is also produced in double-glazed format with an 8 mm air gap achieved with steel spacer bars. Overall thicknesses are 15, 21, 44 and 50 mm, to attain integrity and insulation for up to two hours.

Table 8.2.1   Fire test performance of Pilkington Pyrodur® and Pyrostop® glass.

Glass type	Thickness (mm)	Integrity (minutes)	Insulation (minutes)
Pyrodur®	10	30	N/A
	13	60	N/A
Pyrostop®	15	60	30
	21	60	60
	44	90	90
	50	120	120

GLAZING

Glass must be secured into prepared openings such as doors, windows and partitions, and this is termed glazing. As the area of the glass pane increases, so must its thickness to resist wind load and building usage.

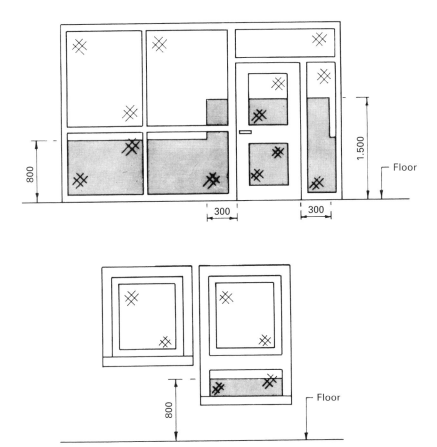

Figure 8.2.1 Zones of safe glazing – shown shaded.

The Building Regulations stipulate suitable standards for the safe use and application of glass through Approved Document N. The position of the glass is very important, and critical locations are shown in Fig. 8.2.1. These are defined as a zone within 800 mm of the finished floor level and 1,500 mm if in a door or adjacent side panel. Glass in these critical locations must be classified as safe, which means it must do one of the following:

- conform to the 'safe break' characteristics defined in BS 6206: *Specification for impact performance requirements for flat safety glass and safety plastics for use in buildings.* BS EN 12600 partially supersedes BS 6206 and defines impact testing;

- be annealed glass within the limitations of Table 8.2.2. Small panes isolated or in groups separated by glazing bars: maximum pane width is 250 mm, maximum pane area is 0.5 m², nominal thickness minimum 6 mm;

- be a robust glass substitute material, such as polycarbonate;

- be permanently protected by a screen, as shown in Fig. 8.2.2.

The effect of wind loading on glazed area and glass thickness can be determined by reference to BS 6262: *Code of practice for glazing for buildings*. Tables and graphs permutate data for topography, ground roughness, life and glass factors to provide the recommended minimum glass thickness.

Table 8.2.2 **Maximum dimensions of annealed glass panels.**

Annealed glass thickness (mm)	Maximum width (m)	Maximum height (m)
8	1.10	1.10
10	2.25	2.25
12	4.50	3.00
15	Any	Any

GLAZING WITHOUT BEADS

This is a suitable method for general domestic window and door panes. The glass is bedded in a compound (normally putty), secured with sprigs, and fronted with a weathered surface putty that slopes away from the glass. Putty is a glazing compound that will require a protective coating of paint as soon as practicable after glazing. This is done because the surface of the putty will dry out and crack. Two kinds of putty are in general use:

■ **linseed oil putty**, for use with primed wood members; made from linseed oil and whiting, usually to the recommendations of BS 544;

■ **acrylic putty,** manufactured from butyl benzyl phthalate; water-based for use with metal and timber glazing installations.

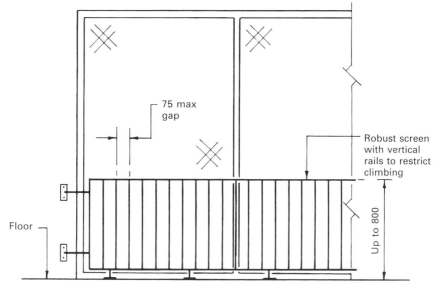

Figure 8.2.2 **Screen protection to glazing.**

The glass pane should be cut to allow a minimum clearance of 2 mm all round for both wood and metal frames, and should be packed to ensure that the gap is equal on all four sides. Sufficient putty is applied to the rebate to give at least 2 mm of back putty when the glass is pressed into the rebate, any surplus putty being stripped off level or at an angle above the rebate. The glass should be secured with sprigs at not more than 440 mm centres and finished off on the front edge with a weathered putty fillet so that the top edge of the fillet is at or just below the sightline (see Fig. 8.2.3). This is done to avoid the rear of the rebate becoming visible from outside the window.

COMPOUNDS AND SEALANTS

Putty has the disadvantage of setting after a period of time, which may be unsuitable for large glazed areas subject to thermal movement. Therefore, several non-setting compounds and sealants are available for this type of application. One-part sealants of polysulphide, silicone and urethane cure by chemical reaction with exposure to the atmosphere. They form a firm, resilient seal with a degree of flexibility that can absorb the thermal movement of the glass. One-part sealants of butyl or acrylic do not cure, but remain soft and pliable. These are preferred for bedding glass and beads only, as they can deteriorate if exposed. Therefore a bead will need fixing to form the external weatherproof element. Two-part polysulphide or polyurethane-based sealants are mixed for application to cure into rubbery material within 14 days. These have excellent adhesion.

GLAZING WITH BEADS

For domestic work, glazing with beads is generally applied to good-class joinery. The beads should be secured with either panel pins or screws; for hardwoods it is standard to use cups and screws. The beads should be secured internally for security to prevent the glass from been removed. The glass is bedded in a compound or a suitable glazing felt, mainly to prevent damage by vibration to the glass. Beads are usually mitred at the corners to give continuity to any moulding features. Beads for metal windows are usually supplied with the surround or frame, and fixing of glass should follow the manufacturer's instructions (see Fig. 8.2.3).

DOUBLE GLAZING

Double glazing is now standard practice in window installation to satisfy Part L of the Building Regulation requirements for energy conservation in external walls through the reduction of heat loss. Factory-made units contain two parallel glass panes sealed with an air gap between the glass. The air gap is formed using a spacer bar, which is filled with desiccant that absorbs any moisture, and the air gap is often filled with argon gas that improves the insulation properties. The spacer bar is sealed to the glass then covered over with a seal to form an airtight unit. This not only reduces heat loss considerably but also improves sound insulation and restricts surface condensation. Typical manufactured units comprise two panes with a dry air gap of between 4 and 20 mm. The greater the air space, the better the thermal insulation, although care must be observed when applied at extremes of atmospheric conditions,

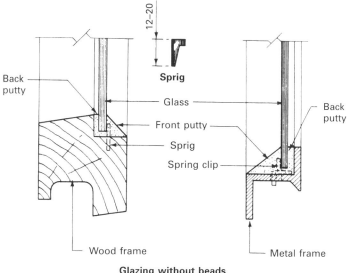

Glazing without beads

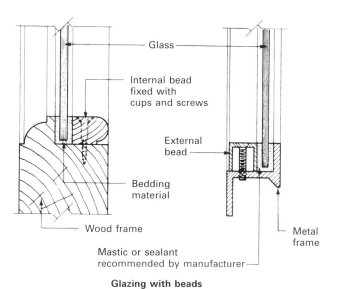

Glazing with beads

Figure 8.2.3   Glazing details.

as explosion or implosion may occur with wider-gapped units. Generally, the insulative effect is to reduce the thermal transmittance coefficient (U-value, a measurement of the amount of heat loss) by about half. Some applications for double-glazed units are shown in Fig. 8.2.4.

LOW-EMISSIVITY GLASS

The use of low-emissivity or 'Low E' glass in double-glazed units can considerably enhance the insulation properties of a double-glazed window. Pilkington UK Ltd manufactures this under the trade name of Pilkington K Glass. Units are manufactured with a standard pane of float glass to the

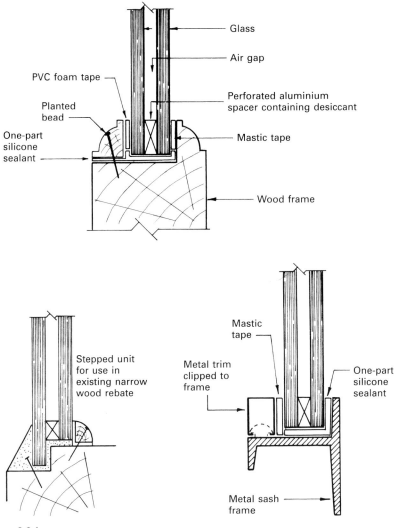

Figure 8.2.4 Double glazing.

outside and a microscopic-metal-coated outer surface to the inner pane of glass, as shown in Fig. 8.2.5.

The effect is to permit short-wave radiation, such as sunlight, to pass through the glazed unit, while reflecting long-wave radiation from internal heating back into the room. Using argon gas, which has lower thermal conductivity than air, between the panes will also reduce heat losses. Table 8.2.3 compares the insulation properties of glazed systems. It is worth noting that only the bottom two would meet the current Building Regulation standards for new buildings if placed within controlled fittings.

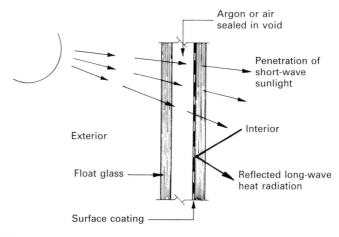

Figure 8.2.5 'Low E' double glazing.

Table 8.2.3 Typical U-values based on 6 mm-thick glass and a 12 mm void in double glazing.

Glazing system	Void	U-value (W/m² K)
Single glazing	–	5.6
Double glazing:		
Float glass × 2	air	2.8
Float glass × 2	argon	2.7
Float glass + 'Low E'	air	1.9
Float glass + 'Low E'	argon	1.6

GLASS BLOCK WALLING

Glass blocks are now produced in a variety of sizes, colour tones and surface finishes. They can be used for aesthetic, non-loadbearing partitioning, particularly in offices where a degree of separation is required while still retaining access for light. They are also acceptable as a light source through external walling, provided they are within a structural support frame. Modern glass-block technology has now produced blocks that have a 60-minute fire integrity, enabling them to be used within fire barrier walls.

Standard units are shown in Fig. 8.2.6. These are manufactured in two halves, rather like square ash trays, of 10 mm-thick glass and then heat-fused together to make a hollow block. The periphery is sprayed with a white adhesive textured finish to bond with mortar. Blocks can be staggered or stretcher-bonded like traditional brickwork. For aesthetic reasons they are usually laid with continuous vertical and horizontal joints (stack-bonded) with steel wire-reinforced bed joints.

EXTERNAL APPLICATION

Blocks are laid and tied to a structural support frame of maximum area 9 m^2 and 3 m in any direction, as shown in Fig. 8.2.7. The lowest course is laid on a bituminous bed and mortar. The mortar to this and subsequent courses comprises 1 part white Portland cement, 0.5 parts lime and 4 parts white quartzite or silver sand, measured by volume. A waterproofing agent is added to the mix. Pointing can be with a weather-resistant silicon sealant. To accommodate expansion and contraction, a polyethylene foam strip is located around the periphery of blocks at the sides and head of the support frame. Nine gauge (3.7 mm) steel reinforcement in ladder pattern is placed in every horizontal course with 300 mm blocks, every other course with 200 mm blocks and every third course with 150 mm blocks (sizes nominal).

SOUND INSULATION AND GLAZING

The main problem with sound insulation and glazing is the requirement for ventilation: opening a window reduces its sound-reduction properties. Windows cannot provide the dual function of insulation against noise and

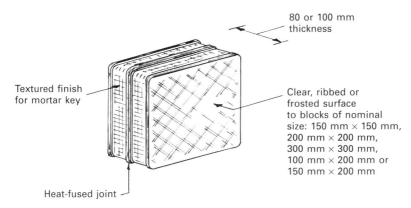

80 or 100 mm thickness

Textured finish for mortar key

Clear, ribbed or frosted surface to blocks of nominal size: 150 mm × 150 mm, 200 mm × 200 mm, 300 mm × 300 mm, 100 mm × 200 mm or 150 mm × 200 mm

Heat-fused joint

Figure 8.2.6   Standard glass block.

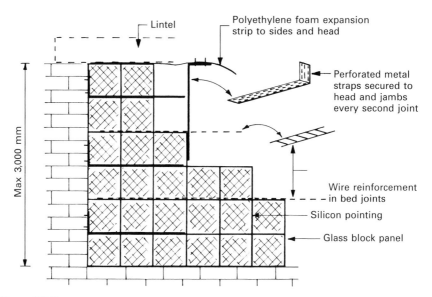

Figure 8.2.7 Typical glass block framing and laying.

ventilation, because letting in air will also let in noise. When opened, any type of window will only give a sound reduction of about 10 dB, as opposed to the 45–50 dB reduction of the traditional cavity wall. A closed window containing single glazing will give a reduction of about 20 dB, or approximately half that of the surrounding wall. It is obvious that the window-to-wall ratio will affect the overall sound reduction of the enclosing structure.

Double glazing can greatly improve the sound insulation properties of windows provided the following points are observed.

- Sound insulation increases with the distance between the glazed units; for a reduction of 40 dB, the airspace should be 150–200 mm wide, which is not practical.

- If the double windows are capable of being opened, they should be weather-stripped.

- Sound insulation increases with glass thickness, particularly if the windows are fixed; this may mean the use of special ventilators with specific performances for ventilation and acoustics.

Double glazing designed specifically to improve the thermal properties of a window has no real value for sound insulation, as it does not really have any mass or density to slow down the passage of air. In some situations, such as around airports, triple glazing may have a better effect in reducing sound transmission.

Windows

8.3

The primary function of a window is to provide a means for admission of natural daylight to the interior of a building and also to serve as a means of providing the necessary background ventilation of dwellings, as required under Building Regulation F1, by including opening lights in the window design or vents within the glass or frames.

Windows, like doors, can be made from a variety of materials, or a combination of these materials, such as timber, metals like aluminium, and plastic. They can also be designed to operate in various ways by arranging the sashes to slide, pivot or swing, or hang as a casement to one of the frame members.

BUILDING REGULATIONS

Approved Document F details the requirements for providing ventilation to habitable rooms, kitchens, utility rooms, bathrooms and sanitary accommodation in domestic buildings. Ventilation is required to remove high humidity levels, which may cause condensation. A habitable room is defined as a room used for dwelling purposes but not a kitchen. A habitable room must have ventilation openings unless it is adequately ventilated by mechanical means, such as extractor fans.

Passive stack ventilation (PSV) is an acceptable alternative in some rooms (see Approved Document F, Section 1). This relies on the 'chimney effect' of drawing air out of a building by negative pressure cause by air movement across a stack externally. This is a simple system of ducts used to convey air by natural movement to external terminals. Non-domestic buildings have different requirements to suit various purposes, with criteria applied to occupied rooms instead of habitable rooms and greater emphasis on mechanical ventilation and floor areas (see Approved Document F, Section 2).

A ventilation opening includes any permanent or closable means of ventilation that opens directly to a supply of external air. Included in this category are openable parts of windows, louvres, airbricks, window trickle ventilators and doors that open directly to the external air. The three general objectives of providing a means of ventilation are:

■ to extract moisture from rooms where it is produced in significant quantities, such as kitchens and bathrooms through cooking and showering;

■ to provide for occasional rapid ventilation where moisture presence is likely to produce condensation, such as over a cooker;

■ to achieve a background ventilation that is adequate without affecting the comfort of the room, as for gas-fire combustion air.

Guidance in Approved Document F gives advice on how ventilation can be provided to meet these objectives. Solutions range from simple trickle-type ventilators (see Fig. 8.3.3, page 379) built into the frames to opening sashes, mechanical extractor fans and fully ducted air ventilation systems.

Unless a mechanical system of air movement is used throughout a dwelling, background ventilation through trickle vents is required in all habitable rooms. Table 5.2a of Part F gives the requirement based on floor area and number of bedrooms for new dwellings.

Intermittent use of mechanical fan extract ventilation is also required for the following specific situations, as detailed in Table 5.1a of Part F:

■ kitchen – minimum 60 litres per second; 30 l/s if adjacent to hob;

■ utility room – 30 l/s;

■ bathroom – 15 l/s;

■ sanitary accommodation – 6 l/s.

If any of the above situations is without a window with an opening sash, a combined light/extract fan switch should be fitted with an over-run facility for the fan. A gap of at least 10 mm should be provided between the door and floor finish.

TRADITIONAL CASEMENT WINDOWS

Fig. 8.3.1 shows a typical arrangement and annotated details for the common component names for this type of window. A wide range of designs can be produced using various combinations of the members, to produce different sizes and combinations of opening sashes and fixed lights. The only limiting factor is the size of glass pane relevant to its thickness.

The general arrangement of the framing is important. Heads and sills always extend across the full width of the frame and in many cases have projecting horns for building into the wall, or projecting the sill beyond the opening.

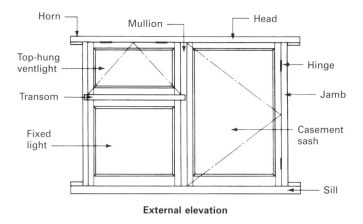

Horn

Mullion

Head

Top-hung
ventlight

Hinge

Transom

Jamb

Fixed
light

Casement
sash

Sill

External elevation

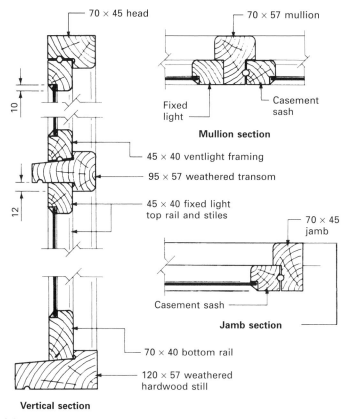

70 × 45 head

70 × 57 mullion

Fixed
light

Casement
sash

Mullion section

45 × 40 ventlight framing

95 × 57 weathered transom

45 × 40 fixed light
top rail and stiles

70 × 45
jamb

Casement sash

Jamb section

70 × 40 bottom rail

120 × 57 weathered
hardwood still

Vertical section

Figure 8.3.1 Traditional timber casement window.

The jambs and mullions span between the head and sill; these are joined to them by a wedged or pinned mortise and tenon joint. This arrangement gives maximum strength, as the vertical members will act as struts; it will also give a simple assembly process. Traditional timber windows are now being re-engineered into high-performance timber windows that are factory-produced and fully compliant with the requirements of the Building Regulations with regard to air permeability and ventilation.

The traditional casement window frame has deep rebates to accommodate the full thickness of the **sash**, which is the term used for the framing of the opening ventilator. If fixed glazing or lights are required, it is necessary to have a sash-frame surround to the glass, because the depth of rebate in the window frame is too great for direct glazing to the frame. This is also done to provide a balanced and aesthetically pleasing aspect to the window. Capillary grooves are cut in specific locations to break the capillary action of water through rain and driving wind.

STANDARD WOOD CASEMENT WINDOWS

BS 644: *Timber windows. Factory assembled windows of various types* provides details of the quality, construction and design of a wide range of wood casement windows. These are constructed using highly engineered timber sections. The frames, sashes and ventlights are made from standard sections of softwood timbers arranged to give a variety in design and size. The fixed and opening sashes are designed so that their edges rebate over the external face of the frame to form a double barrier to the entry of wind and rain. A much deeper capillary groove is formed with this method. The general construction is similar to that described for traditional casement windows, and the fixing of the frame into walls follows that described for door frames.

Most joinery manufacturers produce a range of modified standard casement windows following the basic principles set out in BS 644 but with improved head, sill and sash sections, and the inclusion of draft strips built into the frames. The range produced is based on brick modules to avoid unnecessary cutting and disturbance to the masonry appearance. Window types are identified by a notation of figures and letters. For example, for 212 CV:

   2 = width divided into two units

   12 = 1,200 mm height

   C = casement

   V = ventlight

For typical details see Figs. 8.3.1–8.3.3.

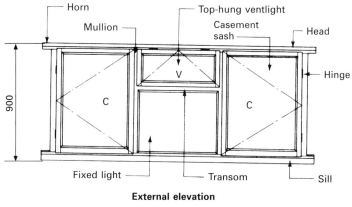

External elevation

Labels (top to bottom, left to right): Horn, Mullion, Top-hung ventlight, Casement sash, Head, Hinge, 900, C, V, C, Fixed light, Transom, Sill

Vertical section

95 × 57 head

10

12

95 × 70 softwood or hardwood sill

Mullion section

70 × 57 mullion

Fixed light

Casement sash

41 × 46 ventlight framing

70 × 57 transom

Direct glazing to fixed light

Jamb section

70 × 57 jamb

41 × 46 casement sash framing

Figure 8.3.2 Typical modified BS casement window (see Fig. 8.3.3 for double-glazed vertical section).

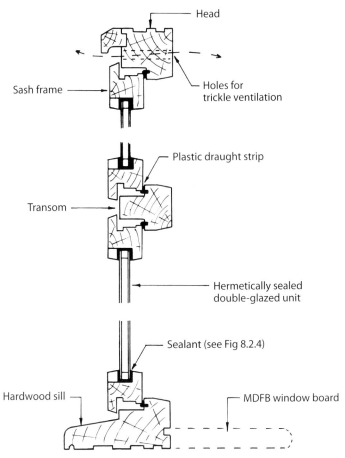

Figure 8.3.3 Double-glazed casement window.

uPVC HOLLOW PROFILE WINDOWS

Due to the wide variety of systems available for constructing a uPVC window they tend to be certified under the BBA scheme. This is the British Board of Agrément Certification Scheme, which gives a European Technical Approval to a manufacturer for their system. Generally BS EN 12608 covers the requirements of unplasticised polyvinylchloride (uPVC) profiles for the fabrication of windows and doors.

uPVC windows are produced from extruded plastic sections, which are mitred and heat-welded together to form a rigid structure. This process is used with the opening sashes, which are fixed using stainless steel pivot hinges into the rebates of the uPVC frames. The double-glazed units are retained into their sash openings using mitred glazing beads with seals already inserted into them. Drainage holes are provided within the uPVC design to allow any water drawn in by capillary action to drain out of the uPVC sections. All opening sashes have draught seals fitted to them. Steel point locking systems make uPVC windows secure because they lock at more than one point into the opposing frame. A key-operated handle is used to open the locking system and the double-glazed units are internally fixed for additional security.

UPVC windows can now be produced in a wood-effect finish which gives grain and colour to the frames as an alternative to the traditional white uPVC. The window systems produced are thermally efficient, lightweight and maintenance-free, and provide an attractive finish to any dwelling.

Typical uPVC sections are shown in Fig. 8.3.4.

ALUMINIUM HOLLOW PROFILE CASEMENT WINDOWS

Extruded aluminium profiles can be designed for use in window and door framing. The main advantages over other materials in these applications are high strength and durability relative to low weight, and very little maintenance except for occasional cleaning. The aluminium is protected from surface oxidisation and corrosive atmospheres with a coloured polyester powder coating. A range of standard colours is available that can be coated to order.

Metals are poor thermal insulators, but the overall thermal performance of aluminium-profiled casements can be considerably improved by specifying a closed-cell foam infill to the hollow sections. Cold or thermal bridging and associated condensation is prevented by incorporating a thermal barrier or break between internal and external components. The thermal break is produced from a high-strength two-part polyurethane resin. Typical section details are shown in Fig. 8.3.5.

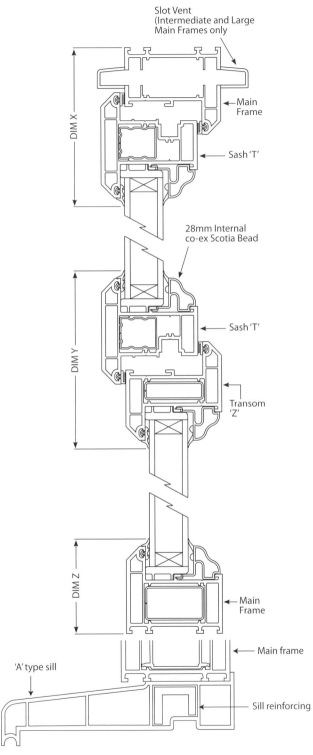

Slot Vent
(Intermediate and Large
Main Frames only

←Main
Frame

Sash 'T'

28mm Internal
co-ex Scotia Bead

Sash 'T'

Transom
'Z'

←Main
Frame

←Main frame

'A' type sill

Sill reinforcing

DIM X

DIM Y

DIM Z

Figure 8.3.4   Typical uPVC sections.

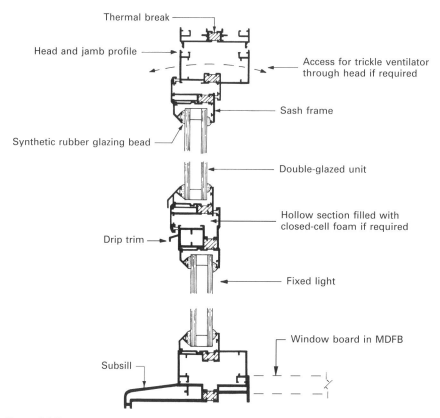

Thermal break

Head and jamb profile

Access for trickle ventilator through head if required

Sash frame

Synthetic rubber glazing bead

Double-glazed unit

Hollow section filled with closed-cell foam if required

Drip trim

Fixed light

Window board in MDFB

Subsill

Figure 8.3.5   Aluminium casement window.

BAY WINDOWS

Any window which projects in front of the main wall line is considered a bay window; various names are, however, given to various plan layouts (see Fig. 8.3.6). Bay windows can be constructed of timber, uPVC and/or metal, and designed with casement or sliding sashes; the main difference in detail is the corner post, which can be made from the solid, jointed or masked in the case of timber and tubular for metal windows (see Fig. 8.3.6). With uPVC windows, an aluminium or steel insert is placed with the corner posts to provide additional strength to the construction.

The bay window can be applied to one floor only or continued over several storeys. The intermediate section can be covered in a variety of materials from tile hanging to lead. Any roof treatment can be used to cover in the projection and weather-seal it to the main wall (see Fig. 8.3.7). No minimum headroom heights for bay windows or habitable rooms are given in the Building Regulations, but 2.000 in bay windows and 2.300 in rooms would be considered reasonable. A bay window that occurs only on upper storeys is generally called an oriel window.

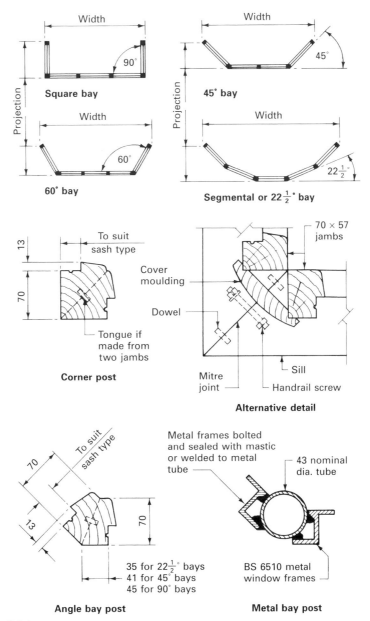

Figure 8.3.6 Bay window types and corner posts.

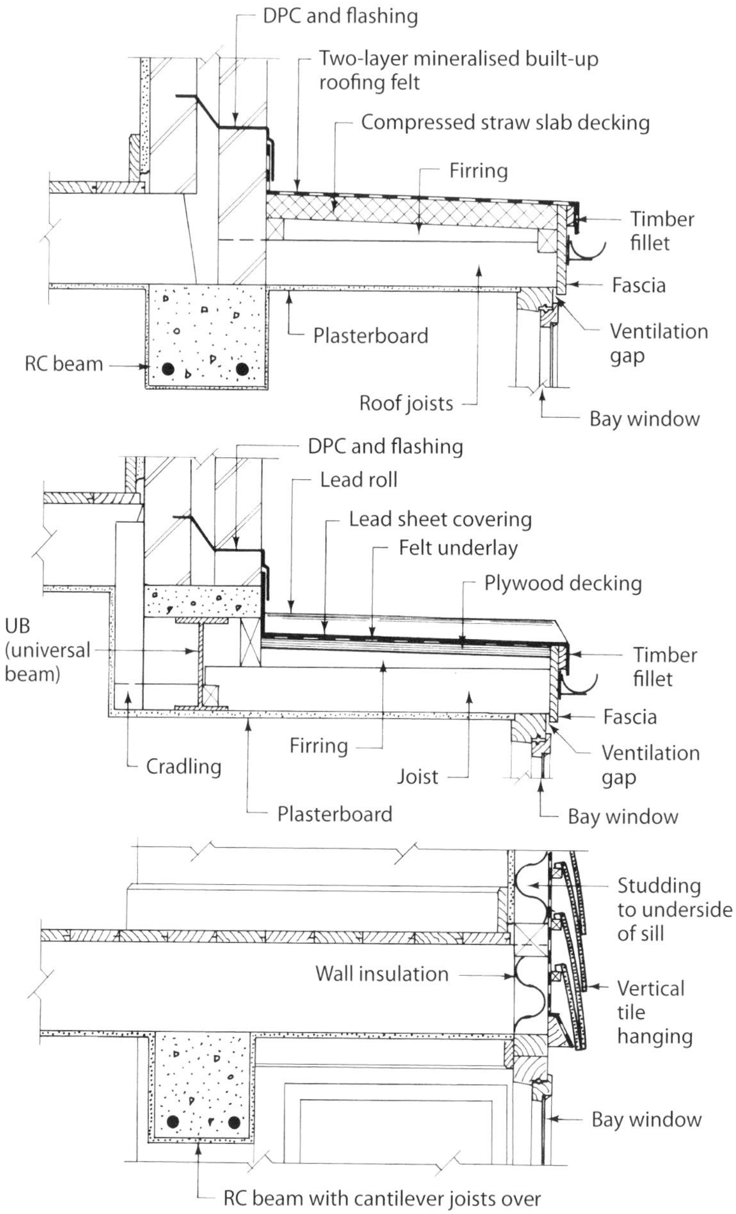
DPC and flashing
Two-layer mineralised built-up roofing felt
Compressed straw slab decking
Firring
Timber fillet
Fascia
Ventilation gap
Plasterboard
RC beam
Roof joists
Bay window
DPC and flashing
Lead roll
Lead sheet covering
Felt underlay
Plywood decking
UB (universal beam)
Timber fillet
Fascia
Ventilation gap
Firring
Cradling
Joist
Plasterboard
Bay window
Studding to underside of sill
Wall insulation
Vertical tile hanging
Bay window
RC beam with cantilever joists over

Note: Current practice is to insulate flat roofs (see Part 6.1) and stud framing (see Part 3.7)

Figure 8.3.7 Typical existing bay window roofs.

DOUBLE-HUNG SASH WINDOWS

These windows are sometimes called **vertical sliding sash windows** and consist of two sashes sliding vertically over one another. They are a traditional window that may be found in listed buildings and conservation areas where they will have to be maintained, repaired or replaced like for like to match the existing material and construction. They are costly to construct, but are considered to be more stable than side-hung sashes and have a better control over the size of ventilation opening, thus reducing the possibility of draughts. Modern uPVC construction methods have produced aesthetically pleasing copies of this type of window, but with modern fittings and sliding mechanisms that do not incorporate lead weights.

In timber, two methods of suspension are possible:

■ weight-balanced type;

■ spring-balanced type.

The former is the older method, in which the counterbalance weights suspended by cords are housed in a boxed framed jamb or mullion. It has been generally superseded by the metal spring balance, which uses a solid frame and needs less maintenance (see Figs. 8.3.8 and 8.3.9).

PIVOT WINDOWS

A pivot window has a central horizontal or vertical pivot. This enables the window to be turned over in order to clean the external glazing without having to access this via ladders, with the associated risks of working at height. The design of a pivot window must take into account the Building Regulations requirement for escape windows in situations where a secondary means of escape is required in case of a fire. The Regulations state that a clear opening of 0.33 m^2 must be achieved with a minimum dimension of 450 mm clear opening.

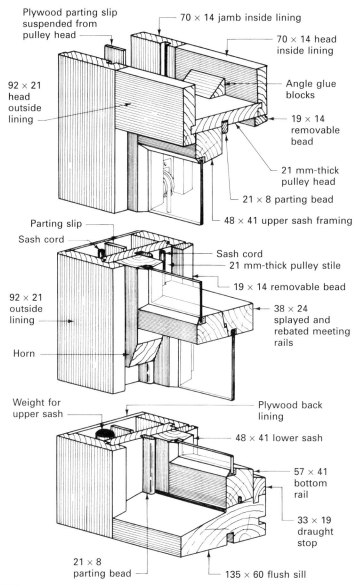

Plywood parting slip suspended from pulley head

70 × 14 jamb inside lining

70 × 14 head inside lining

92 × 21 head outside lining

Angle glue blocks

19 × 14 removable bead

21 mm-thick pulley head

21 × 8 parting bead

48 × 41 upper sash framing

Parting slip

Sash cord

Sash cord

21 mm-thick pulley stile

19 × 14 removable bead

92 × 21 outside lining

38 × 24 splayed and rebated meeting rails

Horn

Weight for upper sash

Plywood back lining

48 × 41 lower sash

57 × 41 bottom rail

33 × 19 draught stop

21 × 8 parting bead

135 × 60 flush sill

Figure 8.3.8   Traditional double-hung weight-balanced sliding sash windows.

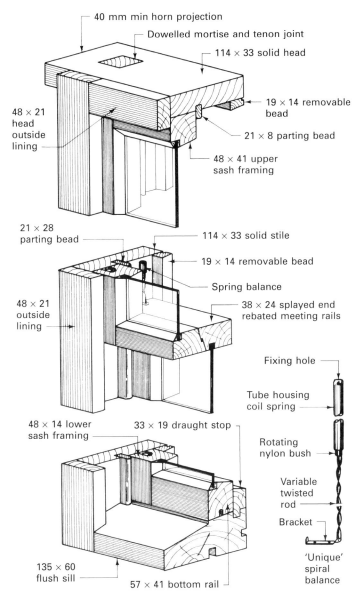

40 mm min horn projection

Dowelled mortise and tenon joint

114 × 33 solid head

19 × 14 removable bead

48 × 21 head outside lining

21 × 8 parting bead

48 × 41 upper sash framing

21 × 28 parting bead

114 × 33 solid stile

19 × 14 removable bead

Spring balance

48 × 21 outside lining

38 × 24 splayed end rebated meeting rails

Fixing hole

Tube housing coil spring

48 × 14 lower sash framing

33 × 19 draught stop

Rotating nylon bush

Variable twisted rod

Bracket

135 × 60 flush sill

57 × 41 bottom rail

'Unique' spiral balance

Note: If 114 × 60 solid stiles are used, balances can be housed in grooves within the stile thickness

Figure 8.3.9 Double-hung spring-balanced sash windows.

Timber stairs 8.4

A stair is a means of providing access from one floor level to another. Modern stairs with their handrails are designed with the main emphasis on simplicity, employing design methods that provide comfortable access and egress from each level safely.

The total rise of a stair – that is, the distance vertically from floor finish to floor finish in any one-storey height – is fixed by the storey heights and floor finishes being used in the building design: the stair designer has only the total vertical going or total horizontal distance with which to vary the stair layout. This in effect alters the pitch of the stairs (the angle at which they are inclined). It is good practice to keep door openings at least 450 mm away from the head or the foot of a stairway (for example, in the case of an attic conversion where space is required at the base of the stairs) and to allow at least the stair width as circulation space at the head or foot of the stairway.

Modern stairs are usually designed as one straight flight between floor levels for domestic installation, which is the simplest and cheapest layout that can be factory-manufactured. Alternatively they can be designed to turn corners by the introduction of quarter space (90°) or half space (180°) intermediate landings. This enables the building designer to accommodate flexibility of space in hallways. Stairs that change direction of travel, without the use of landings, use tapered steps or are based on geometrical curves in plan. Irrespective of the plan layout, the principles of stair construction remain constant, and are best illustrated by studying the construction of simple straight-flight stairs in this volume.

Terminology

- **Stairwell** The space between walls in which the stairs and landings are housed.

- **Stairs** The actual means of ascent or descent from one level to another.

- **Tread** The horizontal surface of a step on which the foot is placed.
- **Nosing** The exposed edge of a tread, usually projecting with a square, rounded or splayed edge.
- **Riser** The vertical member between two consecutive treads.
- **Step** Riser plus tread.
- **Going** The horizontal distance between two consecutive risers or, as defined in Approved Document K, the horizontal dimensions from front to back of a tread less any overlap with the next tread above.
- **Rise** The vertical height between two consecutive treads.
- **Flight** A series of steps without a landing.
- **Newel** Post forming the junction of flights of stairs with landings or carrying the lower end of strings.

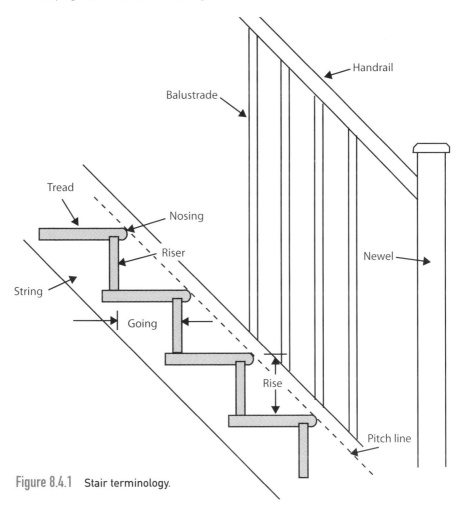

Figure 8.4.1 Stair terminology.

- **Strings** The members receiving the ends of steps, which are generally housed to the string and secured by wedges, called wall or outer strings according to their position.
- **Handrail** Protecting member usually parallel to the string and spanning between newels.
- **Baluster** The vertical infill member between a string and handrail.
- **Pitch line** A line connecting the nosings of all treads in any one flight.

BUILDING REGULATIONS PART K

Approved Document K defines three categories of stairs.

1. **Private**: intended for use solely in connection with one dwelling.
2. **Institutional and assembly**: serves one place in which a substantial number of people will gather.
3. **Other**: serves in buildings other than dwellings, institutional and assembly.

The practical limitations applicable to each of these three categories are shown in Table 8.4.1.

Table 8.4.1 **Stairs (Approved Document K).**

Category	Rise (mm)	Going (mm)
Private	155–220 165–200	245–260 223–300
Institutional and assembly*	135–180	280–340
Other*	150–190	250–320

* Subject to requirements for access for disabled people. See Building Regulations Approved Document M, Section 3.51 (max rise 170 mm, min going 250 mm).

CONSTRUCTION AND DESIGN

The dimensions of the treads and risers need to be constant throughout any flight of steps to reduce the risk of accidents by changing the rhythm of movement up or down the stairway. The height of the individual step rise is calculated by dividing the total rise (for example, ground floor to first floor) by the chosen number of risers. The individual step going is chosen to suit the floor area available that the stairs can be fitted within but still meet the requirements of the Building Regulations (see Table 8.4.1). It is important to note that in any one flight there will be one more riser than treads, as the last tread is in fact the landing.

Stairs are constructed by joining the steps into side rails (known as strings) using housing joints, then glueing and wedging the steps into position to form a complete and rigid unit. Small angle blocks can be glued at the junction of tread and riser in a step to reduce the risk of slight movement giving rise to the annoyance of creaking. The flight can be given extra rigidity by using triangular brackets placed under the steps on the centreline of the flight. A central beam or carriage piece with rough brackets as a support is used only on stairs over 1,200 mm wide, especially where they are intended for use as a common stairway (see Figs. 8.4.2 and 8.4.3).

Stairs can be designed to be fixed to a wall with one outer string, fixed between walls, or free-standing, fixed at the top and base; the majority have one wall string and one outer string. The wall string is fixed directly to the wall along its entire length, or is fixed to timber battens plugged to the wall, the top of the string being cut and hooked over the trimming member of the stairwell. The outer string is supported at both ends by a newel post: in the case of the bottom newel, this rests on the floor; in the case of the upper newel, it is notched over and fixed to the stairwell trimming member. If the upper newel is extended to the ground floor to give extra support, it is called a **storey newel post**. The newel posts provide a fixing point for handrails that span between them and are then infilled with balusters, balustrade or a solid panel to complete the protection to the sides of the stairway (see Fig. 8.4.4). The distance between the balusters must not let a 100 mm-diameter sphere pass between them for safety reasons in accordance with Part K of the Building Regulations. Headroom required when climbing stairs from the pitch line to the ceiling above is 2 m, but 1.8–1.9 m can be accommodated for loft conversions.

If the headroom distance is critical it is possible to construct a bulkhead arrangement over the stairs, as shown in Fig. 8.4.5; this may give the increase in headroom required to comply with the Building Regulations. The raised floor in the room over the stairs can be used as a shelf or can form the floor of a hanging cupboard.

LAYOUT ARRANGEMENTS

A stair flight incorporating landings or tapered steps will enable the designer to economise with the space required to accommodate the stairs. Landings can be quarter space, giving a 90° turn, or half space, giving a 180° turn: for typical arrangements, see Fig. 8.4.6. The construction of the landing is similar to that of a timber upper floor except that, with the reduced span, joist depths can be reduced (see Fig. 8.4.6). The landing area must be level and be provided at the top and base of every flight. The landing can be incorporated in any position up the flight, and if sited near the head may well provide sufficient headroom to enable a cupboard or cloakroom to be constructed below the stairs. A dogleg or string-over-string stair is economical in width, as it will occupy a width less than two flights, but this form has the disadvantage of a discontinuous handrail because this abuts to the underside of the return or upper flight.

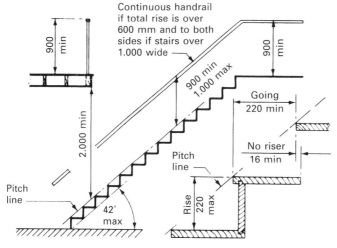

Continuous handrail if total rise is over 600 mm and to both sides if stairs over 1.000 wide

900 min

2.000 min

900 min
1.000 max

Going
220 min

No riser
16 min

Pitch line

Pitch line

42° max

Rise 220 max

900 min

Sum of going + twice rise = 550 min to 700 max in any flight all risers of equal height and all goings of equal width

Private stairways

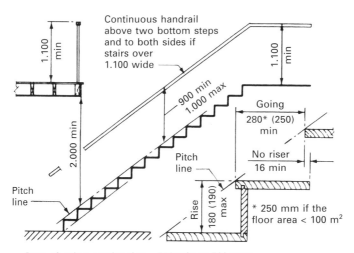

Continuous handrail above two bottom steps and to both sides if stairs over 1.100 wide

1.100 min

2.000 min

900 min
1.000 max

Going
280* (250) min

No riser
16 min

Pitch line

Pitch line

Rise 180 (190) max

1.100 min

* 250 mm if the floor area < 100 m²

Sum of going + twice rise = 550 min to 700 max in any flight all risers of equal height and all goings of equal width

Institutional and assembly and (other) stairways

Figure 8.4.2   Timber stairs and Approved Document K.

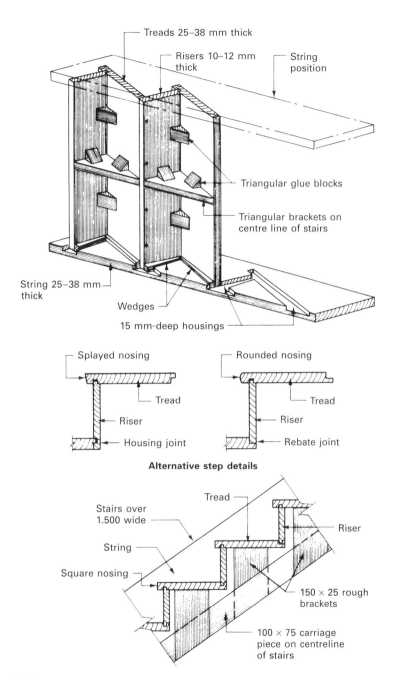

Treads 25–38 mm thick

Risers 10–12 mm thick

String position

Triangular glue blocks

Triangular brackets on centre line of stairs

String 25–38 mm thick

Wedges

15 mm-deep housings

Splayed nosing

Tread

Riser

Housing joint

Rounded nosing

Tread

Riser

Rebate joint

Alternative step details

Tread

Stairs over 1.500 wide

Riser

String

Square nosing

150 × 25 rough brackets

100 × 75 carriage piece on centreline of stairs

Figure 8.4.3   Stair construction details.

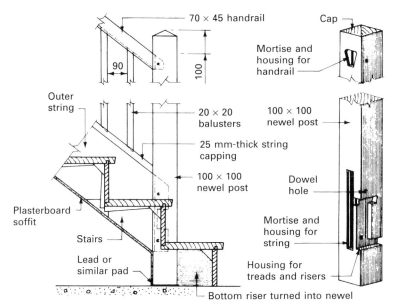

Typical detail at bottom newel

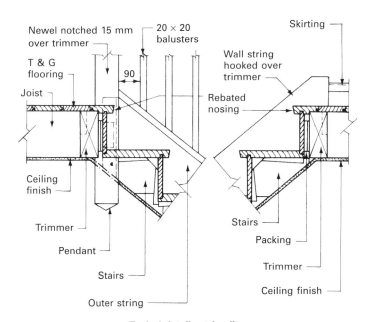

Typical details at landing

Figure 8.4.4 Stair support and fixing details.

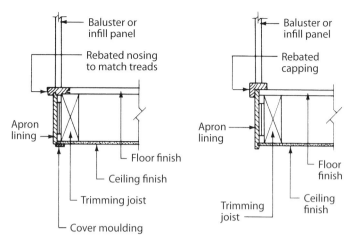

Typical stairwell finishes

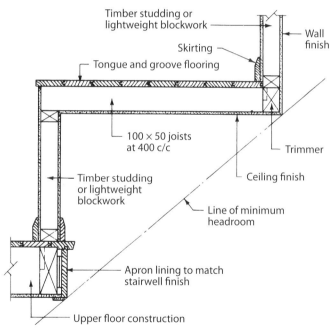

Typical bulkhead over stairs

Figure 8.4.5 Stairwell finishes and bulkhead details.

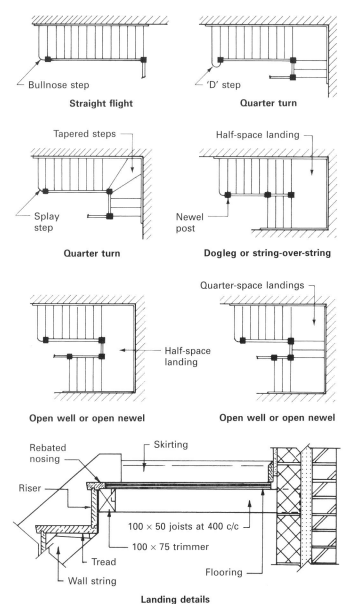

Bullnose step

Straight flight

'D' step

Quarter turn

Tapered steps

Splay step

Quarter turn

Half-space landing

Newel post

Dogleg or string-over-string

Half-space landing

Open well or open newel

Quarter-space landings

Open well or open newel

Rebated nosing

Skirting

Riser

100 × 50 joists at 400 c/c

100 × 75 trimmer

Tread

Flooring

Wall string

Landing details

Figure 8.4.6 Typical stair layouts.

TAPERED STEPS

With the introduction of the Building Regulations special attention has been given to the inclusion of tapered steps in Approved Document K, which makes the use of tapered steps less of an economic proposition and more difficult to design (see Fig. 8.4.7). A tapered step gives a designer the ability to produce a curved stairway that can add substantial architectural features to a hallway.

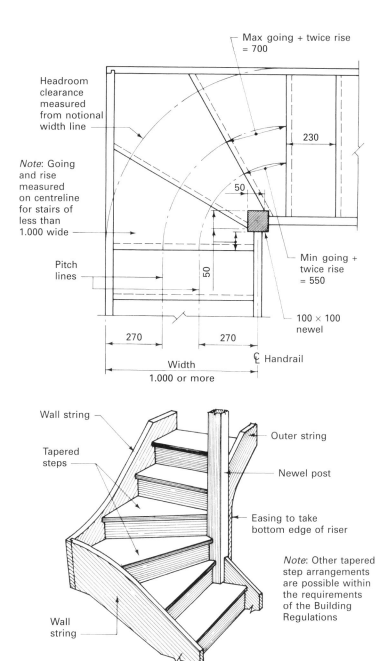

Max going + twice rise = 700

Headroom clearance measured from notional width line

Note: Going and rise measured on centreline for stairs of less than 1.000 wide

Pitch lines

230

50

50

Min going + twice rise = 550

100 × 100 newel

270

270

℄ Handrail

Width
1.000 or more

Wall string

Tapered steps

Outer string

Newel post

Easing to take bottom edge of riser

Note: Other tapered step arrangements are possible within the requirements of the Building Regulations

Wall string

Figure 8.4.7 Tapered steps (or winders) for private stairways.

A popular space-saving variation is the application of continuous tapered steps to create a spiral stair. A spiral stairway has the advantage of a smaller footprint area. The example illustrated in Fig. 8.4.8 can be in hardwood or softwood timber, but contemporary design versions are produced in stainless steel or in any combination of steel, timber and glass. With spiral stairs, every tread has a continuous taper, and the arrangement of these treads to produce a stairway should comply with the safety in use requirements indicated in the Building Regulations. Approved Document K1 of the Regulations gives a specific reference for spiral stairs as conforming to BS 5395-2: *Stairs, ladders and walkways. Code of practice for the design of helical and spiral stairs*. Joinery manufacturers can be expected to adhere to these recommendations to ensure that there is a market for their product. However, before specifying a particular type, the details should be checked with the local authority building control department. Any structural alteration to the floor and landing will also require approval along with the requirements for a loft conversion, which may require that a stairway is enclosed.

Spiral stair units are generally produced as a kit for site assembly of simple-fit components about a central column or continuous newel post. The kit comprises of a central column that is bolted to the floor using a steel fixing plate. From the column are attached treads and spacers to form the stairway. Balusters fit to the periphery of the treads and provide support to the helical spire handrail. On plan, overall diameters from outside of tread to outside of tread are 1,200 and 1,400 mm at 14 treads per 360° turn, and 1,600 and 2,200 mm at 16 treads per 360° turn. Ascent may be clockwise or anti-clockwise. The starting point needs to be carefully considered so the finishing point of the spiral exits the stair at the correct point and headroom.

OPEN-TREAD STAIRS

These are a contemporary form of stairs used in homes, shops and offices based on the simple form of access stair, which has been used for many years in industrial premises. They are designed primarily to save weight and provide visibility through the stairway, as many of the components can be removed, such as nosings, cappings and risers. The open tread or riserless stair must comply fully with Part K of the Building Regulations and in particular Approved Document K, which recommends a minimum tread overlap of 16 mm. Where they are likely to be used by children under five years, such as in dwellings or pre-school nurseries, the opening between treads must contain a barrier sufficient to prevent a 100 mm sphere passing through.

Four basic types of open-tread stairs can be produced: closed string, cut string, mono-carriage and alternating tread.

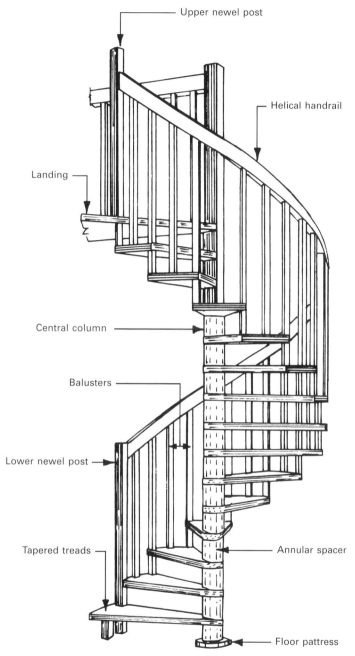

Upper newel post

Helical handrail

Landing

Central column

Balusters

Lower newel post

Tapered treads

Annular spacer

Floor pattress

Figure 8.4.8   Timber spiral stair.

CLOSED STRING

This will terminate at the floor and landing levels and be fixed as for traditional stairs. The treads are tightly housed into the strings, which are tied together with long steel tie bars under the first, last and every fourth tread. This provides the lateral support, while the stair wand holds it together. The nuts and washers can be housed into the strings and covered with timber inserts (see Fig. 8.4.9).

CUT STRINGS OR CARRIAGES

'Cut strings' are so called because the string is cut to accept the treads. These are used to support cantilever treads, and can be worked from the solid or of laminated construction. Because the strings are exposed, the level of quality and detail needs to be precise to produce a quality finished product. The upper end of the carriage can be housed into the stairwell trimming member, with possible additional support from metal brackets. The foot of the carriage is housed in a purpose-made metal shoe or fixed with metal angle brackets (see Fig. 8.4.10). This type of stair is very much an institutional application.

MONO-CARRIAGE

Sometimes called a spine beam, this employs a single central carriage with double cantilever treads. The carriage, which is by necessity large, is of laminated construction and very often of a tapered section to reduce the apparently bulky appearance. The foot of the carriage is secured with a purpose-made metal shoe in conjunction with timber connectors (see Fig. 8.4.11). This type provides a design with an aesthetical feature to the stair.

ALTERNATING TREAD STAIRS

These make economic use of space and are frequently applied to loft conversions, as they have no risers and can be reduced in width to fit within a smaller stairwell space. They have a pitch of about 60° and paddle-shaped treads, which are both controversial issues in the interests of user safety, especially concerning small children. However, the Building Regulations Approved Document K, accepts 'familiarity and regular use' as a reasonable safety argument provided the stair accesses only one habitable room. Additional requirements include a non-slip surface, handrails on both sides, a minimum going of 220 mm with a maximum rise of 220 mm, and a gap between treads of no more than 100 mm if likely to be used by children under five years (see Fig. 8.4.12).

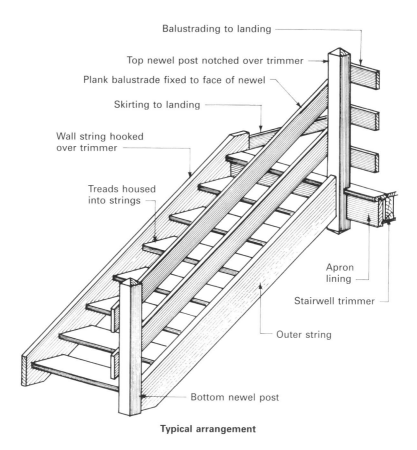

Balustrading to landing

Top newel post notched over trimmer

Plank balustrade fixed to face of newel

Skirting to landing

Wall string hooked over trimmer

Treads housed into strings

Apron lining

Stairwell trimmer

Outer string

Bottom newel post

Typical arrangement

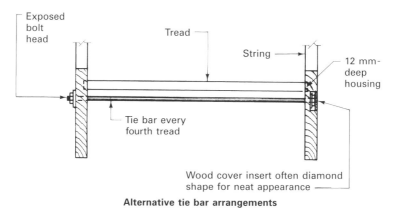

Exposed bolt head

Tread

String

12 mm-deep housing

Tie bar every fourth tread

Wood cover insert often diamond shape for neat appearance

Alternative tie bar arrangements

Figure 8.4.9   Closed string open-tread stairs.

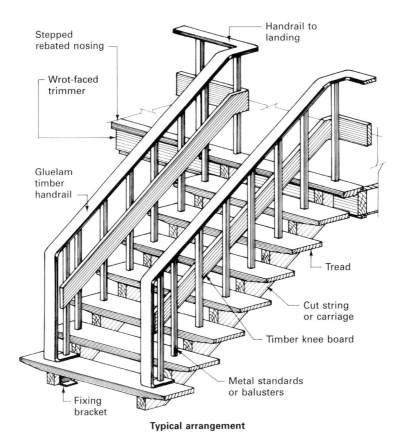

Stepped rebated nosing

Wrot-faced trimmer

Handrail to landing

Gluelam timber handrail

Tread

Cut string or carriage

Timber knee board

Metal standards or balusters

Fixing bracket

Typical arrangement

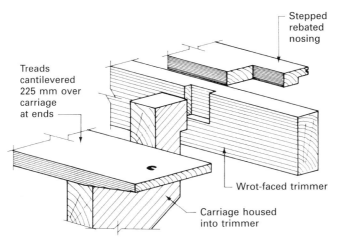

Stepped rebated nosing

Treads cantilevered 225 mm over carriage at ends

Wrot-faced trimmer

Carriage housed into trimmer

Figure 8.4.10   Cut string open-tread stairs.

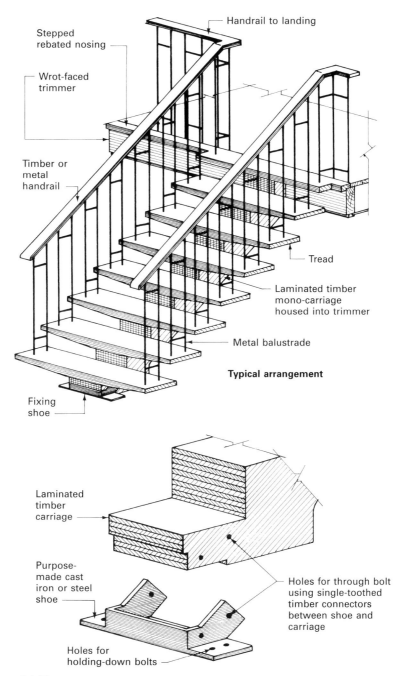

Stepped
rebated nosing

Wrot-faced
trimmer

Handrail to landing

Timber or
metal
handrail

Tread

Laminated timber
mono-carriage
housed into trimmer

Metal balustrade

Typical arrangement

Fixing
shoe

Laminated
timber
carriage

Purpose-
made cast
iron or steel
shoe

Holes for through bolt
using single-toothed
timber connectors
between shoe and
carriage

Holes for
holding-down bolts

Figure 8.4.11 Mono-carriage open-tread stairs.

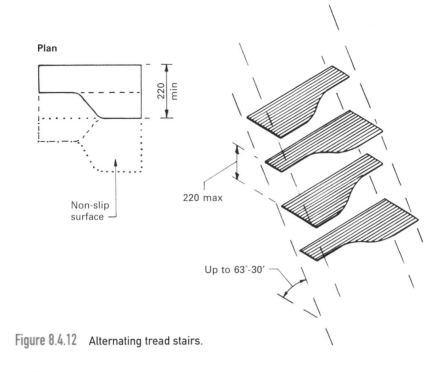

Plan

220 min

220 max

Non-slip
surface

Up to 63°-30'

Figure 8.4.12 Alternating tread stairs.

TREADS

Treads must be of adequate thickness, because there are no risers to give extra support; usual thicknesses are 38 and 50 mm. To give a lighter appearance it is possible to taper the underside of the treads at their cantilever ends for a distance of 225–250 mm. This distance is based on the fact that the average person will ascend or descend a stairway at a distance of about 250 mm in from the handrail. The development of laminated and toughened glass has made this a material that can be used to form stair treads and balustrades that can be combined with timber or stainless steel supporting work to provide an impressive feature to modern house construction.

BALUSTRADING

Together with the handrail, balustrading provides both the visual and the practical safety barrier to the side of the stair that is not adjacent to a wall or other structure. Children present special design problems, because they can and will explore any gap big enough to crawl through. BS 5395 for wood stairs recommends that the infill under handrails should have no openings that would permit the passage of a sphere 90 mm in diameter, a slight reduction on the 100 mm stated in Approved Document K1, Section 1.29, to the Building Regulations. Many variations of balustrading are possible, ranging from simple newels with planks to elaborate metalwork or toughened glass of open design (see Figs. 8.4.9, 8.4.10 and 8.4.11).

Simple precast concrete stairs 8.5

Precast concrete stairs can be designed and constructed to satisfy a number of different requirements. They can be a simple inclined straight-flight slab, a cranked slab with a landing built in, open riser stairs, or constructed from a series of precast steps built into, and if required cantilevered from, a structural wall.

The design considerations for the simple straight flight are the same as those for in-situ stairs of comparable span, width and loading conditions, but exclude the self-weight and any supporting formwork. However, the fixing and support require a different approach: the ends of the precast stairway, which require the physical connection, need careful consideration. Bearings for the ends of the flights must be provided at the floor or landing levels in the form of a haunch, rebate or bracket, and continuity of reinforcement can be achieved by leaving projecting bars and slots in the floor into which they can be grouted (see Fig. 8.5.1).

The delivery of precast stairs should be carefully programmed so that they can be lifted, positioned and fixed direct from the delivery vehicle, thus avoiding double handling. Precast components are usually designed for two conditions:

- lifting and transporting by the inclusion of lifting rings;
- final fixed condition by the inclusion of projecting reinforcing bars.

It is essential that the flights are lifted from the correct lifting points, which may be in the form of loops or hooks projecting from or recessed into the concrete member. These are specific points that have been calculated to avoid any undue tensile and compressive stresses during lifting that would cause cracking within the precast concrete.

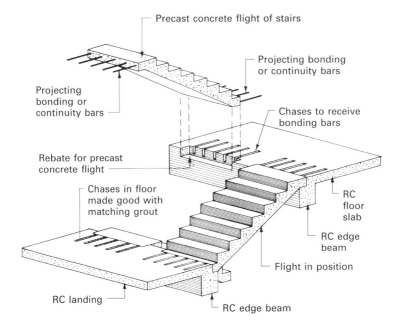

Precast concrete flight of stairs

Projecting bonding
or continuity bars

Projecting
bonding or
continuity bars

Chases to receive
bonding bars

Rebate for precast
concrete flight

Chases in floor
made good with
matching grout

RC
floor
slab

RC edge
beam

Flight in position

RC landing

RC edge beam

Simple precast concrete stairs

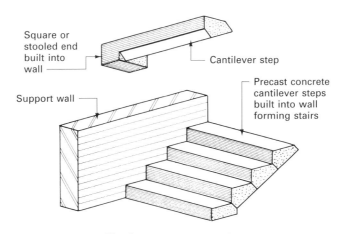

Square or
stooled end
built into
wall

Cantilever step

Precast concrete
cantilever steps
built into wall
forming stairs

Support wall

Simple precast concrete steps

Figure 8.5.1 Precast concrete stairs and steps.

The precast concrete stairs are finished using a sand and cement screed on the landing areas. The whole of the stair surfaces can then be covered by non-slip vinyl floor finishes. This is applied to the treads, risers and landings with aluminium nosings screwed and plugged to the treads, and safety inserts placed into the aluminium nosing. Metal balustrade systems can then be fixed down into the concrete to provide safe access in accordance with Part K of the Building Regulations. Alternatively the internal well of the stairway can be filled using partitioning and the handrail can be fixed to the wall.

The use of precast concrete steps to form a stairway is limited to situations such as short flights between changes in floor level, and external stairs to basements and other areas. They rely on the loadbearing wall for support and, if cantilevered, on the downward load of the wall to provide the equal supporting reaction. The support wall has to provide this necessary load and strength, and at the same time it has to be bonded or cut around the stooled end(s) of the steps. The use of precast concrete stairs in domestic house construction is useful for the application of flats, where the properties of concrete assist with sound reduction. Precast concrete stairs used in commercial buildings have many advantages, namely:

- fire resistance;
- sound reduction;
- faster speed of construction;
- prefinished soffits;
- high-quality factory-produced product;
- instant access to upper floors.

Partitions 8.6

Internal walls that divide the interior of a building into areas of accommodation and circulation are called **partitions**, and these can be classified as loadbearing or non-loadbearing partitions.

LOADBEARING PARTITIONS

These are designed and constructed to receive superimposed loadings and transmit these loads to a foundation. For example, the first floor joists of a domestic dwelling will rest on a loadbearing internal wall and transmit their floor live and dead loads to the partition's foundation. Generally, loadbearing partitions are constructed of 3.5 N/mm^2 solid blocks bonded to the external walls for stability. Openings are formed in the same manner as for external walls; a lintel spans the opening, carrying the load above to the reveals on either side of the opening. Fixings can usually be made direct into block walls. Blockwork partitions are lighter, cheaper and quicker to build, and have excellent thermal and sound insulation properties. For these reasons blockwork is usually specified for loadbearing partitions in domestic work. Loadbearing partitions, because of their method of construction, are considered to be permanently positioned.

NON-LOADBEARING PARTITIONS

These partitions, like loadbearing partitions, must be designed and constructed to carry their own weight and any fittings or fixings that may be attached to them, but they must not, in any circumstances, be used to carry or assist in the transmission of structural loadings. They are designed to provide an economical division between internal spaces and must be able to resist impact loadings on their faces, as well as any vibrations set up by doors being closed.

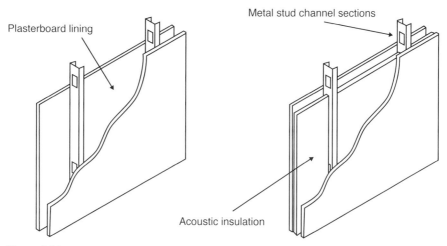

Plasterboard lining

Metal stud channel sections

Acoustic insulation

Figure 8.6.1   Typical metal stud partition wall construction.

Lightweight aerated concrete blocks can be used to construct non-loadbearing partitions. They may be built directly off a concrete floor with a thickened slab or steel reinforcement to support the wall's self-weight. At ceiling level stability is provided by the joists above: they can be built in, or hangers can be secured to the partition wall. Openings are constructed as for loadbearing walls; alternatively a storey-height frame could be used with a fanlight above the door.

Using timber to form stud partitions clad with plasterboard is a suitable alternative to using blockwork (see Fig. 8.6.1). These partitions are lighter in construction than blockwork but are less efficient as sound or thermal insulators. Their insulation properties can be increased by:

■   adding to the internal core of the timber partition sound-reducing mineral or glass fibre insulation;

■   adding additional layers of plasterboard to increase density, to reduce sound transfer.

They are easy to construct and provide a good fixing background, and because of their lightness are suitable for building off a suspended timber floor.

The provision of double joists below the stud partitions should be considered. The basic principle is to construct a simple framed grid of timber to which a dry lining of plasterboard can be attached. The lining material will determine the spacing of the uprights or studs to save undue wastage in cutting the boards to terminate on the centre line of a stud. CLS timber (Canadian Lumber Sizes) is a suitable material for studwork as it has been machined with uniform dimensions and rounded arises.

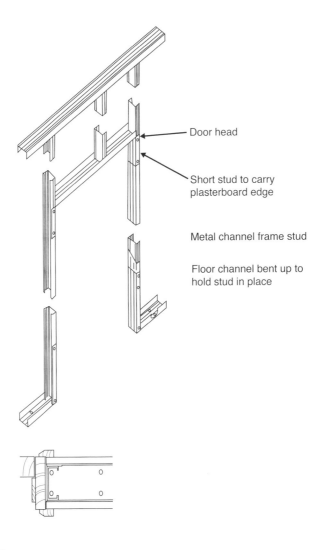

Door head

Short stud to carry
plasterboard edge

Metal channel frame stud

Floor channel bent up to
hold stud in place

Figure 8.6.2 Typical door arrangement using metal studs.

Openings are formed by framing a head between two studs and fixing a lining or door frame into the opening in the stud partition; typical details are shown in Figs. 8.6.2 and 8.6.3. Where plasterboard is the specified facing/lining material, the extent of nailing will depend on the stud spacing, the facing thickness and its width. Table 8.6.1 provides guidance when using galvanised steel taper-head nails at 150 mm spacing. The introduction of battery-operated automatic-feeding drywall screw guns has made the use of nailing redundant. By adding taper-edged plasterboards the wall can be finished using a drylining system, which reduces the need for a final plaster skim coat.

The introduction of cold-rolled steel channel sections into partition systems has created a fast and simple way to construct a partition that provides several benefits, such as:

- fire resistance;
- speed of erection;
- strength and stability;
- resistance to insect and rot attack;
- accommodation for service runs within the partition void;
- lightweight.

Ceiling and floor channels are normally placed first, then in between these a series of C-shaped studs are placed at fixing centres to suit the plasterboard widths being fixed. These vertical steel channel studs are normally fixed using a crimping tool that secures one section to another. The system relies on the final layer of plasterboard, which is fixed using drywall screws to the steel studs. The internal core can be filled with a sound-reduction quilt where this is a requirement for Building Regulations.

Various configurations are available to satisfy the function of the wall, be it:

- a separating wall between two properties;
- an internal non-loadbearing wall;
- a loadbearing wall;
- an acoustic reduction wall.

Table 8.6.1   Guidance when using galvanised steel taper nails.

Plasterboard thickness	Plasterboard width	Stud or batten spacing	Nails
9.5	900	450	2 × 30 long
	1,200	400	2 × 30 long
12.5	600	600	2 × 40 long
	900	450	2 × 40 long
19.0	1,200	600	2.6 × 40 long
	600	600	2.6 × 40 long

Note: All dimensions in millimetres.

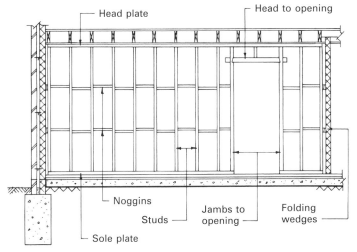

Head plate — Head to opening

Noggins
Studs — Jambs to opening — Folding wedges
Sole plate

Typical arrangement

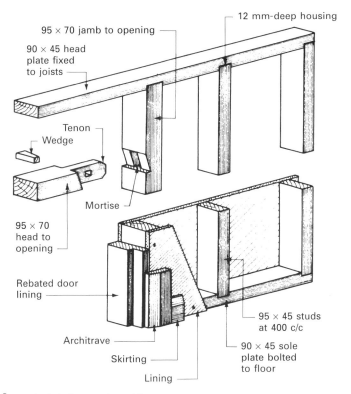

95 × 70 jamb to opening — 12 mm-deep housing
90 × 45 head plate fixed to joists
Tenon
Wedge
Mortise
95 × 70 head to opening
Rebated door lining
95 × 45 studs at 400 c/c
90 × 45 sole plate bolted to floor
Architrave
Skirting
Lining

Figure 8.6.3 Typical timber stud partition.

Finishes: floor, wall and ceiling 8.7

FLOOR FINISHES

The type of floor finish to be applied to a floor will depend upon a number of factors such as type of base, function of the room, degree of comfort required, maintenance problems, cost, appearance, safety and individual preference.

Floor finishes can be considered under three main headings:

- in-situ floor finishes – those finishes that are mixed on site, laid in a fluid state, allowed to dry, and set to form a hard jointless surface;
- applied floor finishes – those finishes that are supplied in tile or sheet form and are laid onto a suitably prepared base;
- timber floor finishes – boards, sheets and blocks of timber laid on or attached to a suitable structural frame or base.

IN-SITU FLOOR FINISHES

MASTIC ASPHALT – BS 6925 AND BS 8204-5

This is a naturally occurring bituminous material obtained from asphalt lakes like those in Trinidad; it can also be derived from crude oil residues. Trinidad lake asphalt is used as a matrix or cement to bind a suitably graded mineral aggregate together to form mastic asphalt as a material suitable for floor finishing. It must be laid by hand on a solid base, such as screed or concrete. This material would be used for commercial and industrial applications such as wet rooms, external balconies or process areas. When set, the asphalt is totally impervious to water and can be used in areas subjected to regular cleaning and washing. It also forms a very good surface on which to apply thin tile and sheet finishes (for example, PVC), and will at the same time fulfil the function of a damp-proof membrane or waterproof decking.

Mastic asphalt is a thermoplastic material that is only pliable when heated, so it has to be melted before it can be applied to form a jointless floor finish. Hot mastic asphalt is applied by hand using a float at a temperature of between 180 and 210 °C; it is applied in a single 13 mm coat as a base for applied finishes, or in a 16 mm single coat for a self-finish. As long as the subfloor is sound, the mastic asphalt will cover any imperfections. A thermal movement layer of black sheathing felt should be included between the subfloor and mastic asphalt to overcome the problems caused by any differential movement. The finish obtained is smooth and hard-wearing, but the colour range is limited to dark colours such as red, brown and black. A matt surface can be produced by giving the top surface a dusting of sand or powdered stone.

PITCH MASTIC

Pitch mastic is a similar material to mastic asphalt but is produced from a mixture of calcareous and/or siliceous aggregates bonded with coal-tar pitch. It is laid to a similar thickness and in a similar manner to mastic asphalt with a polished or matt finish. Pitch mastic floors are resistant to water but have a better resistance to oil and fats than mastic asphalt, and are therefore suitable for washrooms and kitchens.

GRANOLITHIC

This involves a mixture of Portland cement and granite chippings. Introducing granite chippings makes the finished surface extremely hard-wearing and ideal for high-traffic areas. The mixture can be applied as a granolithic screed or can be applied to a 'green concrete' subfloor or to a cured concrete subfloor. **Green concrete** is a term used to describe newly laid concrete that is not more than three hours old, and can take the application of the trowelling of the cement and chippings into the surface. A typical mix for granolithic is one part cement: one part sand: two parts granite chippings (5–10 mm free from dust) by volume. Although the finish obtained is very hard-wearing, it is also noisy underfoot and cold to touch due to its density; it is used mainly in situations where easy maintenance and durability are paramount, such as a common entrance hall to a block of flats.

If granolithic is being applied to a green concrete subfloor as a topping, it is applied in a single layer approximately 20 mm thick, in bay sizes not exceeding 28 m², and trowelled to a smooth surface. This method will result in a monolithic floor and finish construction.

The surface of mature and fully cured concrete will need to be prepared. The entire area will need to be hacked to provide a key for the screed, and brushed well to remove all the laitance before laying a single layer of granolithic, which should be at least 40 mm thick. The finish should be laid on a wet cement slurry coating or PVA bonding agent to improve the bond, in bay sizes not exceeding 14 m².

APPLIED FLOOR FINISHES

Many of the applied floor finishes are thin, flexible materials, and should be laid on a subfloor with a smooth finish. This is achieved by laying a cement/sand bed or screed with a steel float finish to the concrete subfloor. The usual screed mix is 1:3 cement to sand, and great care must be taken to ensure that a good bond is obtained with the subfloor. Screeds are normally laid after the concrete subfloor has cured and should be laid with a thickness of 40 mm. A mature concrete subfloor must be clean, free from dust and contaminants, and dampened with water to reduce the suction before applying the bonding agent to receive the screed. To reduce the possibility of drying shrinkage cracks, screeds should not be laid in bays exceeding 15 m² in area. Screeds laid directly over rigid insulation should be at least 50 mm in thickness, and reinforced with light steel mesh (chicken wire). The thickness of a screed is determined by several factors including loading, traffic, applied floor finish and the rigidity of the insulation it is covering.

FLEXIBLE PVC TILES AND SHEET – BS EN 649 AND BS 8203

Flexible polyvinyl chloride (PVC) is a popular, hard-wearing floor finish produced by a mixture of polyvinyl chloride resin, pigments, mineral fillers and plasticisers to control flexibility. It is produced as 300 mm × 300 mm square tiles or in sheet form up to 2,400 mm wide, with a range of thicknesses from 1.5 to 4.5 mm. The surface can be produced in a variety of different colours, and can have textures to aid its non-slip properties. The floor tiles and sheet are fixed with an adhesive in accordance with the manufacturer's instructions, and produce a surface suitable for most situations.

PVC tiles, like all other small unit coverings, should be laid from the centre of the area towards the edges so that, if the area is not an exact tile module, an even border of cut tiles is obtained. PVC tiles are an economical, easily fixed, easily repaired and hard-wearing floor finish.

Modern sheet vinyl flooring is usually heat-sealed at its joints using a welding gun and a coloured sealant. This is then trimmed smooth to provide a uniform and water-resistant covering. This makes it ideal for applications in hospitals and kitchens, where high standards of hygiene are required.

LINOLEUM – BS EN 12104

Linoleum is produced in sheet or tile form from a mixture of drying oils, resins, fillers and pigments, pressed onto a hessian or bitumen-saturated felt paper backing. Good-quality linoleum has the pattern inlaid or continuing through the thickness, whereas the cheaper quality has only a printed surface pattern. Linoleum gives a quiet, resilient and hard-wearing surface suitable for most domestic floors. Thicknesses vary from 2 to 6.5 mm for a standard sheet width of 1,800 mm; tiles are usually 300 mm square with the same range of thicknesses. Fixing of linoleum tiles and sheet is by adhesive to any dry, smooth surface, although the adhesive is sometimes omitted with the thicker sheets.

CARPET – BS 4223, BS 5808 AND BS 5325

The chief materials used in the production of carpets are nylon, acrylics and wool, or mixtures of these materials. Carpet is mass-produced in a vast range of styles, types, patterns, colours, qualities and sizes for general domestic use in dry situations, as the resistance of carpets to dampness is generally poor. A carpet should be laid over an underlay, which removes any imperfections in the subfloor that would transmit through the surface of the carpet. Carpet-edge grippers are normally fixed to the perimeter of a room and the carpet is stretched over these. Carpet is supplied in narrow or wide rolls and is also available as carpet squares (600 mm × 600 mm) which must be bonded by the use of adhesives to a latexed subfloor.

CORK TILES – BS EN 12104

Cork tiles are cut from baked blocks of granulated cork; the natural resins act as the binder. The tiles are generally 300 mm square with thicknesses of 5 mm upwards according to the wearing quality required, and are supplied in three natural shades. They are hard-wearing, warm to the touch, quiet and resilient, but may collect dirt and grit unless they are treated with a surface sealant. Fixing is by manufacturer's adhesive.

QUARRY TILES – BS EN ISO 10545

The term 'quarry' is derived from the Italian word *quadro* meaning square; it does not mean that the tiles are quarried from an open excavation or quarry. Quarry tiles are made from ordinary or unrefined clays worked into a plastic form, pressed into shape, and hard-burnt. This makes the finished product very hard-wearing and with a good resistance to water, they are suitable for kitchens and entrance halls, but they tend to be noisy and cold. Quarry tiles are produced as square tiles in sizes ranging from 100 mm × 100 mm × 20 mm thick to 225 mm × 225 mm × 32 mm thick. The application of quarry tiles is much in fashion with their use in kitchens, toilets, washrooms and high-wearing applications.

With the advent of modern cement-based adhesives, quarry tiles can be thin-bed jointed directly onto the cured sand and cement screed. These special adhesives cure quickly overnight so the floor can be grouted the next day. Alternatively the quarry tiles can be thick-bedded by laying onto a sand and cement screed directly. As tiling work proceeds the screed can be advanced and tiles laid and tapped into position to bond. With any solid subfloor, care should be taken to ensure that it contains a damp-proof membrane to prevent moisture penetration into the subfloor. With this method the concrete subfloor should be dampened to reduce suction and then covered with a semi-dry 1:4 cement: sand mix to a thickness of approximately 40 mm. The top surface of the compacted semi-dry thick bed should be treated with a 1:1 cement:sand grout before tapping the tiles into the grout with a wood beater.

Clay tiles have a tendency to expand, probably as a result of physical absorption of water and chemical hydration, and for this reason an expansion joint of compressible material should be incorporated around the perimeter of the floor (see Fig. 8.7.1). In no circumstances should the length or width of a quarry tile floor exceed 7,500 mm without an expansion joint. The joints between quarry tiles are usually finished by grouting with a 1:1 cement:sand grout. The surface of the tile after laying must be sealed in accordance with the manufacturer's instructions to prevent discoloration from use. There is a wide variety of tread patterns available to provide a non-slip surface; also a wide range of fittings is available to form skirtings and edge coves (see Fig. 8.7.1).

PLAIN CLAY OR CERAMIC FLOOR TILES – BS EN ISO 10545

These are similar to quarry tiles but are produced from refined natural clays, which are pressed after grinding and tempering into the desired shape before being fired at a high temperature. Plain clay floor tiles, being denser than quarry tiles, are made as smaller and thinner units ranging from 50 mm × 50 mm to 300 mm × 300 mm, in thicknesses of 9.5–13 mm. Laying, finishes and fittings available are all as described for quarry tiles.

TIMBER FLOOR FINISHES

Timber is a very popular floor finish with both designer and user because of its natural appearance, sustainable properties, resilience and warmth. It is available as a board, strip, sheet or block finish. If attached to joists, as in the case of a suspended timber floor, it also acts as the structural decking. Solid timber flooring is available as hardwoods and softwoods: the latter is usually covered with a carpet or secondary finish; the former is usually sealed to enhance its decorative grain.

TIMBER BOARDS

Softwood timber floorboards are joined together by machined, tongued and grooved joints along their edges, and are fixed by nailing to the support joists

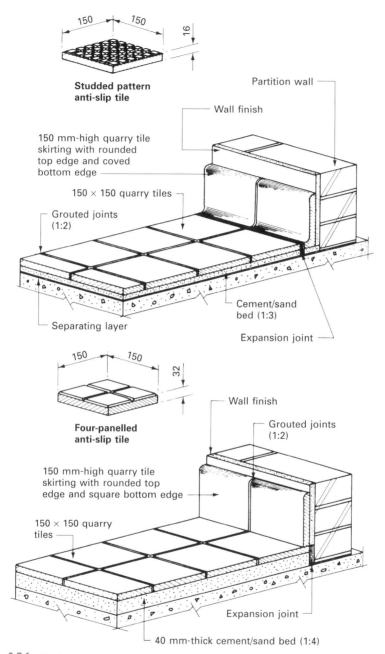

150

150

16

Studded pattern anti-slip tile

Partition wall

Wall finish

150 mm-high quarry tile skirting with rounded top edge and coved bottom edge

150 × 150 quarry tiles

Grouted joints (1:2)

Cement/sand bed (1:3)

Separating layer

Expansion joint

150

150

32

Four-panelled anti-slip tile

Wall finish

Grouted joints (1:2)

150 mm-high quarry tile skirting with rounded top edge and square bottom edge

150 × 150 quarry tiles

Expansion joint

40 mm-thick cement/sand bed (1:4)

Figure 8.7.1   Typical quarry tile floors.

or fillets attached to a solid floor. The boards are butt-jointed in their running length, the joints being positioned centrally over a joist so that both edges are supported. Joints should be staggered so that butt joints do not occur in the same position in consecutive lengths, which would weaken the floor's surface. The support spacing will be governed by the spanning properties of the board (which is controlled by its thickness); floor joist supports placed at 400 mm centres are usual for boards 19 and 22 mm thick. The tongue is positioned slightly off centre, and the boards should be laid with the narrow shoulder on the underside to give maximum wear to the top surface. It is essential that the boards are well cramped together before being fixed to form a tight joint, to avoid any gaps through shrinkage, and that they are laid in a position where they will not be affected by dampness. Timber is a hygroscopic material and so it will swell and shrink as its moisture content varies; ideally this should be maintained at around 12 per cent.

TIMBER STRIP

These are narrow boards, being under 100 mm wide to reduce the amount of shrinkage and consequent opening of the joints. Timber strip can be supplied in softwood or hardwood, and is considered to be a superior floor finish to boards. Jointing and laying is as described for boards, except that hardwood strip is very often laid one strip at a time and secret-nailed (see Fig. 8.7.2).

PARTICLEBOARD FLOOR FINISH – BS EN 312

Chipboard is manufactured from wood chips or shavings bonded together with thermosetting synthetic resins, forming rigid sheets 18 and 22 mm thick, which are suitable as a floor finish. They are an economical alternative to softwood floorboards and can cover larger areas quickly and efficiently. The sheets are fixed by nailing or screwing to support joists or fillets; often the tongue and groove are glued for added stability. If the sheet is used, as an exposed finish a coating of sealer should be used. Alternatively chipboard can be used as a decking to which a thin tile or carpet finish can be applied. Tongued and grooved boards of 600 mm width are also available as a floor-decking material that does not need to be jointed over a joist.

WOOD BLOCKS

These are small blocks of timber, usually of hardwood, which are designed to be laid in set patterns, producing a decorative finish to the completed floor. Lengths range from 150 to 300 mm, with widths up to 89 mm; the width is proportional to the length to enable the various patterns, such as herringbone, to be created. Block thicknesses range from 20 to 30 mm, and the final thickness after sanding and polishing is about 5–10 mm less.

The blocks are jointed along their edges with a tongued and grooved joint, and have a rebate, chamfer or dovetail along the bottom longitudinal edges to take up any surplus adhesive used for fixing. Two methods can be used for fixing wood blocks: the first uses hot bitumen and the second a cold latex bitumen emulsion. If hot bitumen is used, the upper surface of the subfloor is first primed with black varnish to improve adhesion and then, before laying, the bottom face of the block is dipped into the hot bitumen. The cold adhesive does not require a priming coat to the subfloor. Blocks, like tiled floors, should be laid from the centre of the floor towards the perimeter, which is generally terminated with a margin border. To allow for moisture movement, a cork expansion strip should be placed around the entire edge of the block floor (see Fig. 8.7.2).

PARQUET

This is a superior form of wood-block flooring made from specially selected hardwoods chosen mainly for their decorative appearance. Parquet blocks are generally smaller and thinner than hardwood blocks, and are usually fixed to a timber subfloor that is level and smooth. Fixing can be by adhesives or secret-nailing; alternatively parquet can be supplied as a patterned panel fixed to a suitable backing sheet, in panel sizes from 300 to 600 mm square.

WOOD VENEER AND LAMINATE FLOORING

This modern material consists of a series of strips generally 1,200 mm long that interlock with each other to provide a finished floor surface. The surface finish of the laminate is available in a variety of colours and designs many that mimic a more expensive hardwood floor. Many laminates contain a snap-lock system, where each board clips into the other and is then pressed down to finally secure it. Some systems may require a glued joint along the tongue and groove. End jointing is staggered between adjacent strips to avoid continuity of joints.

Veneered floors have a thin, hardwood facing bonded to a high density fiber core (HDFB) core with a paper underlayer. Laminate floors comprise a thin but dense aluminium oxide clear, protective surface over a photographic image of wood, an HDFB core and a paper backing. If damaged, veneers can be sanded to some extent, but laminated artificial veneers are less easily repaired.

- **Underlay** A suitable underlay may be required for acoustic function or to reduce damp penetration.

- **Sublayer** Structural concrete or screed must dry out and be level; suspended timber must be stable and level; floorboards must be screwed to joists and surface irregularities must be sanded.

- **Movement** An 8–10 mm expansion gap is required between strips and the room perimeter. With a new build this gap can be accommodated under the skirting board; where there is an existing skirting, a moulding is used to cover the gap.

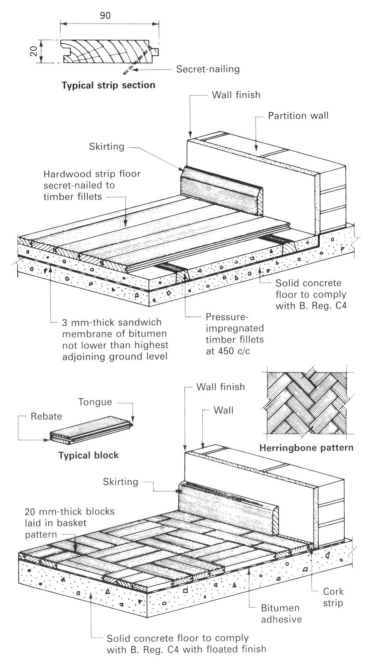

90

20

Secret-nailing

Typical strip section

Wall finish

Partition wall

Skirting

Hardwood strip floor
secret-nailed to
timber fillets

3 mm-thick sandwich
membrane of bitumen
not lower than highest
adjoining ground level

Pressure-
impregnated
timber fillets
at 450 c/c

Solid concrete
floor to comply
with B. Reg. C4

Tongue

Rebate

Typical block

Wall finish

Wall

Herringbone pattern

Skirting

20 mm-thick blocks
laid in basket
pattern

Solid concrete floor to comply
with B. Reg. C4 with floated finish

Bitumen
adhesive

Cork
strip

Figure 8.7.2   Typical timber floor finishes.

Facing brickwork can be used internally to provide a decorative finish. By using a coloured mortar and a recessed joint, many visual affects can be produced by the variety of colours available. High-quality blockwork that gives an acceptable face for painting offers a suitable and hard-wearing finish for many applications, such as sports halls and corridor areas. Internal walls or partitions can be built with a fair face of natural materials such as bricks or stone, but generally it is cheaper to use a material such as blocks with an applied finish, plaster, drylining or glazed tiles.

EXTERNAL RENDERING – BS EN 13914

This is a finish that is similar to plastering and uses a mixture of cement and sand, or cement, lime and sand, applied to the face of a building in several coats. External rendering can give extra protection against the penetration of moisture or can provide a desired trowelled texture. It can also be used in the dual capacity of providing protection and appearance.

The rendering must have the properties of durability, moisture resistance and an acceptable appearance, as modern construction often uses blockwork outer skins, which the render then covers. The factors to be taken into account in achieving the above requirements are mix design, bond to the backing material, texture of surface, degree of exposure of the building, and the standard of workmanship in applying the rendering.

Cement and sand mixes will produce a strong moisture-resistant rendering, but one that is subject to cracking because of high-drying shrinkage. These mixes are used mainly on members that may be vulnerable to impact damage, such as columns. Cement, lime and sand mixes have a lower drying shrinkage but are more absorbent than cement and sand mixes; they will, however, dry out rapidly after periods of rain and are, therefore, the mix recommended for general use.

The two factors that govern the proportions to be used in a mix are:

- the type of background to which the rendering is to be applied;
- the degree of exposure of the building to weathering.

The two common volume mix ratios are outlined in Table 8.7.1. Class I is the strongest and Class V the weakest mix.

The final coat is usually a weaker mix than the undercoats: for example, when using an undercoat mix 1:1:6 (1:5–6 cement:sand with plasticiser), the final coat mix ratio should be 1:2:9 (1:7–8 cement:sand with plasticiser), to reduce the amount of shrinkage in the final coat. The number of coats required will depend upon the surface condition of the background and the degree of exposure. Generally a two-coat application is acceptable, except where the

Table 8.7.1 Common volume mix ratios for external rendering.

| Mix Designation | Mix proportions by volume basd on damp sand | | | | |
| | Cement: lime:sand[a] | Cement:ready-mixed lime:sand[a] | | Cement:sand[a] (using plasticiser) | Masonry cement:sand[a] |
		Ready-mixed lime:sand	Cement:ready-mixed material		
I	1:¼:3	1:12	1:3	—	—
II	1:½:4 to 4½	1:9	1:4 to 4½	1:3 to 4	1:2½ to 3½
III	1:1:4 to 6	1:6	1:5 to 6	1:5 to 6	1:4 to 5
IV	1:2:8 to 9	1:4½	1:8 to 9	1:7 to 8	1:5½ to 6:½
V	1:3:10 to 12	1:4	1:10 to 12	—	—

Note: In special circumstances, e.g. where soluble salts in the background are likely to cause problems, mixes based on sulphate-resisting Portland cement should be employed.
[a] With fine or poorly graded sands, the lower volume of sand should be used.

background is very irregular or the building is in a position of severe exposure, when a three-coat application would be specified. The thickness of any one coat should not exceed 15 mm, and each subsequent coat thickness is reduced by approximately 3 mm to give a final coat thickness of 6–10 mm. Wood-float treatment produces a smooth, flat surface, which may be treated with a felt-coated float or lambswool roller to raise the surface texture and provide better resistance to cracking.

Various textured surfaces can be obtained on renderings.

- **Roughcast** A wet plaster mix of 1 part cement: ½ part lime: 1½ parts shingle: 3 parts sand is thrown onto a porous coat of rendering to give an even distribution.

- **Pebbledash or dry dash** Selected aggregate, such as pea shingle, is dashed or thrown onto a rendering background before it has set, and is tamped into the surface with a wood float to obtain a good bond.

- **Spattered finishes** These are finishes applied by a machine (which can be hand-operated), guns or sprays using special mixes prepared by the machine manufacturers.

When the render has set and sufficiently dried out, it can be finished with a masonry paint to provide a range of colours to the final finish.

PLASTERING

Plastering – like brickwork – is one of the oldest, established crafts in this country, having been introduced by the Romans. The plaster used was a lime plaster, which generally now has been superseded by gypsum. Any plaster finish must smooth out irregularities in the backing wall, provide a continuous and flat surface that is suitable for direct decoration, and be sufficiently hard to resist damage by impact upon its surface; gypsum plasters fulfil these requirements.

Gypsum is a crystalline combination of calcium sulphate and water. Deposits of suitable raw material are found in several parts of England. After crushing and screening the gypsum is heated to dehydrate the material. The amount of water remaining at the end of this process defines its class under BS EN 13279, which covers the definition and requirements for gypsum building plasters. If powdered gypsum is heated to about 170 °C it loses around three-quarters of its combined water and is called hemihydrate gypsum plaster, but is probably better known as **Plaster of Paris**. If a retarder is added to the hemihydrate plaster, a new class of finishing plaster is formed to which the addition of expanded perlite and other additives will form a one-coat or universal plaster, a renovating-grade plaster or a spray plaster for application by a spray machine.

BS EN 13279 establishes three types of plaster:

- Type A – plaster binders;
- Type B – gypsum plaster (see Table 8.7.2);
- Type C – gypsum plaster for special purposes.

Type C covers the special plasters. Here are some examples.

- **Thin-wall plasters** – for skimming and filling. These may contain organic binders that could be incompatible with some backgrounds. Check with manufacturer before application.
- **Projection plasters** – for spray machine application.
- **X-ray plasters** – contain barium sulphate aggregate. This is a heavy metal with salts that provide a degree of exposure protection.
- **Acoustic plaster** – contain porous gaps, which will absorb sound.
- **Resinous plaster** – 'Proderite' formula S, a cement-based plaster specially formulated for squash courts.

Gypsum plasters are not suitable for use in temperatures exceeding 43 °C, and should not be applied to frozen backgrounds. However, plasters can be applied under frosty conditions provided the surfaces are adequately protected from freezing after application.

Gypsum plasters are supplied in multi-walled paper sacks, which require careful handling. Damage to the bags will admit moisture and start off the setting process. Plaster in an advanced state of set will have a short working time and may lack considerable strength. Characteristics are manifested in lack of bond and surface irregularities. Plaster should be stored in a dry location on pallets or other means of avoiding the ground. The dry, bagged plaster in store will be unaffected by low temperatures.

Table 8.7.2 Types of gypsum binders and gypsum plasters.

Designation	Notation
Gypsum binders e.g.:	A
■ gypsum binders for direct use or further processing (dry powder products);	A1
■ gypsum binders for direct use on site;	A2
■ gypsum binders for further processing (e.g. for gypsum blocks, gypsum plasterboards, gypsum elements for suspended ceilings, gypsum boards with fibrous reinforcement).	A3
Gypsum plaster:	B
■ gypsum building plaster;	B1
■ gypsum-based building plaster;	B2
■ gypsum-lime building plaster;	B3
■ lightweight gypsum-building plaster;	B4
■ lightweight gypsum-based building plaster;	B5
■ lightweight gypsum-lime building plaster;	B6
■ gypsum plaster for plasterwork with enhanced surface hardness.	B7
Gypsum plaster for special purposes:	C
■ gypsum plaster for fibrous plasterwork;	C1
■ gypsum mortar;	C2
■ acoustic plaster;	C3
■ thermal insulation plaster;	C4
■ fire protection plaster;	C5
■ thin coat plaster, finishing product;	C6
■ finishing product.	C7

Deterioration with age must also be avoided by regular checks on the manufacturer's date stamps, and by applying a system of strict rotation when using the product.

The plaster should be mixed in a clean plastic or rubber bucket using clean water only. Cleanliness is imperative, because any set plaster left in the mixing bucket from a previous mixing will shorten the setting time, which may reduce the strength of the plaster when set.

Pre-mixed plasters incorporate lightweight aggregates such as expanded perlite and exfoliated vermiculite. The density of a lightweight plaster is about one-third that of a comparable sanded plaster, and it has a thermal value of about three times that of sanded plasters, resulting in a reduction of heat loss, less condensation and a reduction in the risk of pattern staining. It also has superior adhesion properties to all backgrounds including smooth, dry, clean concrete.

The choice of plaster mix, type and number of coats will depend upon the background or surface to which the plaster is to be applied. Roughness and suction properties are two of the major considerations.

Generally all lightweight concrete blocks provide a suitable key for the direct application of plasters. Bonding agents in the form of resin emulsions, such as PVA, are available for smooth surfaces; these must be applied in strict accordance with the manufacturer's instructions to achieve satisfactory results.

The suction properties of the background can affect the drying rate of the plaster by absorbing the moisture of the mix. Too much suction can result in the plaster failing to set properly, thus losing its adhesion to its background; too little suction can give rise to drying shrinkage cracks due to the retention of excess water in the plaster.

Undercoat plasters are applied by means of a wooden float or rule worked between dots or runs of plaster to give a true and level surface. The runs or rules and dots are of the same mix as the backing coat and are positioned over the background at suitable intervals to an accurate level, so that the main application of plaster can be worked around the guide points. The upper surface of the undercoat plaster should be scored or scratched to provide a suitable key for the finishing coat. The thin finishing coat of plaster is applied to a depth of approximately 3 mm and finished with a steel float to give a smooth surface. In hot and dry conditions, care must be taken to ensure that water is not allowed to evaporate from the mix. Water is very important in the setting process: if applied plaster is allowed to dry too quickly, loss of strength, loss of surface adhesion and surface irregularities will become apparent.

Most paints and wall coverings are compatible with plaster finishes, provided that the plaster is thoroughly dried out prior to application, which normally takes several months. As an interim measure, permeable water-based emulsion paint could be used.

Guidance in application and standards of workmanship can be found in BS EN 13914-2: *Design, preparation and application of external rendering and internal plastering. Design considerations and essential principles for internal plastering.*

DRYLINING TECHNIQUES

External skins of walls or internal walls and partitions can be drylined with a variety of materials, which can:

■ be self-finished, ready for direct decoration;

■ have a surface suitable for a single final coat of board finish plaster; or

■ have the joints drylined.

The main advantages of drylining techniques are: speed of construction; reduction in the amount of water used in the construction of buildings, thus reducing the drying out period; and, in some cases, increased thermal insulation.

Suitable materials are veneered finished plywoods, which are available in many different decorative finishes and plasterboard. The background will require timber battens to fix the plywood to. Finishing can be a direct application of paint, varnish or wallpaper, but masking the fixings and joints may present problems. As an alternative the joints can be made a feature of the design by the use of edge chamfers or moulded cover fillets.

Plasterboard consists of an aerated gypsum core that is encased in and bonded to specially prepared bonded paper liners. The grey-coloured liner is intended for a skim coat of plaster, and the ivory-coloured liner is for direct decoration.

Gypsum plasterboards are manufactured in a variety of specifications to suit many applications. They are generally available in metric co-ordinated widths of 900 and 1,200 mm, with co-ordinated lengths from 1,800 to 3,000 mm, in thicknesses of 9.5 and 12.5 mm. Boards can be obtained with a tapered edge for a seamless joint or a square edge for a final coat of skim plaster. Some variations on this are listed in Table 8.7.3.

Boards are normally fixed at the required centres to joists for ceilings or directly to battened-out walls. The joints on square-edged wall boards are covered with a scrim material that resists cracking at this weak spot. (see Fig. 8.7.3).

The fixing of drylinings is usually by nails or drywall screws to timber battens suitably spaced, plumbed and levelled to overcome any irregularities in the background. Fixing battens are placed vertically between the horizontal battens, and fixed at the junctions of the ceiling and floor with the wall. It is advisable to treat all wooden fixing battens with an insecticide and a fungicide to lessen the risk of beetle infestation and fungal attack, although this is not necessary with metal battens. The spacing of the battens will be governed by the spanning properties of the lining material: fixing battens at 450 mm centres are required for 9.5 mm-thick plasterboards and at 600 mm centres for 12.5 mm-thick plasterboards, placed so that they coincide with the board joints.

Plasterboard of 12.5 mm can also be fixed to brick or block masonry with dabs of plaster. This is known as the 'dot and dab' method of dry lining. Dabs of board-finish plaster about 50–75 mm wide by 250 mm long are close-spaced

Table 8.7.3 Common types of plasterboard.

Type	Thickness (mm)	Width (mm)	Application
Plank	19	600	Sound insulation
Moisture-resistant	12 and 15	1,200	Bathroom/kitchen
Contoured	6	1,200	Curved surfaces
Vapour check	9.5 and 12.5	900 and 1,200	External wall/roof
Fireline	12.5 and 15	900 and 1,200	Fire cladding
Lath	9.5 and 12.5	400 and 600	Ceiling lining
Thermal*	22, 30 and 40 mm	1,200	Thermal insulation
Sound reduction	12.5–15	1,200	Sound reduction

* Overall thickness, including laminates of expanded polystyrene, extruded polystyrene or phenolic foam.

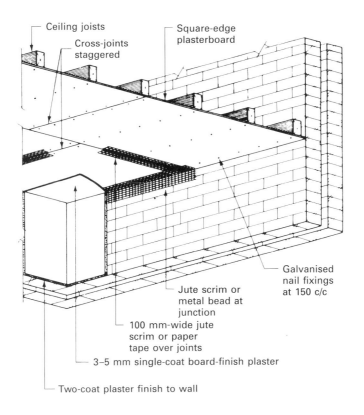

Ceiling joists

Cross-joints staggered

Square-edge plasterboard

Galvanised nail fixings at 150 c/c

Jute scrim or metal bead at junction

100 mm-wide jute scrim or paper tape over joints

3–5 mm single-coat board-finish plaster

Two-coat plaster finish to wall

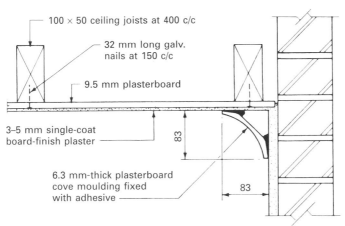

100 × 50 ceiling joists at 400 c/c

32 mm long galv. nails at 150 c/c

9.5 mm plasterboard

3–5 mm single-coat board-finish plaster

83

6.3 mm-thick plasterboard cove moulding fixed with adhesive

83

Figure 8.7.3   Plasterboard ceilings.

vertically at a horizontal spacing (max 600 mm) to suit the board width. Intermediate dabs are similarly spaced with a continuous spread at ceiling and ground levels. The plasterboard is then placed in position, tapped and horizontally aligned with a spirit level until it is firmly in contact with the dabs. Double-headed (duplex) nails can be used to secure the boards temporarily while the dabs set, after which they are removed and joints are made good. It is recommended that tapered-edge boards are used for this method of fixing.

GLAZED WALL TILES

Two processes are used in the manufacture of these hard glazed tiles. First, the body or 'biscuit' tile is made from materials such as china clay, ball clay, flint and limestone, which are mixed by careful processes into a fine powder before being heavily pressed into the required shape and size. The tile is then fired at a temperature of up to 1,150 °C. Next, glazing is applied in the form of a mixed liquid consisting of fine particles of glaze and water. The coated tiles are then fired for a second time at a temperature of approximately 1,050 °C, when the glazing coating fuses to the surface of the tile. By applying a secondary finish and refiring, the dimensions of the tile can be controlled and a vast range of colours and patterns can be produced. Gloss, matt and eggshell finishes are available.

A range of fittings with rounded edges and mitres is produced to blend with the standard 150 mm × 150 mm square tiles, 5 or 6 mm thick; a similar range of fittings is made for the 108 mm × 108 mm, 4 or 6 mm thick, square tiles. The appearance and easily cleaned surface of glazed tiles makes them suitable for the complete or partial tiling of bathrooms and as splashbacks for sinks and basins.

Tiles are fixed with a suitable adhesive, which can be of a thin bed of mastic adhesive or a bed of a cement-based adhesive; the former requires a flat surface such as a mortar screed. Ceramic or glazed tiles are considered to be practically inert: therefore careful selection of the right adhesive to suit the backing and final condition is essential. Manufacturers' instructions or the recommendations of BS 5385 should always be carefully followed.

Glazed tiles can be cut easily using the same method as employed for glass. The tile is scored on its upper face with a glass-cutting tool; this is followed by tapping the back of the tile behind the scored line over a rigid and sharp angle such as a flat, straight edge.

CEILING FINISHES

The usual method of finishing ceilings is to use a plasterboard base with a skim coat of plaster. The plasterboards are fastened to the underside of the floor or ceiling joists with galvanised plasterboard nails or drywall screws to reduce the risk of corrosion to the fixings. Concrete beam and block floors will require timber or metal battens to be fixed to the soffit to enable the boards to be secured. At the abutment of boards, a 3–5 mm gap is filled with jointing plaster and bridged with jute scrim or special jointing tape. This is done to reinforce the point in a ceiling that is most vulnerable to cracking, which is the junction between the ceiling and wall (see Fig. 8.7.3); alternatively the junction can be masked with a decorative plasterboard or polystyrene cove moulding, which adds a decorative feature to a room.

The cove moulding is made in a similar manner to plasterboard and is intended for direct decoration. Plasterboard cove moulding is jointed at internal and external angles with a mitred joint, and with a butt joint in the running length. Any clean, dry and rigid background is suitable for the attachment of plasterboard cove, which can be fixed by using a special water-mixed adhesive applied to the contact edges of the moulding, which is pressed into position; any surplus adhesive should be removed from the edges before it sets. A typical plasterboard cove detail is shown in Fig. 8.7.3.

The introduction of metal channel support systems enables the soffit of any beam and block flooring system to be covered and at the same time function as a services void. A primary grid is spanned across a room from metal hangers attached to the concrete beams. Secondary channels run at 90° to this and it is these that the plasterboards are fixed to (see Fig. 8.7.4). All joints are scrimmed and a finishing coat of plaster applied over the plasterboards. This method is economical and fast to construct, and produces a flat soffit to the underside of the beam and block floors; it can also be used effectively with timber suspended floors. The void can also be used to introduce thermal or sound insulation materials into the construction.

Other forms of finish that may be applied to ceilings are sprayed plasters, which can be of a thick- or thin-coat variety. Spray plasters are usually of a proprietary mixture, applied by spraying apparatus directly on to the soffit, giving a coarse texture which can be trowelled smooth, patterned with combs or stippled with a stiff brush (Artexing). Artexing is generally applied onto a plasterboard ceiling. Application can be with a brush if preferred.

Wallpapers can be applied to ceilings to give a patterned ceiling with a textured finish. These papers are applied directly to the plasterboard soffit or over a stout lining paper. Some ceiling papers are designed to be a self-finish but others require one or more coats of emulsion paint (see page 449).

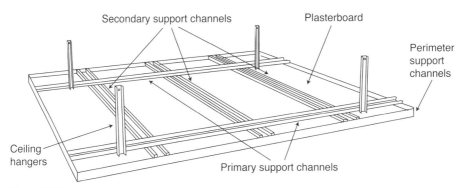

Figure 8.7.4 Construction of a plasterboard and metal channel suspended ceiling.

Internal fixtures and fittings 8.8

Fixtures consist of trims, in the form of skirtings, dado rails, frieze or picture rails, architraves and cornices, whereas fittings would include such things as cupboards, shelves and third-fix items, which have to be screwed and plugged to the building.

SKIRTINGS

A skirting is a horizontal member fixed around the skirt or base of a wall, which is used primarily to mask the junction between the wall finish and the floor. Plasterwork is often not taken down to the floor as this may be below the level of any existing DPC, and it is difficult to finish plasterwork against a floor without a crack forming. In essence this is the same problem that plaster coving solves. Skirting can be an integral part of the floor finish, such as quarry tile or PVC skirtings, or it can be made from softwood timber or MDF. Timber is the most common material used, and this is fixed by nails direct to the background, but a dense material that will not accept nails, such as engineering bricks, may have to be fixed using plugs and screws or an impact contact adhesive. External angles in skirtings are formed by mitres, but internal angles are usually scribed (see Fig. 8.8.1).

ARCHITRAVES

These are mouldings cut and fixed around door and window openings to mask the joint between the wall finish and the frame, which is subject to movement between timber and plaster finishes. As with skirtings, the usual material is timber, but MDF and plastic mouldings are available. Architraves are fixed with nails to the frame or lining, and to the wall in a similar manner to skirtings

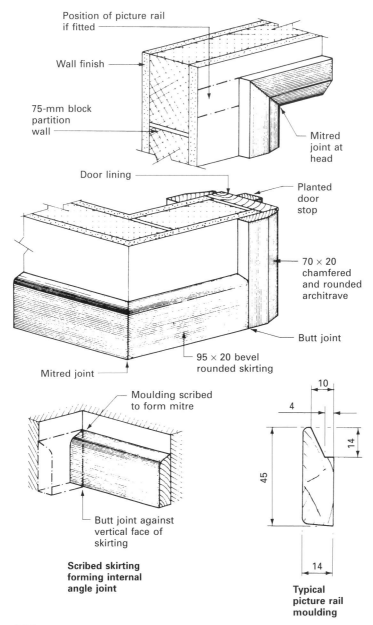

Position of picture rail if fitted

Wall finish

75-mm block partition wall

Mitred joint at head

Door lining

Planted door stop

70 × 20 chamfered and rounded architrave

Butt joint

95 × 20 bevel rounded skirting

Mitred joint

Moulding scribed to form mitre

Butt joint against vertical face of skirting

Scribed skirting forming internal angle joint

10

4

14

45

14

Typical picture rail moulding

Figure 8.8.1   Typical wood trim details.

if the architrave section is large (see Fig. 8.8.1). Mitres are used at the corners on the heads of the frames.

DADO RAILS

These are horizontal mouldings fixed in a position to prevent the walls from being damaged by the backs of chairs pushed against them. They are used today as a decorative feature to break up large walls and provide a two-tone wall finish using emulsion paints. If used, they are fixed by nails directly to the wall or to plugs inserted in the wall. Once again, mitred joints are used for external angles and scribed joints for internal.

PICTURE RAILS

These are moulded rails fixed horizontally around the walls of a room, from which pictures may be suspended using hooks and wires. Picture rails are usually positioned in line with the top edges of the door architrave. Like dado rails, picture rails are becoming popular once again as a decorative feature to a room. They can be manufactured from timber or MDF, and can be fixed by nails in the same manner as dado rails and skirtings. Fig. 8.8.1 shows a typical picture rail moulding.

CORNICES

Cornices are timber or plaster ornate mouldings used to mask the junction between the wall and ceiling. These are very seldom used today, having been superseded by the cove mouldings. However, in conservation work cornices have to be reproduced to match any work during replacement.

CUPBOARD FITTINGS

These are usually supplied as a complete fitting and only require positioning on site; they can be free-standing or plugged and screwed to the wall to aid stability when loaded with items. Built-in cupboards can be formed by placing a cupboard front in the form of a frame across a recess (usually the chimney breast) and then hanging suitable doors to the frame. Another method of forming built-in cupboards is to use a recessed partition wall to serve as cupboard walls and room divider, and attach to this partition suitable cupboard fronts or a standard door.

SHELVES

Shelves can be part of a cupboard fitting or can be fixed to wall brackets or bearers that have been plugged and screwed to the wall. Timber or MDF cores are usually used for shelving, faced with any of a large variety of modern plastic finishes. Shelves are classed as solid or slat: the latter are made of 45 mm × 20 mm slats, spaced about 20 mm apart, and are used mainly in airing and drying cupboards. Proprietary adjustable shelf brackets and uprights are available to allow the spacing between shelves to be adjusted. Typical shelf details are shown in Fig. 8.8.2.

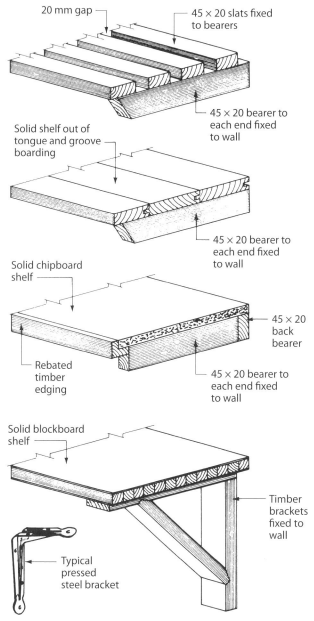

Figure 8.8.2 Shelves and supports.

KITCHEN FITTINGS

Kitchen units are manufactured in preassembled or flat-pack form. Flat pack is more convenient for storage and delivery, but does require additional time for site assembly before fitting. Materials vary; they may be selected timber or its less expensive derivatives and composites of plywood, chipboard or MDF. The carcass base tends to be made from chipboard, which is laminated in Formica® plastic, while the doors, drawer fronts and worktops tend to be manufactured from more expensive quality materials. Purpose-made plastic and metal brackets are used to assemble the components.

The design of a kitchen is based around the system of modular standardised metric units as shown in Fig. 8.8.3. Any remaining gaps in a run of units are normally filled using plain panels to match the kitchen design. Special components, such as corner pieces and purpose-made base units like wine racks, are not included. These vary in style and measurements between manufacturers, but retain overall dimensions to co-ordinate with the standard range.

Heights of floor-supported units are nominal, with adjustable legs fitted as standard, which are then covered with plinths. Surfaces or worktops are secured to the base units with small angle brackets. The worktops are available in laminates of solid timber (usually beech), although the less expensive patterned or grain-effect plastic-faced particleboard is most often used. Stone granite and other natural materials are now increasingly being used for worktop surfaces instead of the traditional laminated chipboard worktops. Thickness varies between 28 and 40 mm, with the option of a rounded edge or wood trimming.

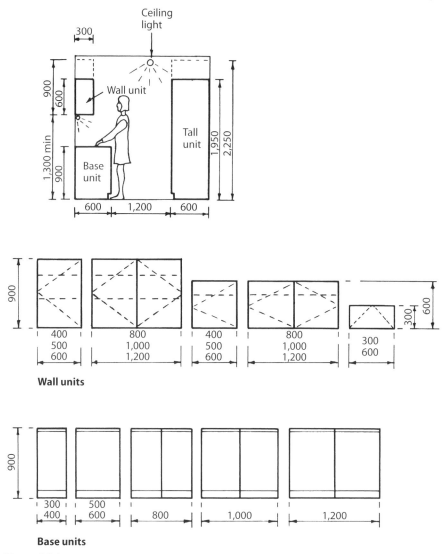

Wall units

Base units

Figure 8.8.3 Standard kitchen units (dimensions in mm).

Ironmongery 8.9

Ironmongery is a general term that is applied to hardware fitted to any doors, windows and other fixtures, and includes such items as nails, screws, bolts, hinges, locks, catches, window and door closers.

NAILS

A nail is a fixing device that relies on the grip around its shank and the shear strength of its cross-section to fulfil its function when it is being driven. It is therefore important to select the right type and size of nail for any particular situation, as too large a nail can split the item being fastened. Nails are specified by their type, length and diameter, with the range given in millimetres. The complete range of nails is detailed in BS 1202. Steel is the main material used in the manufacture of nails but other metals used are copper and aluminium alloy where roofing applications require corrosion resistance.

These are some of the nails in common use.

- **Cut clasp nails** These are made from black rolled steel and are used for general carcassing work.
- **Cut floor brads** These are made from black rolled steel and are used mainly for fixing floorboards, because their rectangular section reduces the tendency of thin boards to split.
- **Round wire** These are made in a wide variety of lengths. They are used mainly for general carpentry work, but have a tendency to split thin members.
- **Oval brad head** These are made with an oval cross-section to lessen the risk of splitting the timber members. They are used for the same purpose as round wire nails but have the advantage of being able to be driven below the surface of the timber. A similar nail with a smaller head is also produced, called an oval lost head nail.

- **Clout nails** Also called slate nails, these have a large flat head and are used for fixing tiles and slates. The felt nail is similar but has a larger head and is only available in lengths up to 38 mm.

- **Panel pins** These are fine wire nails with a small head that can be driven below the surface. They are used mainly for fixing plywood and hardboard.

- **Plasterboard nails** The holding power of these nails, with their countersunk head and jagged shank, makes them suitable for fixing ceiling and similar boards.

WOOD SCREWS

A wood screw is a fixing device used mainly in joinery, and relies upon its length, thread and head for holding power and resistance to direct extraction. For a screw to function properly it must be inserted by rotation using a screwdriver or power drill with screw bit. It is usually necessary to drill pilot holes for the shank and/or core of the screw depending on the quality of the finish required and the density of the objects being fastened.

Wood screws are manufactured from cold-drawn wire, steel, brass, stainless steel, aluminium alloy, silica bronze and nickel–copper alloy. In addition to the many different materials, a wide range of painted and plated finishes is available, such as an enamelled finish known as black japanned. Plated screws are used mainly to match the fitting that they are being used to fix, and include such finishes as galvanised steel, copper-plated, nickel-plated and bronze metal antique. Mirror screws are the only common speciality: the head has a threaded hole within it which accepts a chromed metal dome that neatly covers the screw head, leaving a quality finish to the mirror being fastened.

Screws are specified by their material, type, length and gauge. The screw gauge is the diameter of the shank, and is designated by a number; however, unlike the gauge used for nails, the larger the screw gauge number, the greater the diameter of the shank. Various head designs are available for all types of wood screw, each having a specific purpose.

- **Countersunk head** This is the most common form of screw head, and is used for a flush finish, the receiving component being countersunk to receive the screw head.

- **Raised head** This is used mainly with good-quality fixtures; the rounded portion of the head with its mill slot remains above the surface of the fixture, ensuring that the driving tool does not come into contact with the surface, causing damage to the finish.

- **Round head** The head, being fully exposed, makes these screws suitable for fittings of material that is too thin for the screw to be countersunk.

- **Recessed head** Screws with a countersunk head and a recessed cruciform slot, giving a positive drive with a specially shaped screwdriver.

- **Coach screws** These are made of mild steel with a square head for driving in with the aid of a spanner, and are used mainly for heavy carpentry work.

CAVITY FIXINGS

Various fixing devices are available for fixing components to thin materials of low structural strength, such as plasterboard and hardboard. Cavity fixings are designed to spread the load over a wide area of the board. Here are some typical examples.

- **Steel spring toggles** Spring-actuated wings open out when the toggle fixing has been inserted through a hole in the board and spread out on the reverse side of the board. Spring toggles are specially suited to suspending fixtures from a ceiling.

- **Steel gravity toggles** When inserted horizontally into a hole in the board, the long end of the toggle drops and is pulled against the reverse side of the board as the screw is tightened.

- **Hollow wall anchors** These contain a threaded screw and a steel sleeve with the nut at the end of the sleeve. A setting tool is required to squeeze the steel sleeve by extracting the bolt within it, which crushes the sleeve forming a fastening within the cavity. The bolt is extracted and the fitting then fastened using the bolt until tight.

- **Plasterboard screws** These have a self-drilling point and a large diameter screw which embeds itself into the plasterboard. A central screw fits into the middle of the plasterboard screw and tightens normally.

Typical examples of nails, screws and cavity fixings are shown in Fig. 8.9.1.

HINGES

Hinges are devices used to attach doors, windows and gates to a frame, lining or post so that they are able to pivot about one edge. It is of the utmost importance to specify and use the correct number and type of hinges in any particular situation to ensure correct operation of the door, window or gate. Hinges are classified by their function, length of flap and material used, and sometimes by the method of manufacture. Materials used for hinges are steel, brass, cast iron, aluminium and nylon with metal pins. Here are some typical examples of hinges in common use.

- **Steel butt hinge** These are the most common type in general use, and are made from steel strip, which is cut to form the flaps and is pressed around a steel pin.

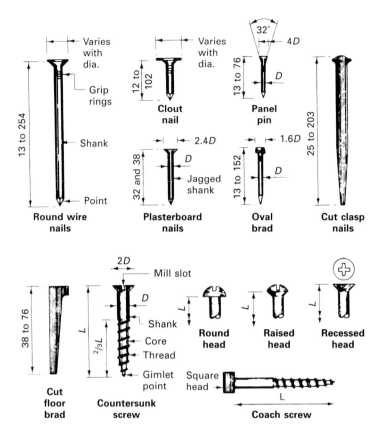

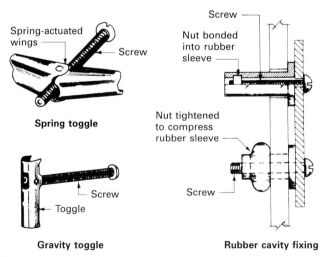

Figure 8.9.1 Nails, screws and cavity fixings.

- **Steel double-flap butt hinge** Similar to the steel butt hinge, but made from two steel strips to give extra strength.

- **Rising butt hinge** This is used to make the door level rise when opened, to clear floor coverings. It will also act as a gravity self-closing door when fitted with these butts, sometimes called skew butt hinges.

- **Tee hinge** Sometimes called a cross garnet, this type of hinge is used mainly for hanging matchboarded doors, where the weight is distributed over a large area.

- **Band and hook** A stronger type of tee hinge made from wrought steel and used for heavy doors and gates. A similar hinge is produced with a pin that projects from the top and bottom of the band, and is secured with two retaining cups screwed to the post; these are called reversible hinges.

Examples of typical hinges are shown in Fig. 8.9.2.

LOCKS AND LATCHES

Any device used to keep a door in the closed position can be classed as a lock or latch. A lock is activated by means of a key, a set of levers and a bolt, whereas a latch is operated by a lever or bar. Latches used on lightweight cupboard doors are usually referred to as catches. Locks can be obtained with a latch bolt so that the door can be kept in the closed position without using a key; these are known as **deadlocks**.

Locks and latches that can be fixed to the face of the door are termed **rim locks or latches**. Alternatively, they can be fixed within the construction of the door, when they are called **mortise locks or latches**. When this form of lock or latch is used, the bolts are retained in mortises cut behind the striking plate, which is fixed to the frame (see Fig. 8.9.3).

Cylindrical night latches are fitted to the stile of a door, and a connecting bar withdraws the latch when the key is turned. Most night latches have an internal device to stop the bolt from being activated from outside with a key.

Door handles, cover plates and escutcheons used to complete lock or latch fittings are collectively called **door furniture** and are supplied in a wide range of patterns and materials.

DOOR BOLTS

Door bolts are security devices fixed to the inside faces of doors, and consist of a slide or bolt, operated by hand, located in a keep fixed to the frame. Two general patterns are produced: the **tower bolt**, which is the cheaper form; and the stronger but more expensive **barrel bolt**. The bolt of a tower bolt is retained with staples or straps along its length, whereas in a barrel bolt it is completely enclosed along its length (see Fig. 8.9.3).

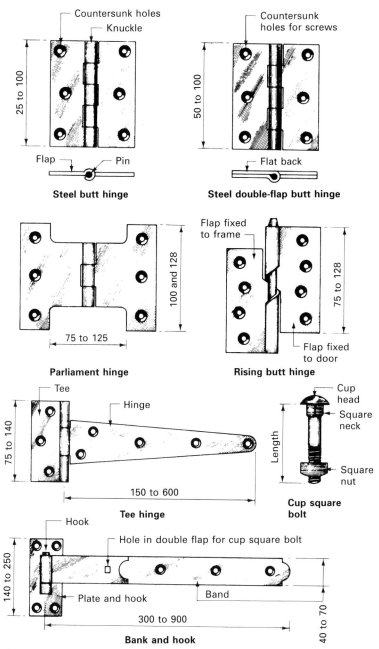

Figure 8.9.2 Typical hinges.

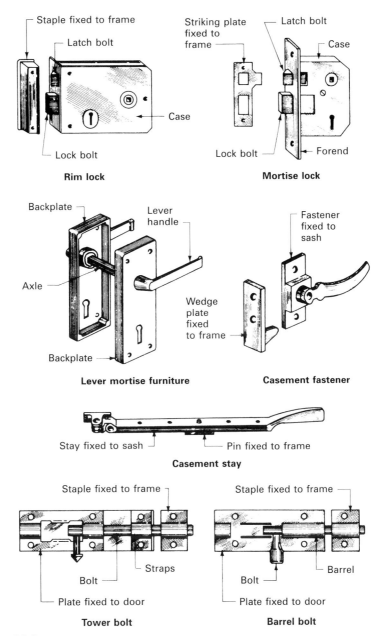

Figure 8.9.3   Door and window ironmongery.

CASEMENT WINDOW FURNITURE

Two fittings are required for opening sashes: the **fastener**, which is the security device; and the **stay**, which holds the sash in the opened position. Fasteners operate by the blade being secured in a mortise cut into the frame, or by the blade locating over a projecting wedge or pin fixed to the frame (see Fig. 8.9.3). Casement stays can be obtained to hold the sash open in a number of set positions by using a pin fixed to the frame and having a series of locating holes in the stay, or they can be fully adjustable by the stay sliding through a screw-down stop fixed to the frame (see Fig. 8.9.3). Additional security is now provided by the use of a locking butt that fits over the casement pin, which can be closed when the window is shut tight, and a security screw fixing through the casement handle that locks it into position by the use of an Allen key.

Shootbolt multi-point espagnolette locking systems provide improved security for window casements. This type of security is common on uPVC doors and frames where a four-point locking system is employed. They have two or three bolt functions (the number depends on the casement size) operating through one application of the casement fastener or lock mechanism. The bolts slide within a recess in the casement sash and lock into corresponding keeps secured to the casement frame. Fig. 8.9.4 illustrates the principle.

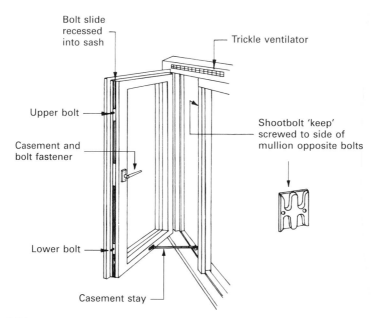

Figure 8.9.4 Shootbolt multi-point espagnolette casement fastener.

LETTER PLATES

These are the hinged covers attached to the outside of a door, which cover the opening made to enable letters to be delivered through the door. The minimum opening size for letter plates recommended by the postal services is 200 mm x 45 mm; the bottom of the opening should be sited not lower than 750 mm from the bottom edge of the door and not higher than 1,450 mm to the upper edge of the opening. A wide range of designs is available in steel, aluminium alloy and plastic, some of which have a postal knocker incorporated in the face design (see BS EN 13724) and a built-in draught excluder.

BS EN 1906 and BS EN 12209 cover builders' hardware for housing, and give recommendations for materials, finishes and dimensions to a wide range of ironmongery items not covered by a specific standard. These include such fittings as finger plates, cabin hooks, gate latches, cupboard catches and drawer pulls.

Painting and decorating 8.10

The application of coats of paint to the wall and ceiling components, woodwork and fittings of a building has two functions: the paint will provide colour and, at the same time, give a protective coating that will increase the durability of the element to which it has been applied. The covering of wall and ceiling surfaces with wallpaper or fabric is basically to provide colour, aesthetics, contrast and atmosphere. To achieve a good, durable finish with any form of decoration, the preparation of the surface and the correct application of the paint or paper are of the utmost importance. The more time spent on this element, the better the quality of finish.

PAINT

Paint is a mixture of a liquid or medium and a colouring or pigment. Mediums used in paint manufacture range from thin liquids to stiff jellies, and can be composed of linseed oil, drying oils, synthetic resins and water. The latter now meets with the requirement of environmental legislation. The various combinations of these materials form the type or class of paint. The medium's function is to provide the means of spreading and binding the pigment over the surface to be painted and holding it in position while it dries and bonds to the substrate; the pigment provides the body, colour and durability of the paint. The general pigment used for finishing paint is titanium dioxide, which gives good obliteration of the undercoating and is not poisonous, as were the lead pigments that were once used.

ACRYLIC PAINTS

Because of the VOC (volatile organic compounds) content reduction legislation introduced through the European Parliament, oil-based paints are being phased out in favour of water-based acrylic paints.

PRIMING PAINTS

These are first-coat paints used to seal the surface, protect the surface against damp air, act as a barrier to prevent any chemical action between the surface and the finishing coats, and give a smooth surface for the subsequent coats. Priming paints are produced for application to wood, metal and plastered surfaces. After application the priming coat should be sanded down and washed, ready for undercoating.

UNDERCOATING PAINTS

These are used to build up the protective coating's thickness and to provide the correct surface for the finishing coat(s). Undercoat paints contain a greater percentage of pigment than finishing paints, and as a result have a matt or flat finish. To obtain a good finishing colour it is essential to use an undercoat of the type and colour recommended by the manufacturer. Each subsequent coat will require rubbing down and cleaning before the next coat is applied.

FINISHING PAINTS

A wide range of colours and finishes including matt, semi-matt, eggshell, satin, gloss and enamel is available. These paints usually contain an acrylic polymer emulsion that enables them to be easily applied, be quick to dry and have good adhesive properties. Gloss paints have less pigment than the matt finishes and consequently less obliterating power.

POLYURETHANE PAINTS

These are quick-drying paints based on polyurethane resins, which give a hard, heat-resisting surface. They can be used on timber surfaces as a primer and undercoat, but metal surfaces will require a base coat of metal primer; the matt finish with its higher pigment content is best for this 'one paint for all coats' treatment. Other finishes available are gloss and eggshell.

EMULSION PAINTS

Most of the commonly used water-based paints come under a general classification of emulsion paints: they are quick-drying and can be obtained in matt, eggshell, semi-gloss and gloss finishes. The water medium has additives such as polyvinyl acetate and alkyd resin to produce the various finishes. Their general use is for large, flat areas such as ceilings and walls.

VARNISHES AND STAINS

Varnishes form a clear, tough film over a surface, and are available in a gloss or satin finish. They are a solution of water and acrylic co-polymer emulsion, their application being similar to that of acrylic paints. Varnish is best applied over a stain, which colours the wood, highlighting any grains; the varnish will seal in this feature.

PAINT SUPPLY

Paints are supplied in metal and plastic containers of 5 litre, 2.5 litre, 1 litre, 500 millilitre and 250 millilitre capacity. They are usually in one of the 100 colours recommended for building purposes in BS 4800, although any colour can now be matched and mixed for a client. BS 381C is also a useful reference, giving 91 standard colours for special purposes, such as determining colours used by Her Majesty's (HM) services. The colour range in BS 381C can be used for purposes other than paint identification.

The RAL standard notation colour system is widely used within European countries for specifying paints and coatings. It sets out a standard range of colours in charts that can be specified by a designer for an interior design range.

KNOTTING

Knots or streaks in softwood timber may exude resins, which can soften and discolour paint finishes; generally an effective barrier is provided by two coats of knotting applied before the priming coats of paint. Knotting is a uniform dispersion of shellac in a natural or synthetic resin, in a suitable solvent, such as methylated spirit. It should be noted that the higher the grade of timber specified, with its lower proportion of knots, the lower will be the risk of resin disfigurement of paint finishes.

WALLPAPERS

Most wallpapers consist of a base paper of 80 per cent mechanical wood pulp and 20 per cent unbleached sulphite pulp, printed on one side with a colour and/or design. Papers are also available with vinyl plastic coatings making them washable, together with a wide range of fabrics suitably coated with vinyl, which gives a raised, textured pattern for the decoration of walls and ceilings.

The preparation of the surface to receive wallpaper is very important; ideally a smooth, clean and consistent surface is required. Walls that are in a poor condition can be lined with a lightweight lining paper (see following page) for traditional wallpaper, and a heavier esparto lining paper for the heavier classes of wallpaper. Lining papers should be applied horizontally and the final paper covering vertically.

Wallpapers are attached to the surface by means of a modern adhesive, which must be mixed with water to the manufacturer's recommendations and proportions. Heavy washable papers and woven fabrics should always be hung with an adhesive recommended by the paper manufacturer. New plaster and lined surfaces should be treated to provide the necessary 'slip' required for successful paper hanging. The application of a weak solution of wallpaper paste will be sufficient to seal the new plasterwork and provide a surface that wallpaper will adhere to.

Standard wallpapers are supplied in rolls with a printed width of 530 mm and a length of 10 m, giving an approximate coverage of 4.5 m^2; the actual coverage will be governed by the 'drop' of the pattern, which may be as much as 2 m. Decorative borders and motifs are also available for use in conjunction with standard wallpapers.

TYPES OF WALLPAPER

Lining paper is used to provide uniform suction on uneven surfaces and to reinforce cracked plaster before applying decorative wallpaper finishes. It also smoothes out some of the background imperfections. The quality varies from a light gauge to a thicker 2000 grade. A foil-backed lining paper is also available for damp walls, but it is not intended as a substitute for remedial treatment to a source of dampness. Lining papers are manufactured 11 m long by 560 mm or 760 mm in width.

FINISHES

- **Flocks** These are manufactured with a raised pile pattern produced by blowing coloured wool, cotton or nylon fibres onto an adhesive printed pattern.

- **Metallic finishes** Paper coated with metal powders or a metal foil sheet material.

- **Moires** A watered silk effect finish.

- **Ingrained** An effect produced by incorporating sawdust or other fibres in the paper surface (woodchip).

- **Textured effect** An imitation texture achieved by embossing the surface.

- **Washable papers** These include varnish- and plastic-coated paper, in addition to expanded PVC foam and textured PVC sheet.

- **Relief papers** These are in two categories: high relief (Anaglypta) and low relief (Lincrusta). Anaglypta is a quality wallpaper moulded while wet to provide embossed patterns that give the paper strength and resilience. The depressions on the back of the paper can be filled with a paste/sawdust mix to resist impact damage. Lincrusta is a paper composed of whiting, wood flour, lithopone, wax, resin and linseed oil, which is superimposed hot on craft paper. The result is a relatively heavy paper requiring a dense paste for adhesion to the wall.

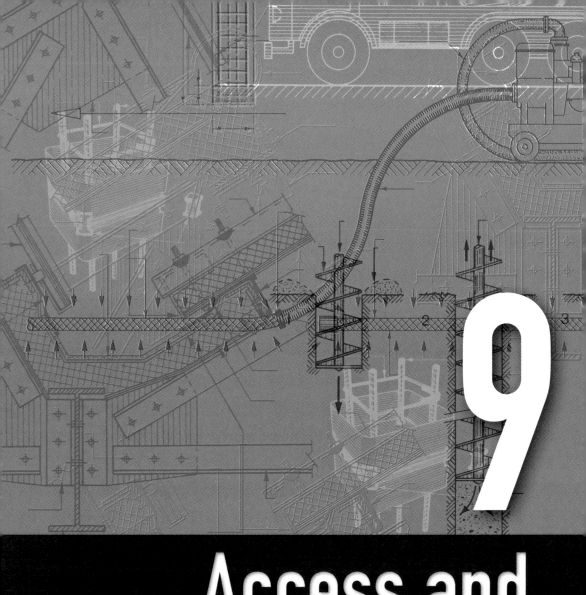

9

Access and facilities for disabled people

Accessibility

9.1

The legislation concerning providing access to buildings for the disabled is very detailed. The two primary pieces of legislation are the Building Regulations and the Equality Act. These both place conditions on premises, ensuring that they are fit for disabled use and access. All new, refurbished and adapted dwellings must have reasonable provisions to allow accessibility for disabled people. The objective is to provide wheelchair users with an easily negotiable approach from the point of alighting a vehicle to the dwelling entrance.

BUILDING REGULATIONS PART M – 2004

Approved Document M: *Access to and use of buildings* has been revised in support of disability legislation. The Regulations provide design and practical guidance for disabled-user convenience in new and refurbished buildings. All new homes (houses and flats) are required to be constructed with sufficient accessibility and use of facilities for disabled people. The objective is to allow disabled and elderly people greater freedom of use and independence in their own homes for a longer period of time. The Regulations also allow for internal and external accessibility for wheelchair users in buildings other than dwellings. Movement in and around all buildings should offer people with disabilities the same means of access as others enjoy when visiting friends, relatives, shops, entertainments and other conveniences, without being impeded. The Regulations are divided into sections that deal with domestic access and building other than dwellings.

The main features of the Approved Document are:

- access approach from car parking area to the main entrance of a building to be level or ramped;

- main entrance threshold to be level, not stepped;

- main entrance door wide enough for a wheelchair;
- WC facilities at ground or entrance floor;
- WC compartment with sufficient space for a wheelchair user to manoeuvre;
- WC facilities at accessible levels for ambulant and disabled people;
- switches for lighting, power sockets, heating control and so on at convenient heights above floor level;
- in flats, lifts provided to access all floors;
- in houses, the structure about a stair to be strong enough to support a stair lift.

BS 8300

A further piece of guidance is British Standard BS 8300: *Design of buildings and their approaches to meet the needs of disabled people. Code of Practice*, which covers many aspects of designing for disability.

This British Standard complements the Building Regulations and the Disability Discrimination Act. It was produced from Government-commissioned research into the ergonomics of modern buildings. The Standard presents the built environment from the perspective of the disabled user, with particular regard to residential buildings as occupier and visitor, and public buildings as spectator, customer and employee. The Standard also considers the disabled as participants in sports events, conferences and performances.

LEVEL APPROACH FOR BUILDINGS OTHER THAN DWELLINGS

A level approach from the boundary of the site and car parking should satisfy the following criteria:

- access should be level from the boundary of the site and from any designated disabled car parking;
- all surfaces must be free from any trip hazards and be an easily travelled surface;
- where the gradient of the approach is 1:20 or steeper, this part of the approach must be designed as a ramped access.

ON-SITE CAR PARKING AND SETTING DOWN

The Regulations make provision for the space around a parked vehicle and the provision of setting-down points, which must satisfy the following criteria:

- the surface of the parking bay and the route to the access entrance must be free of any obstructions;
- provision should be made for access to ticket machines;
- setting-down areas need to be convenient to the entrance of the building;

- at least one parking bay must be provided as close as possible to the entrance;
- the size of bay must be configured as shown in Fig. 9.1.1;
- the surface of the accessibility zone must be non-slip, durable and firm;
- setting-down points must be signposted.

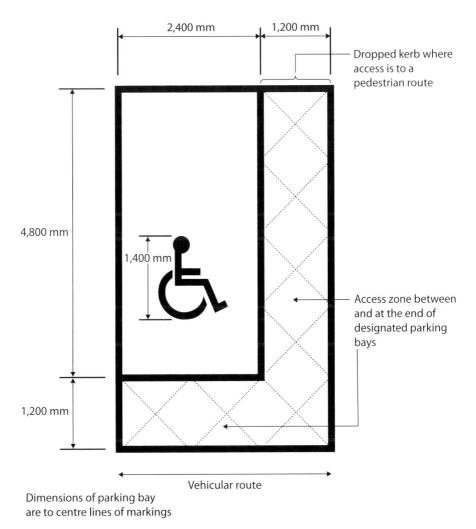

Dimensions of parking bay
are to centre lines of markings

Figure 9.1.1   Disabled parking bay.

RAMPED ACCESS TO BUILDINGS OTHER THAN DWELLINGS

A ramped approach from a parked vehicle is defined as having a measured slope greater than 1 in 20, but not greater than 1 in 15 overall. A ramp should have:

- no going greater than 10 m or rise of more than 500 mm;
- a surface that is firm, even and slip-resistant when wet;
- alternative means of access provided where the rise is greater than 2 m;
- unobstructed width not less than 1,500 mm;
- 1,200 mm-long landings at top and bottom;
- intermediate landings 1,500 mm long;
- a kerb on any open side of a ramp;
- a signposted approach.

STEPPED APPROACHES FOR BUILDINGS OTHER THAN DWELLINGS

Steps are to be avoided if at all possible. However, if the site constraints have a gradient where a ramp cannot be accommodated, steps designed to suit

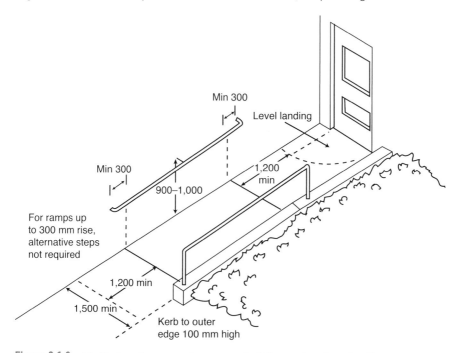

Figure 9.1.2 Limitations for ramped access to buildings other than dwellings. Source: *Designing for Accessibility*, 2004 Edition, copyright Centre for Accessible Environments/RIBA Publishing.

ambulant disabled people may be acceptable, provided they satisfy the requirements of M1 and M2, if they meet the following conditions on site.

A stepped access will satisfy the requirements for Part M if:

- a level landing is provided at the top and bottom of each flight;
- the length of each landing is 1,200 mm, unobstructed;
- change in level hazard warning surfaces are provided at the top and bottom landings;
- no doors swing across landings;
- the surface width of flights between walls, strings or upstands is not less than 1,200 mm;
- there are no single steps;
- the provisions shown in Fig. 9.1.3 are made;
- all nosings are visible using a 55 mm-wide contrast strip on tread and riser;
- the nosing projection over the step is less than 25 mm;
- the rise is between 150 mm and 170 mm, and is not open;
- the going is between 280 mm and 425 mm;
- there is a continuous handrail on each side of a flight and landing.

Figs. 9.1.3 and 9.1.4 show step designs that satisfy the Building Regulations.

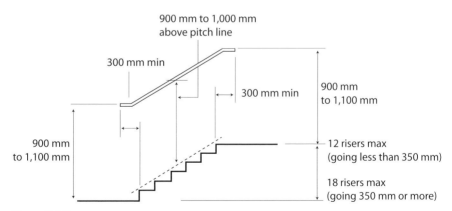

Figure 9.1.3 **Provisions for steps under Part M.**

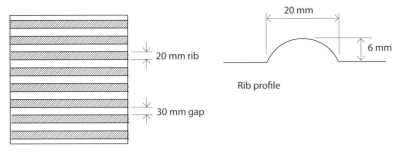

20 mm rib

20 mm

6 mm

Rib profile

30 mm gap

'Corduroy' hazard warning surface (with 8 mm ribs)

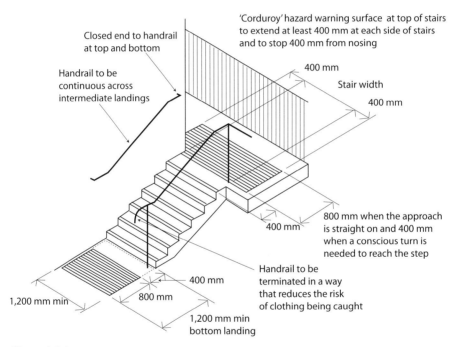

Closed end to handrail at top and bottom

Handrail to be continuous across intermediate landings

'Corduroy' hazard warning surface at top of stairs to extend at least 400 mm at each side of stairs and to stop 400 mm from nosing

400 mm

Stair width

400 mm

800 mm when the approach is straight on and 400 mm when a conscious turn is needed to reach the step

400 mm

Handrail to be terminated in a way that reduces the risk of clothing being caught

400 mm

800 mm

1,200 mm min

1,200 mm min bottom landing

Figure 9.1.4 Acceptable external step design elements.

HANDRAILS TO EXTERNAL STEPPED AND RAMPED ACCESS

The provision of a handrail to provide additional support for disabled access is part of the Regulations. A handrail must be comfortable to touch, non-slip, easily gripped and provide good forearm support. A handrail should be placed at a set height and extend beyond the top and bottom step. Fig. 9.1.5 illustrates the provisions required for handrails.

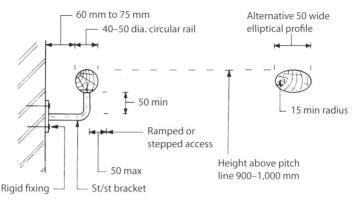

60 mm to 75 mm
40–50 dia. circular rail
Alternative 50 wide
elliptical profile

50 min

15 min radius

Ramped or
stepped access

50 max

Height above pitch
line 900–1,000 mm

Rigid fixing — St/st bracket

Figure 9.1.5 Handrail design.

DWELLING ACCESS

The approach to the dwelling is the best means of providing a level access to the entrance door for a disabled user. An accessible threshold at the entrance must be provided. This means the removal of a step and the provision of a ramped level up to the threshold. Care must be taken with the design of the threshold to allow access but prevent moisture and driving rain from entering the dwelling.

DWELLING ENTRANCE DOOR

The entrance door must allow a wheelchair user to manoeuvre into the building easily. The minimum acceptable width of door opening for wheelchairs is 775 mm. As shown in Fig. 9.1.6, this is measured from the face of the door stop, latch side of the frame, to the face of the door opened at 90° to the frame. It is therefore the clear opening width.

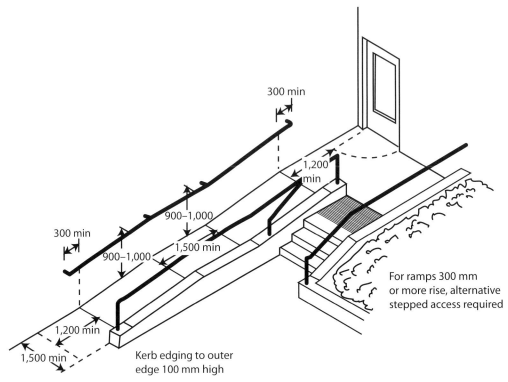

300 min

1,200 min

900–1,000

300 min

900–1,000

1,500 min

1,200 min

1,500 min

Kerb edging to outer
edge 100 mm high

For ramps 300 mm
or more rise, alternative
stepped access required

Figure 9.1.6   Entrance door.
Source: *Designing for Accessibility*, 2004 Edition, copyright Centre for Accessible
Environments/RIBA Publishing.

Circulation space: entrance to dwelling 9.2

With the entrance to a dwelling, the objective is to make it, and any rooms off the entrance corridor, easily accessible for a wheelchair user. A room containing a WC, which may be a bathroom, needs to be on that level and accessible by a wheelchair corridor.

An adequate means of passage is required within the entrance storey, which should be wide enough to permit unobstructed circulation by a wheelchair user. Allowance for normal projections, such as shelves and radiators, should be considered when designing for minimum acceptable corridor width. Where a necessary permanent obstruction such as a radiator cannot be avoided, the passageway widths given in Table 9.2.1 can be reduced to 750 mm for a distance of no more than 2 m. This relaxation is not acceptable where it would impede a wheelchair user from turning into or out of a room, as shown in Fig. 9.2.1.

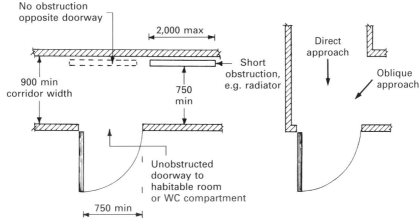

Figure 9.2.1 Corridors and passageways.

Rooms and facilities off a room must also be unobstructed and wheelchair accessible. Internal door openings should have sufficient width to accommodate the manoeuvring of a wheelchair from corridors and passageways. Table 9.2.1 indicates the spatial requirements for internal doors relative to corridor width and direction of approach.

Table 9.2.1   Passageway and door width dimensions

Internal doorway width (mm) – see Fig. 9.2.1	Unobstructed passageway width (mm) minimum
≥ 750	900 (direct approach)
750	1,200 (oblique approach)
775	1,050 (oblique approach)
800	900 (oblique approach)

STEPS

Where the slope of a site is considerable, it may be necessary to incorporate a stepped change of level into the entrance storey. A slight slope is preferred but, where steps are unavoidable, a stair should have a clear width of at least 900 mm. If the stair has three or more rises, a continuous handrail is required both sides of the flight. Step rise and going should satisfy the private stair provisions in Approved Document K of the Building Regulations (see Part 8.4 on page 390).

WC facilities for the disabled 9.3

In this section, the provision of WCs is divided into those for dwellings and those for buildings other than dwellings.

WC PROVISION FOR DWELLINGS

A WC is required in the entrance storey of the dwelling. There should be no need to reach it via stairs from the habitable room. If there are no habitable rooms in the entrance storey, a WC can be still provided here as well as within the principal storey. The WC does not have to be independent; it may be part of a bathroom suite.

To maximise space for a wheelchair user, an outward-opening door will provide clear access to the WC. The minimum clear widths of door openings were defined in Table 9.2.1. A basin of the hand-wash type should be of minimal dimensions so as not to impede access to the WC. Clear space requirements in the vicinity of a WC are indicated in Fig. 9.3.1.

WC PROVISION FOR BUILDINGS OTHER THAN DWELLINGS

The design of sanitary accommodation must take into account the needs of the visually impared, the deaf, people with learning difficulties and those with tactile sensitivity to hot surfaces. Taps and other fittings should be able to be opened and operated with the minimum of strength. Provision should always be made to release cubicle doors from the outside in the event of an occupant collapsing. Sanitary accommodation will satisfy the requirements of Part M if the following points are observed.

- Any tap is controlled automatically or can be operated with a closed fist.
- Toilet doors are fitted with light-action privacy bolts.

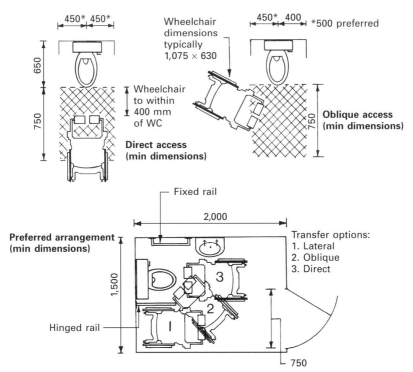

Figure 9.3.1 Clear space requirements in the vicinity of a WC.

- Toilet doors have an external release for use in case of emergency.
- Doors when open do not obstruct escape routes.
- An audible and visual fire alarm is fitted.
- Any heating device has its surfaces limited to 43 °C.
- Grab rails and bars contrast with the background colour.
- Wall and floor finishes contrast.
- At least one wheelchair-access unisex toilet is provided within a building.

WHEELCHAIR-ACCESSIBLE UNISEX TOILETS

A person in a wheelchair must be able to use the WC facilities provided within the building. These are the provisions that must be supplied in buildings other than dwellings.

- The WC must be located near to the building entrance.
- When it is the only toilet facility in a building, the width of the toilet between walls is increased to 2 m.
- The route to the toilet is obstruction-free.

- Doors open outward, preferably fitted with a horizontal closing bar.
- Dimensions comply with those shown in Fig. 9.3.2.
- Heights and fitting comply with those shown in Fig. 9.3.2.
- An emergency assistance alarm is fitted, with identified pull cord.

REFERENCES

- BS 8300: *Design of buildings and their approaches to meet the needs of disabled people. Code of practice.*
- Building Regulations Part M and Approved Document: *Access to and use of buildings.*
- Equality legislation.

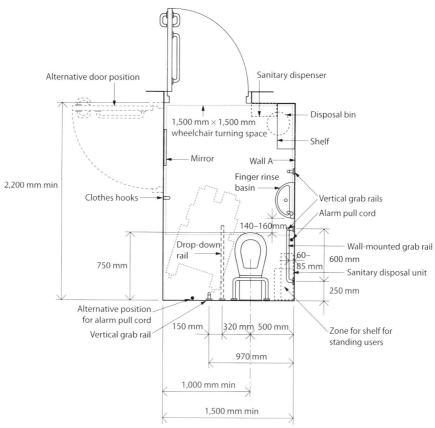

Note: Layout for right-hand transfer to WC   (excluding any projecting heat emitters)

Figure 9.3.2   Provisions for wheelchair-accessible unisex toilets.

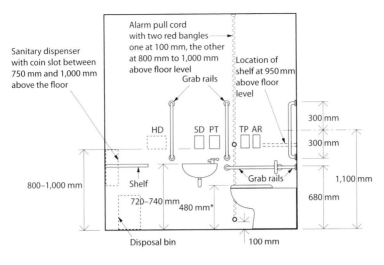

Sanitary dispenser with coin slot between 750 mm and 1,000 mm above the floor

Alarm pull cord with two red bangles one at 100 mm, the other at 800 mm to 1,000 mm above floor level

Grab rails

Location of shelf at 950 mm above floor level

HD   SD PT   TP AR

300 mm

300 mm

1,100 mm

680 mm

800–1,000 mm   Shelf

Grab rails

720–740 mm   480 mm*

Disposal bin   100 mm

* Height subject to manufacturing tolerance of WC pan

HD: Possible position for automatic hand dryer (see also below)
SD: Soap dispenser
PT: Paper towel dispenser
AR: Alarm reset button
TP: Toilet paper dispenser

Height of drop-down rails to be the same as the other horizontal grab rails

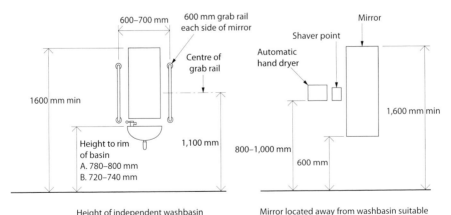

600–700 mm

600 mm grab rail each side of mirror

Mirror

Centre of grab rail

Shaver point

Automatic hand dryer

1600 mm min

1,600 mm min

Height to rim of basin
A. 780–800 mm
B. 720–740 mm

1,100 mm   800–1,000 mm

600 mm

Height of independent washbasin and location of associated fittings, for wheelchair users and standing people

A. For people standing
B. For use from WC

Mirror located away from washbasin suitable for seated and standing people (mirror and associated fittings used within a WC compartment or serving a range of compartments)

Figure 9.3.2   **Provisions for wheelchair-accessible unisex toilets (cont.).**

Accessibility in flats, common stairs and lifts 9.4

Here the objective is to provide disabled people with a reasonable means of access to visit any storey in a block of flats. A lift is the most desirable and suitable means for vertical transportation, but this will not always be provided. A common stairway is essential for general use and as a means of fire escape. It should be designed with regard for the ambulant disabled and in particular for people with limited or impaired vision.

STAIRS

Here are some guidelines on what needs to be incorporated into the design of any common stair.

- Landings provided at the top and bottom of a stair flight. Minimum length not less than stair width and without permanent obstruction.

- Handrail on both sides of the flight and landings, where there are two or more rises (see Fig. 9.4.1).

- Any door that opens onto a landing must leave at least 400 mm clear space across the width of the landing.

- Uniform rise of step, maximum 170 mm.

- Risers full, i.e. no gaps under tread.

- Uniform going of tread, minimum 250 mm.

- Tapered treads to be avoided, but if used, going measured 270 mm from inside of tread.

- Tread nosing profiles as shown in Fig. 9.4.1.

- Nosings made prominent, with contrasting colour and brightness.

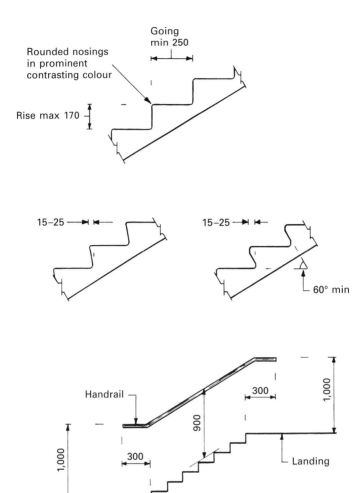

Going
min 250

Rounded nosings
in prominent
contrasting colour

Rise max 170

15–25

15–25

60° min

Handrail

1,000

300

900

300

1,000

Landing

Figure 9.4.1 Step profiles for common stairs.

Lifts offer the best method of providing disabled access for people in wheelchairs. They must be able to accommodate an unaccompanied wheelchair user. Control facilities should also be designed to be accessible and usable from a wheelchair and by people with a visual impairment. Here is some design guidance.

- Lift door opening and closing time adequate for a wheelchair to access without contacting closing doors. Door fully open for at least three seconds.

- Minimum car-load capacity of 400 kg.

- Landing space in front of lift doors, minimum 1,500 mm x 1,500 mm.

- Door opening width, minimum 800 mm.

- Lift car dimensions, minimum 900 mm wide x 1,250 mm deep.

- Controls at landings and in car, minimum 900 mm, maximum 1,200 mm above floor levels. Within car, minimum 400 mm from the front wall.

- Controls prominent/projecting to provide a tactile facility: for example, floor numbers profiled.

- Visual indication that a lift is responding to a call.

- Over three storeys, visual and audible indication of floor level reached.

LIFTS FOR BUILDINGS OTHER THAN DWELLINGS

This section covers the installation and design considerations for lifts that are used in buildings other than dwellings and flats, such as in a hospital, school or college establishment. They satisfy the requirements of Part M of the Building Regulations by complying with a number of criteria.

- Internal dimensions of 1,100 mm wide x 1,400 mm deep.

- 800 mm clear opening at the door.

- Controls between 900 mm and 1,200 mm from car floor level.

- Lift doors and landing doors contrasted against surroundings.

- Visual and audible indication of arrival.

- Any glass areas identifiable by visually impaired.

Switches, sockets and general controls

9.5

This requirement covers a range of control devices normally installed within a dwelling: light switches, power sockets, telephone sockets, entry telephones, heating thermostats and radiator valves. It should also include door furniture and ironmongery. The objective here is to provide these means of control at a level readily accessible for people with reach limitations. In addition to the registered disabled, this category will also include the elderly and infirm.

Design of controls should facilitate ease of operation and prominent visibility: for example, by using large, rocker-type electrical switches and contrasting colours. Interior design and layout of premises should ensure that control facilities are not impeded by obstructions.

A suitable wall-mounting height for controls is recommended as between 450 and 1,200 mm above finished floor level. Ceiling pull-switch pendants should also terminate within this range. Door handles should align with switch controls as shown in Fig. 9.5.1.

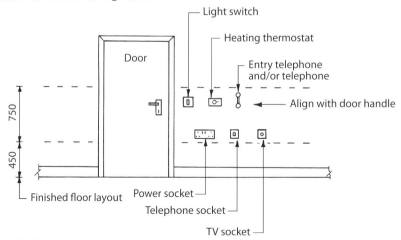

Figure 9.5.1 Recommended accessibility for domestic controls.

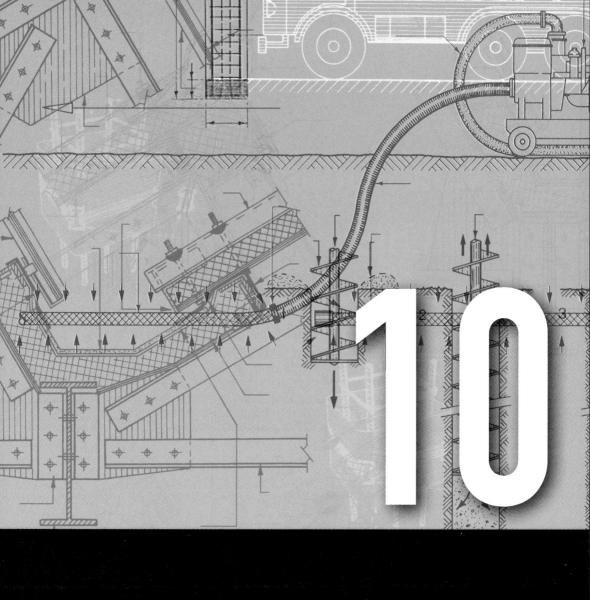

10

Framed buildings

Simple framed buildings 10.1

The purpose of any framed building is to transfer the loads of the structure plus any imposed loads through the members of the frame to a suitable foundation. This form of construction can be clad externally with lightweight, non-loadbearing walls to provide the necessary protection from the elements and to give the required degree of comfort in terms of sound and thermal insulation. Framed buildings are particularly suitable for medium- and high-rise structures and for industrialised low-rise buildings such as single-storey factory buildings.

Frames can be considered under four headings:

- plane frames;
- portal frames;
- space frames;
- skeleton frames.

PLANE FRAMES

These are fabricated in a flat plane and are usually called **trusses** or lattice **girders**, according to their elevation shape. They are designed as a series of connected rigid triangles, which gives a lightweight structural member using the minimum amount of material; main uses are in roof construction and long-span beams of light loading. Plane frames do need to be laterally braced and tied together to resist wind-loading, with additional horizontal braces such as purlins and binders, in a similar way to trussed-rafter construction found in domestic buildings.

PORTAL FRAMES

These are similar to plane frames in that they are two-dimensional structural elements; however, unlike plain frames, they combine a vertical column and a horizontal or inclined beam into a rigid strong unit. For short spans they rely on the rigid welded or bolted connections between the column and the foundation, and between the column and the beam; for longer spans moveable pin connections can be introduced at the ridge or the column base to help redistribute the stresses evenly around the whole frame. As with plane frames, portal frames need to be braced horizontally to resist wind-loadings.

SPACE FRAMES

These are similar in conception to a plane frame but are designed to span in two directions, as opposed to the one-direction spanning of the plane frame. A variation of the space frame is the **space deck**, which consists of a series of linked pyramid frames forming a lightweight roof structure. With the development of curved cladding and glazing systems, it is now possible to achieve curved double-layer space decks in the form of a geodesic dome and its variations. For details of these forms of framing, refer to *Advanced Construction Technology*.

SKELETON FRAMES

Basically, these are a series of interconnected horizontal and vertical members placed typically at right angles to one another, so that the loads are transmitted from member to member until they are transferred through the foundations to the subsoil. They support all the loads of the structure and form a framework to carry the external cladding systems, internal partitions, floors, roofs and services. Skeleton frames can be economically constructed of concrete or steel, or a combination of the two. Timber skeleton frames, although possible, are generally considered to be uneconomic in this form, although buildings up to ten storeys high have been achieved. The choice of material for a framed structure can be the result of a number of factors, such as:

- site conditions;
- proposed occupancy (for example, space requirements, flexibility of use);
- material properties (for example, sound resistance, thermal mass, achievement of fire resistance, integration with building services);
- cost of material (for example, global supply and demand);
- speed of construction and buildability;
- availability of labour and specialist plant;
- maintenance costs during life of building;
- aesthetics and personal preference.

Glossary of terms for skeleton frame members

- **Main beams** Span between columns and transfer the live and imposed loads placed upon them to the columns.

- **Secondary beams** Span between and transfer their loadings to the main beams. Primary function is to reduce the spans of the floors or roof being supported by the frame.

- **Tie beams** Internal beams spanning between columns at right angles to the direction of the main beams and having the same function as a main beam.

- **Edge beams** As tie beams but spanning between external columns.

- **Columns** Vertical members that carry the loads transferred by the beams to the foundations.

- **Foundation** The base(s) to which the columns are connected and which serve to transfer the loadings to a suitable loadbearing subsoil.

- **Floors** May or may not be an integral part of the frame; they provide the platform on which equipment can be placed and on which people can circulate. Besides transmitting these live loads to the supporting beams they may also be required to provide a specific fire resistance, together with a degree of sound and thermal insulation.

- **Roof** Similar to floors but its main function is to provide a weather-resistant covering to the uppermost floor.

- **Cladding walls** The envelope of the structure, which provides the resistance to the elements, entry of daylight, natural ventilation, fire resistance, thermal insulation and sound insulation.

The three major materials used in the construction of skeleton frames – reinforced concrete, precast concrete and structural steelwork – are considered in detail in the following chapters.

CHOICE OF FRAMING SYSTEMS

All the above systems are currently used today, and this means that a comparison must be made before any particular framing material is chosen. The main factors to be considered in making this choice are:

- site costs; - construction costs; - maintenance costs.

SITE COSTS

A building owner will want to obtain a financial return on his or her capital investment as soon as possible; therefore speed of construction is of paramount importance. The use of a steel or precast concrete frame will permit the maximum amount of prefabrication off site, during which time the groundworks contractor can be constructing the foundations in preparation for the erection of the frame. To make the best use of a site, the structure needs to be designed so that the maximum amount of rentable floor area is achieved. Generally, prefabricated section sizes are smaller than comparable in-situ concrete members, mainly because of the greater control over manufacture obtainable under factory conditions, and thus these will occupy less floor area.

Site conditions regarding storage space, fabrication areas and manoeuvrability around and over the site can well influence the framing method chosen. For example, precast concrete or prefabricated steel frames can be delivered to site, lifted into position and erected with minimal site storage facilities required. This is particularly useful on cramped urban or high-rise construction projects. However, all lifts of preformed elements need to be planned and executed carefully to minimise risks.

CONSTRUCTION COSTS

The main factors are design considerations, availability of labour, availability of materials and site conditions. Concrete is an adaptable material, which allows the designer to be more creative than when working within the rigid confines of standard steel sections. However, as the complexity of shape and size increases, so too does the cost of formwork; also, for the erection of a steel structure, skilled labour is required, whereas activities involved with precast concrete structures can be carried out by the more readily available semi-skilled labour, working under the direction of a competent person.

Economy of scale can also be a dominant factor: for example, if there is a great deal of concrete required, it may be cheaper and more convenient to mix and batch concrete on site using raw materials of aggregate and cement, if there is sufficient space to set up a batching plant and good transport links to bring in the materials. The availability of materials fluctuates, and only a study of current market trends can give an accurate answer to this problem.

MAINTENANCE COSTS

These can be considered in the short or long term, but it is fair to say that in most framed buildings the costs are generally negligible if the design and workmanship are sound. Steelwork, because of its corrosion properties, will need some form of protective treatment but, as most steel structures have to be given a degree of fire resistance, the fire protective method may well perform the dual function. Concrete elements, both precast or in-situ, have an intrinsic fire resistance, but careful detailing at the junctions needs to be properly maintained.

Reinforced concrete frames 10.2

CONCRETE

Plain concrete is a mixture of cement, fine aggregate, coarse aggregate and water. Concrete sets to a rock-like mass because of a chemical reaction that takes place between the cement and water, resulting in a paste or matrix that binds the other constituents together. Concrete gradually increases its strength during the curing or hardening period to obtain its working strength in about 28 days if ordinary Portland cement is used. The specification of concrete has been covered earlier in this book on pages 80–81.

REINFORCEMENT

To construct an economic structural member, any material specified for use as a reinforcement to concrete must fulfill certain requirements.

- It must have tensile strength.
- It must be capable of achieving this tensile strength without undue strain.
- It must be of a material that can be easily bent to any required shape.
- Its surface must be capable of developing an adequate bond between the concrete and the reinforcement to ensure that the required design tensile strength is obtained.
- A similar coefficient of thermal expansion is required to prevent unwanted stresses being developed within the member owing to temperature changes.
- It must be available at a reasonable cost, which must be acceptable to the overall design concept.

The material that meets all the above requirements is steel in the form of bars, and is supplied in two basic types: mild steel and high-yield steel.

Hot-rolled steel bars are covered by BS 4449, which specifies a characteristic strength of 250 N/mm^2 for mild steel and 460 N/mm^2 for high-yield steel. The surface of mild steel provides adequate bond, but the bond of high-yield bars, being more critical with the higher stresses developed, is generally increased by rolling longitudinal or transverse ribs on to the surface of the bar. In the current Eurocodes mild steel is not specified for use in reinforced concrete. Grade 460A can be identified by having at least two rows of parallel transverse ribs; grade 460B is similar but has at least one row of ribs contrary to the others. As an alternative to hot-rolled steel bars, cold-worked steel bars can be used. When bars are cold-worked they become harder and stiffer, and develop a higher tensile strength.

The range of nominal sizes available for both round and deformed bars is 6, 8, 10, 12, 16, 20, 25, 32 and 40 mm with length specified by the purchaser. For pricing purposes the 16 and 20 mm bars are taken as basic, with the diameters on either side becoming more expensive as the size increases or decreases. A good design will limit the range of diameters used, together with the type of steel chosen, to achieve an economic structure. Good design also eases the site processes of handling, storage, buying and general confusion that can arise when the contractor is faced with a wide variety of similar materials.

The bending of reinforcement can be carried out on site using a bending machine, which shapes the cold bars by pulling them round a mandrel. Small diameters can also be bent round a simple jig, such as a board with dowels fixed to give the required profile; large diameters may need a power-assisted bending machine. It is more common now for bars to be supplied ready bent and labelled, so that only the fabrication processes take place on site. This helps speed up the fixing process and also frees up valuable storage and circulation space on site.

The bent reinforcement should be fabricated into cages for columns and beams, and into mats for slabs and walls. Where the bars cross or intersect one another they should be tied with soft iron wire, fixed with special wire clips or tack-welded to maintain their relative positions. Structural members that require only small areas of reinforcement can be reinforced with steel fabric, which can be supplied in sheets or rolls. Where exposed ends of fixed steelwork are present, such as with vertical starter bars, plastic mushroom caps are fitted as a temporary measure to prevent operatives from injuring or impaling themselves.

Steel fabric for the reinforcement of concrete is covered by BS 4483, which gives four basic preferred types:

- 'A' for square mesh (floor slabs);
- 'B' for rectangular structural mesh (also for slabs);
- 'C' for long mesh (roads and pavements); and
- 'D', a small-diameter wrapping mesh (binding concrete fireproofing).

The fabric is factory-made by welding or interweaving wires complying with the requirements of BS 4482 to form sheets with a length of 4.800 m and a width of 2.400 m, or rolls 48.000 and 72.000 m long with a common width of 2.400 m. Each type has a letter prefix, which is followed by a reference number, which is the total cross-sectional area of main bars in mm² per metre width. Typical examples of reinforcing bars and fabric are shown in Fig. 10.2.1.

Before placing reinforcement into the formwork it should be brushed free of all loose rust and mill scale, and be free of grease or oil, as the presence of any

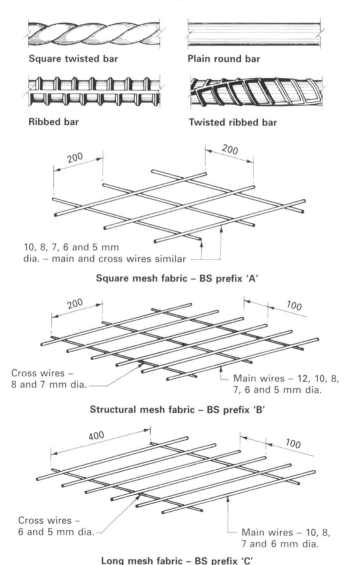

Square twisted bar

Plain round bar

Ribbed bar

Twisted ribbed bar

200

200

10, 8, 7, 6 and 5 mm
dia. – main and cross wires similar

Square mesh fabric – BS prefix 'A'

200

100

Cross wires –
8 and 7 mm dia.

Main wires – 12, 10, 8,
7, 6 and 5 mm dia.

Structural mesh fabric – BS prefix 'B'

400

100

Cross wires –
6 and 5 mm dia.

Main wires – 10, 8,
7 and 6 mm dia.

Long mesh fabric – BS prefix 'C'

Figure 10.2.1 Typical reinforcing bars and welded fabric.

of these on the surface could reduce the bond and hence the strength of the reinforced concrete. Reinforcement must have a minimum cover of concrete to give the steel protection from corrosion due to contact with moisture, and to give the structural member a certain degree of fire resistance. The cover to the steel needs to be maintained to the minimum value set out on the reinforced concrete detailing drawings. The value of the cover varies according to the degree of exposure of the finished concrete (see Part 3 Substructure on pages 82–83). The Building Regulations, Approved Document B, establishes the minimum fire resistance for various groups of buildings. Note that the dimensions of structural members are also of importance, to avoid the failure of the concrete due to the high temperatures encountered during a fire before the reinforcement reaches its critical temperature.

To maintain the right amount of concrete cover during construction, small blocks of concrete may be placed between the reinforcement and the formwork; alternatively, plastic clips or spacer rings can be used. Where top reinforcement has to be retained in position (as in a slab), cradles or chairs made from reinforcing bar may have to be used. All forms of spacers must be of a material that will not lead to corrosion of the reinforcement or cause spalling of the hardened concrete.

MIXING, BATCHING, HANDLING AND PLACING CONCRETE

In simple DIY projects and small domestic projects involving small quantities, like shed bases and post holes, batching is done 'by volume'. However, if the fine aggregate is damp or wet, its volume will increase by up to 25 per cent. This increase in volume, called **bulking**, makes the batching and hence the water/cement ratio unreliable. In most professional projects, batching is done by 'by mass' as this is the far more accurate method of achieving the correct strength. This method involves the use of a balance to give the exact mass of the materials as they are placed in the scales. This method offers greater accuracy, and the balance can be attached to the mixing machine.

HAND MIXING

This should be carried out on a clean, hard surface. The materials should be thoroughly mixed in the dry state before the water is added. The water should be added slowly, preferably using a rose head, until a uniform colour is obtained.

MACHINE MIXING

The mix should be turned over in the mixer for at least two minutes after adding the water. The first batch from the mixer tends to be harsh, because some of the mix will adhere to the sides of the drum. This batch should be used for some less important work, such as filling in weak pockets in the bottom of the excavation.

READY-MIXED

This is used for large batches, with lorry transporters of up to 6 m³ capacity. It has the advantage of eliminating site storage of materials and mixing plant, with the guarantee of concrete manufactured to quality-controlled standards. Placement is usually direct from the lorry via a fold-out chute that can extend up to 6 m from the back of the lorry. When using ready-mixed, site-handling facilities must be co-ordinated with deliveries, which may include the use of concrete skips manoeuvred by crane or concrete pumps facilitated by stout hydraulic hoses.

HANDLING AND PLACING

If concrete is to be transported for some distance over rough ground, the runs should be kept as short as possible, because vibrations of this nature can cause segregation of the materials in the mix. For the same reason, concrete should not be dropped from a height of more than 1 m. If this is unavoidable, a chute should be used.

If the concrete is to be placed in a foundation trench, it will be levelled from peg to peg; if it is to be used as an oversite slab, the external walls could act as a levelling guide. The levelling is carried out by tamping with a straight-edge board; this tamping serves the dual purpose of compacting the concrete and bringing the excess water to the surface so that it can evaporate. Concrete must not be over-tamped, as this will bring not only the water to the surface but also the cement paste that is required to act as the matrix. Concrete should be placed as soon as possible after mixing to ensure that the setting action has not commenced. Concrete that dries out too quickly will not develop its full strength: therefore new concrete should be protected from the drying winds and sun by being covered with damp canvas or polythene sheeting, or being regularly sprayed with water. This protection should be continued for at least three days, because concrete takes about 28 days to obtain its working strength.

OTHER CONSIDERATIONS

DESIGN

The design of reinforced concrete is the prerogative of the structural engineer, and is a subject for special study, but the technologist should have an understanding of the principles involved. By assessing the possible dead and live loads on a structural member, the designer can calculate the reactions and effects that such loadings will have on the member. This will indicate areas of tensile stress and critical locations for reinforcement. Calculations are based on the recommendations and design tables in Eurocode 2: Design of concrete structures BS EN 1992, together with established formulae for determing bending moments, shear forces and the area of steel required. Fig. 10.2.2 shows typical examples of these forces for simple situations.

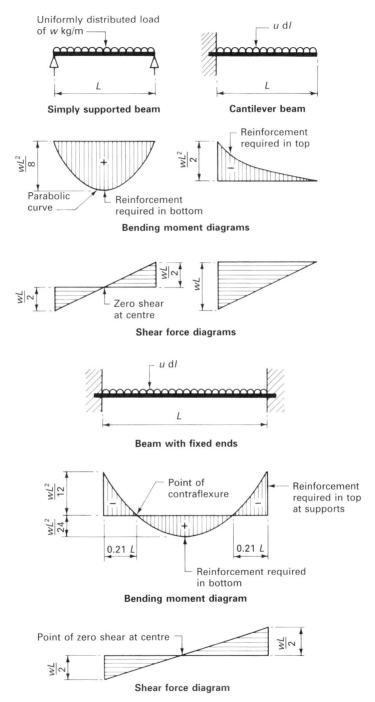

Figure 10.2.2 Bending moment and shear forces.

REINFORCEMENT SCHEDULES AND DETAILS

Once the engineer has determined the reinforcement required, detail drawings can be prepared to give the contractor the information required to construct the structure. The drawings should give the following information:

- sufficient cross-reference to identify the member in relationship to the whole structure;
- all the necessary dimensions for design and fabrication of formwork;
- details of the reinforcement;
- minimum cover of concrete over reinforcement;
- concrete grade required if not already covered in the specification.

Reinforced concrete details should be prepared so that there is a clear distinction between the lines representing the outline of the member and those representing the reinforcement. Bars of a common diameter and shape are normally grouped together with the same reference number when included in the same member. To simplify the reading of reinforced concrete details it is common practice to show only one bar of each group in full, together with the last bar position (see Fig. 10.2.3).

The bars are normally bent and scheduled in accordance with the recommendations of BS 8666. These give details of the common bending shapes, the method of setting out the bending dimensions, the method of calculating the total length of the bar required, and a shape code for use with data-processing routines. A preferred form of bar schedule is also given, which has been designed to give easy cross-reference to the detail drawing.

Reinforcement on detail drawings is annotated by a coding system to simplify preparation and reading of the details: for example, 9 H 12-01-300, which can be translated as:

9 = total number of bars in the group
H = high-yield steel ribbed bar (460 N/mm^2)
12 = diameter in mm
01 = bar mark number
300 = spacing centre to centre.

A typical reinforced concrete beam detail and schedule is shown in Fig. 10.2.3. All other reinforced concrete details shown in this volume are intended to show patterns of reinforcement rather than detailing practice, and are therefore shown in full without reference to bar diameters and types.

HOOKS, BENDS AND LAPS

To prevent bond failure, bars should be extended beyond the section where there is no stress in the bar. The length of bar required will depend upon such

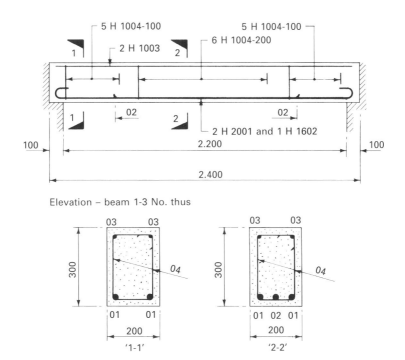

Elevation – beam 1-3 No. thus

Note: Cover to main bars 25 mm

Member	Bar mark	Type and size	No. of mbrs	No. in each	Total no.	Length of each bar[†]	Shape. All dimensions* are in accordance with BS 8666
Beam 1	1	H20	3	2	6	2.660	2.300
	2	H16	3	1	3	1.400	Straight
	3	H10	3	2	6	2.300	Straight
	4	H10	3	16	48	1.000	250 / 150

[†] Specified to nearest 25 mm   * Specified to nearest 5 mm

Figure 10.2.3   Typical reinforced concrete beam details and schedule.

factors as grade of concrete, whether the bar is in tension or compression, and whether the bar is deformed or plain. Hooks and bends can be used to reduce this anchorage length at the ends of bars, and should be formed in accordance with the recommendations of BS 8666: *Scheduling, dimensioning, bending and cutting of steel reinforcement for concrete. Specification* (see Fig. 10.2.4).

Where a transfer of stress is required at the end of a bar, the bars may be welded or lapped.

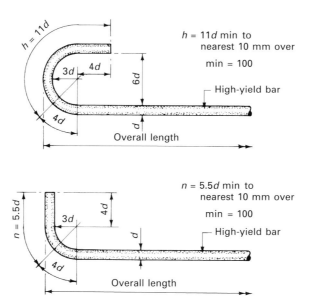

Figure 10.2.4   **Standard hooks and bends.**

REINFORCED CONCRETE BEAMS

Beams can vary in the complexity of their design and reinforcement, from the very simple beam formed over an isolated opening, such as those shown in Figs. 10.2.3 and 10.2.5, to the more common form encountered in frames, where the beams transfer their loadings to the columns (see Fig. 10.2.6).

When tension is induced in a beam the fibres will lengthen until the ultimate tensile strength is reached, when cracking and subsequent failure will occur. With a uniformly distributed load the position and value of tensile stress can easily be calculated by the structural engineer, but the problem becomes more complex when heavy point loads are encountered.

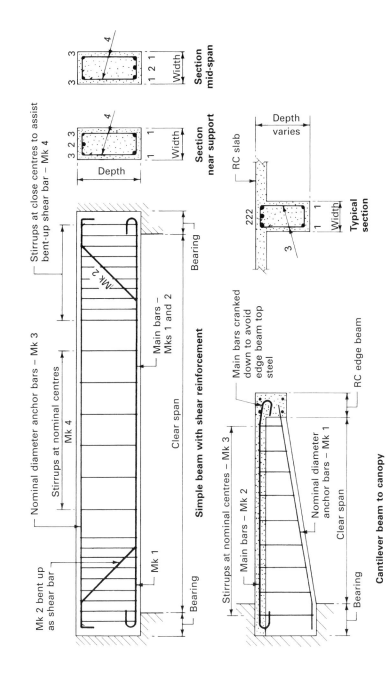

Section mid-span

Section near support

Stirrups at close centres to assist bent-up shear bar – Mk 4

Nominal diameter anchor bars – Mk 3

Stirrups at nominal centres Mk 4

Mk 2 bent up as shear bar

Mk 1

Bearing

Mk 2

Clear span

Main bars – Mks 1 and 2

Bearing

Simple beam with shear reinforcement

Depth varies

RC slab

222

Typical section

Main bars cranked down to avoid edge beam top steel

RC edge beam

Stirrups at nominal centres – Mk 3

Main bars – Mk 2

Nominal diameter anchor bars – Mk 1

Clear span

Bearing

Cantilever beam to canopy

Figure 10.2.5   Simple reinforced concrete beams.

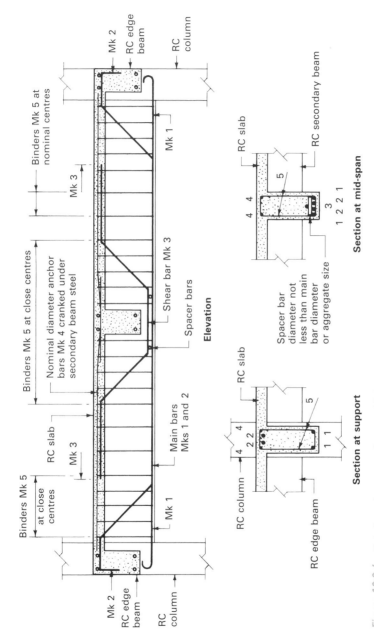

Elevation

Binders Mk 5 at close centres

Nominal diameter anchor bars Mk 4 cranked under secondary beam steel

Binders Mk 5 at nominal centres

Mk 2
RC edge beam

RC column

Mk 3

Mk 1

Shear bar Mk 3

Spacer bars

Binders Mk 5 at close centres

RC slab

Mk 3

Main bars Mks 1 and 2

Mk 1

Mk 2
RC edge beam

RC column

Section at support

RC slab

RC column

RC edge beam

Spacer bar diameter not less than main bar diameter or aggregate size

Section at mid-span

RC slab

RC secondary beam

Figure 10.2.6   Reinforced concrete beam with heavy reinforcement.

The correct design of a reinforced concrete beam will ensure that it has sufficient strength to resist both the compression and the tensile forces encountered in the outer fibres, but it can still fail in the 'web' connecting the compression and tension areas. This form of failure is called **shear failure** and is in fact diagonal tension. Concrete has a limited amount of resistance to shear failure, and if this is exceeded, reinforcement must be added to provide extra resistance. Shear occurs at or near the supports as a diagonal failure line, at an angle of approximately 45° to the horizontal and sloping downwards towards the support. A useful fact to remember is that zero shear occurs at the point of maximum bending (see Fig. 10.2.2).

Reinforcement to resist shearing force may be either links, stirrups or inclined bars, or all of these. Links wrap around the main bars in the reinforcement cage, shown as 'bar mark 4' in the table in Fig. 10.2.3; stirrups are essentially 'U'-shaped and have an open top to the cage to allow more main bars to be dropped down inside the main cage. The total shearing resistance is the sum of the shearing resistances of the inclined bars and the stirrups, calculated separately if both are provided. Inclined or bent-up bars should be at 45° to the horizontal and positioned to cut the anticipated shear failure plane at right angles. These may be separate bars, or main bars from the bottom of the beam that are no longer required to resist tension, and that can be bent up and carried over or on to the support to provide the shear resistance (see Figs. 10.2.5 and 10.2.6). Stirrups or links are provided in beams, even where not required for shear resistance, to minimise shrinkage cracking and to form a cage for easy handling. The nominal spacing for stirrups or links must be such that the spacing dimension used is not greater than the lever arm of the section, which is 0.75 times the effective depth of the beam, measured from the top of the beam to the centre of the tension reinforcement in the bottom of the beam. The spacings of the stirrups or links are closed up in areas where there are high shear stresses, which is usually close to the beam supports (see Figs. 10.2.5 and 10.2.6).

REINFORCED CONCRETE COLUMNS

A column is a vertical member carrying the beam and floor loadings to the foundation, and is a compression member. As concrete is strong in compression it may be concluded that, provided the compressive strength of the concrete is not exceeded, no reinforcement will be required. For this condition to be true, the loading must be truly axial and the column must be short: that is, there must be no vertical bending or buckling in the column as defined by BS EN 1992 Eurocode 2: *Design of concrete structures*.

These conditions rarely occur in framed buildings as beams always connect eccentrically to the exact centre of the column: consequently bending is induced and the need for reinforcement to provide tensile strength is apparent. Bending in columns may be induced by one or more of the following conditions:

- load coupled with the slenderness of the column;
- reaction to beams upon the columns – as the beam deflects, it tends to pull the column towards itself, thus inducing tension in the far face;
- the reaction of the frame to wind-loadings, both positive and negative.

All bars in compression should be tied by a link passing around the bar in such a way that it tends to move the bar towards the centre of the column; typical arrangements are shown in Fig. 10.2.7.

Where the junction between beams and columns occurs there could be a clash of steel, as bars from the beam may well be in the same plane as bars in the columns. To avoid this situation, one group of bars must be offset or cranked into another plane. It is generally considered that the best practical solution is to crank the column bars to avoid the beam steel: typical examples of this situation, together with a method using straight bars, are shown in Fig. 10.2.8. A similar situation can occur where beams of similar depth intersect: see the cantilever beam example in Fig. 10.2.5.

REINFORCED CONCRETE SLABS

A reinforced concrete slab will behave in exactly the same manner as a reinforced concrete beam, and it is therefore designed in the same manner. The designer will analyse the loadings, bending moments, shear forces and reinforcement requirements on a slab strip 1.000 m wide. In practice, the reinforcement will be fabricated to form a continuous mat. For light loadings a mat of welded fabric could be used.

There are three basic forms of reinforced concrete slab:

- flat slab floors or roofs;
- beam and slab floors or roofs;
- pre- or post-tensioned slabs – these will be dealt with in *Advanced Construction Technology*.

Flat slabs

These are basically slabs contained between two plain surfaces, and can be either simple or complex. The design of the complex form is based on the slab acting as a plate in which the slab is divided into middle and column strips, the reinforcement being concentrated in the latter strips.

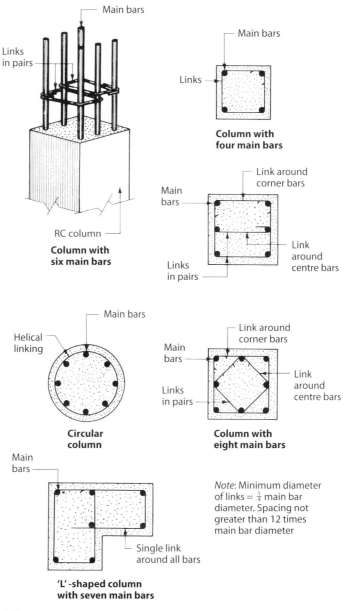

Main bars

Links
in pairs

RC column

**Column with
six main bars**

Main bars

Links

**Column with
four main bars**

Link around
corner bars

Main
bars

Link
around
centre bars

Links
in pairs

Helical
linking

Main bars

Link around
corner bars

Main
bars

Link
around
centre bars

Links
in pairs

**Circular
column**

**Column with
eight main bars**

Main
bars

Single link
around all bars

**'L'-shaped column
with seven main bars**

Note: Minimum diameter
of links $= \frac{1}{4}$ main bar
diameter. Spacing not
greater than 12 times
main bar diameter

Figure 10.2.7   Typical reinforced concrete column binding arrangements.

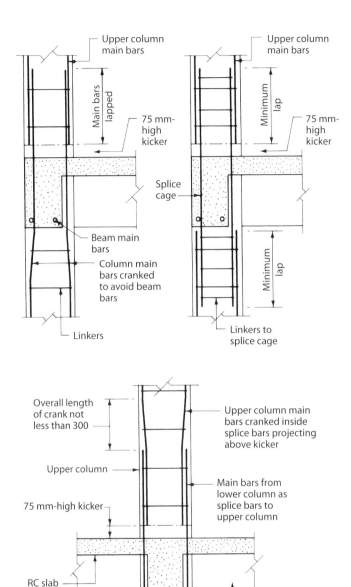

Figure 10.2.8   Reinforced concrete column and beam junctions.

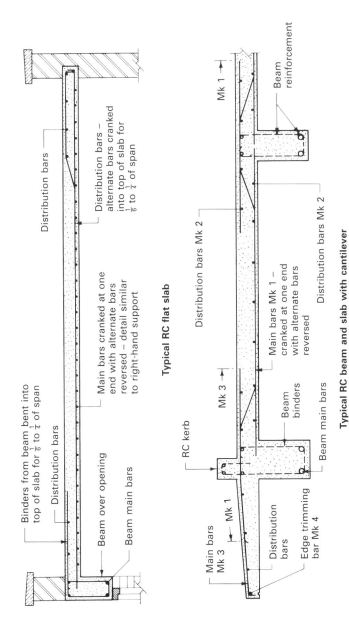

Distribution bars

Distribution bars – alternate bars cranked into top of slab for $\frac{1}{6}$ to $\frac{1}{4}$ of span

Main bars cranked at one end with alternate bars reversed – detail similar to right-hand support

Typical RC flat slab

Binders from beam bent into top of slab for $\frac{1}{6}$ to $\frac{1}{4}$ of span

Distribution bars

Beam over opening

Beam main bars

Beam reinforcement

Mk 1

Distribution bars Mk 2

Main bars Mk 1 – cranked at one end with alternate bars reversed

Distribution bars Mk 2

RC kerb

Mk 3

Beam binders

Beam main bars

Main bars Mk 3

Mk 1

Distribution bars

Edge trimming bar Mk 4

Typical RC beam and slab with cantilever

Figure 10.2.9 Typical reinforced concrete slab details.

Simple flat slabs can be thick and heavy, but have the advantage of giving clear ceiling heights as there are no internal beams. They are generally economic up to spans of approximately 9.000 m, and can be designed to span one way – that is, across the shortest span – or to span in two directions. These basic slabs are generally designed to be simply supported: that is, there is no theoretical restraint at the edges, and therefore tension is not induced and reinforcement is not required. However, it is common practice to provide some top reinforcement at the supports as anti-crack steel should there, in practice, be a small degree of restraint. Generally this steel is 50 per cent of the main steel requirement and extends into the slab for 0.2 m of the span. An economic method is to crank up 50 per cent of the main steel or every alternate bar over the support, as the bending moment will have reduced to such a degree at this point that it is no longer required in the bottom of the slab. If there is an edge beam, the top steel can also be provided by extending the beam binders into the slab (see Fig. 10.2.9).

Beam and slab

By adopting this method of design, large spans are possible and the reinforcement is generally uncomplicated. A negative moment will occur over the internal supports, necessitating top reinforcement; as with the flat slabs, this can be provided by cranked bars (see Fig. 10.2.9). Each bar is in fact cranked, but alternate bars are reversed, thus simplifying bending and identification of the bars. Alternatively, a separate mat of reinforcement supported on chairs can be used over the supports.

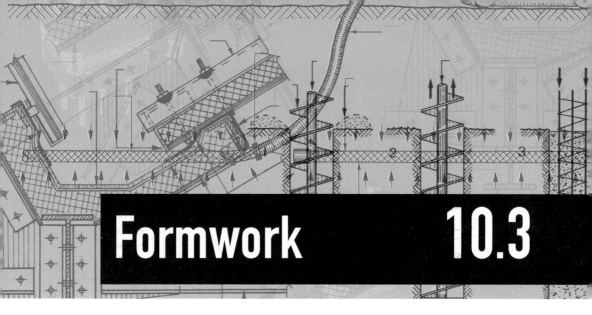

Formwork 10.3

Formwork for in-situ concrete work may be described as a mould or box into which wet concrete can be poured and compacted so that it will flow and finally set to the inner profile of the box or mould. It is important to remember that the inner profile must be opposite to that required for the finished concrete: so if, for example, a chamfer is required on the edge, a triangle fillet must be inserted into the formwork.

To be successful in its function, formwork must fulfil certain requirements.

- It should be strong enough to support the load of wet concrete, which is generally considered to be approximately 2,400 kg/m^3.

- It must not be able to deflect under load, which would include the loading of wet concrete, self-weight, and any superimposed loads, such as operatives and barrow runs over the formwork.

- It must be accurately set out; as concrete is a fluid when placed, it will take up the shape of the formwork, which must therefore be of the correct shape and size and be in the right position.

- It must have grout-tight joints. Grout leakage can cause honeycombing of the surface, or produce fins, which have to be removed. The making-good of defective concrete surfaces is both time-consuming and costly. Grout leakage can be prevented by using sheet materials and sealing the joints with flexible, foamed-polyurethane strip or by using a special self-adhesive tape.

- Form sizes should be designed so that they are the maximum size that can easily be handled, by hand or by a mechanical lifting device.

- Material must be chosen so that it can be easily fixed using double-headed nails, round-wire nails or wood screws. The common method is to use nails, and these should be of a length at least two-and-a-half times the thickness of the timber being nailed.

- The design of the formwork units should be such that they can easily be assembled and dismantled without any members being trapped.

The requirements for formwork described above make timber the most suitable material for general formwork. It can be of board form, either wrot or unwrot depending on whether a smooth or rough surface is required. Formwork is the material that makes up the mould of the cross-section, while falsework is the array of temporary props and braces that supports the forms.

Softwood boards used to form panels for beam and column sides should be joined together by cross-members over their backs, at centres not exceeding 24 times the board's thickness.

The moisture content of the timber should be between 15 and 20 per cent, so that the moisture movement of the formwork is reduced to a minimum. If the timber is dry it will absorb moisture from the wet concrete, which could weaken the resultant concrete member. It will also cause the formwork to swell and bulge, which could give an unwanted profile to the finished concrete. If timber with a high moisture content is used it will shrink and cup, which could result in open joints and a leakage of grout.

Plywood is extensively used to construct formwork units as it is strong, light and supplied in sheets of 1.200 m wide with standard lengths of 2.400, 2.700 and 3.000 m. The quality selected should be an exterior grade, and the thickness should be related to the anticipated pressures so that the minimum number of strengthening cleats on the back is required.

Chipboard can also be used as a formwork material but, because of its lower strength, will require more supports and stiffeners. The number of uses that can be obtained from chipboard forms is generally less than plywood, softwood boarding or steel.

Steel forms are generally based on a manufacturer's patent system, and within the constraints of that system are an excellent material, which can be installed and dismantled much faster than traditional timber formwork elements. A great deal of technological development has gone into making these systems lightweight, flexible and adaptable, and most major reinforced concrete projects benefit greatly from the repetition and reuse of such forms. Timber can often only be reused up to a dozen times; however, steel, if treated with care, will give 30 or 40 uses, which is more than double that of similar timber forms. In modern formwork construction, system formwork or shuttering is primarily used, which consists of standard modular components. The formwork lining panels or forms can be marine-quality plywood, aluminum or steel-faced shutters, which can be fixed together to form a variety of shapes and sizes; depending on the type of system, these forms are clipped and bolted together to form a vast range of reinforced concrete cross-sections. Specialist companies will often provide a full structural design service to ensure the safety and stability of the formwork throughout the concreting process, and to maximise the reuse of forms without the need to reconfigure them. Another major advantage of using these systems is improved health and safety for operatives, as they can combine safe access and egress ladders, provision of working platforms and edge protection as part of the formwork and falsework system.

RELEASE AGENTS

The face of the formwork in contact with the concrete needs to be treated so that, when it is eventually removed, the concrete does not adhere to the formwork and damage the finish of the cast concrete. Using a release agent also improves the finished look of the concrete surface in terms of even colour-matching and the presence of tiny but noticeable air or blowholes in the surface.

There are four types of release agent:

- neat oils with surfactants – used on steel forms but can also be used on timber and plywood;
- mould cream emulsions – can be used on a variety of timber-based forms but should not be used on steel due to its water content, which may cause rust staining;
- chemical release agents – specially formulated compounds recommended for high-quality finished work for all types of steel and timber forms;
- vegetable oil-based release agents (VERAs) – used on timber forms, and more environmentally friendly because they are biodegradable, water-based products.

Two common defects that can occur on the surface of finished concrete for untreated formwork are blowholes and uneven colour.

- **Blowholes** These are small holes, less than 15 mm in diameter, caused by air being trapped between the formwork and the concrete face.
- **Uneven colour** This is caused by irregular absorption of water from the wet concrete by the formwork material. A mixture of old and new material very often accentuates this particular defect.

Generally oils will encourage blowholes but will discourage uneven colour, whereas an emulsion cream release agent will discourage blowholes and reduce uneven colouring. Chemical release agents are preferred where both an even colour and reduced blowholes are specified, but these can be more costly.

Release agents are prepared by the manufacturer to suit various conditions and applications. It is important that operatives follow the application instructions on the product health and safety data sheets, particularly with regard to COSHH Regulations for the product.

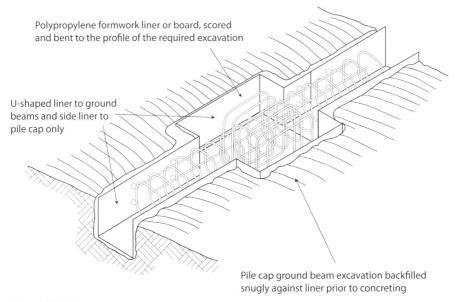

Polypropylene formwork liner or board, scored
and bent to the profile of the required excavation

U-shaped liner to ground
beams and side liner to
pile cap only

Pile cap ground beam excavation backfilled
snugly against liner prior to concreting

Figure 10.3.1   **Typical permanent formwork liner for excavations.**

FORMWORK LININGS

To obtain smooth, patterned or textured surfaces the inside of a form can be lined with various materials, such as oil-tempered hardboard, moulded rubber, moulded PVC or glass-fibre-reinforced polyester; the latter is also available as a complete form mould. When using any form of lining, the manufacturer's instructions regarding sealing, fixing and the use of mould oils must be strictly followed to achieve a satisfactory result.

TYPES OF FORMWORK

FOUNDATION FORMWORK

Foundations to a framed building generally consist of a series of isolated bases or pads, although if these pads are close together it may be more practicable to merge them together to form a strip. Where the soil is poor, deep piles with pile caps and ground beams may be specified, but whatever foundation is needed, the requirement of formwork needs to be considered.

If the subsoil is firm and hard, it may be possible to excavate the trench or pit for the foundations to the size and depth required and cast the concrete against the excavated faces. Where this method is not practicable, formwork will be required. There are two methods that can be used:

■ permanent formwork liners or boards;

■ traditional reusable timber or steel formwork.

Permanent formwork liners

There are various proprietary permanent formwork systems for reducing the amount of overdig and minimising concrete wastage for pad, strip, pile caps and ground beam foundations. A typical system uses strong 10 mm-thick twin-walled rigid corrugated polypropylene liner (see Fig. 10.3.1). Once the liner is in place the reinforcement cage is inserted, with clips to provide the cover to the steel reinforcing bars. When the liner and reinforcement is in the correct position, the soil is backfilled to counteract the pressure exerted by the wet concrete pour. After the concrete has been poured, the liner stays in place. The benefits of this type of system are:

- it reduces concrete waste;
- it requires less excavation than traditional shuttering;
- it is light but rigid so easily delivered, stored and handled;
- installation needs only a sharp knife and assembly does not need skilled labour;
- no shutter stripping, cleaning or release agents are needed;
- it can be adapted to suit pipes, cables and other services passing through the foundation.

Traditional formwork

If traditional reusable timber or steel forms are chosen then side and end panels will be required for the foundation. These can constructed of plywood panels with timber studs or runners to provide framing around the edges, or by using proprietary steel forms bolted together. The formwork panels should be firmly strutted against the excavation faces, to resist the horizontal pressures of the wet concrete and to retain the formwork in the correct position. Ties will be required across the top of the form as a top restraint, and these can be used to form the kicker for a reinforced concrete column or as a template for casting in the holding-down bolts for precast concrete or structural steel columns (see Fig. 10.3.2).

COLUMN FORMWORK

A column form or box consists of a vertical mould that has to resist considerable horizontal pressures in the early stages of casting. The column box should be located against a 75 mm-high plinth or kicker that has been cast monolithic with the base or floor. The kicker not only positions the formwork accurately but also prevents loss of grout from the bottom edge of the form.

Two column formwork systems are available:

- traditional timber formwork;
- proprietary steel formwork systems.

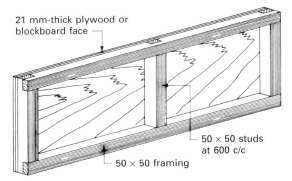

21 mm-thick plywood or
blockboard face

50 × 50 studs
at 600 c/c

50 × 50 framing

Typical framed formwork panel

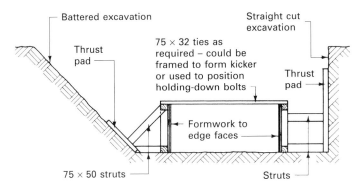

Battered excavation

Straight cut
excavation

Thrust
pad

75 × 32 ties as
required – could be
framed to form kicker
or used to position
holding-down bolts

Thrust
pad

Formwork to
edge faces

75 × 50 struts

Struts

Typical foundation formwork

Figure 10.3.2 **Formwork to foundations.**

For traditional timber formwork, the panels forming the column sides can be strengthened using horizontal cleats or vertical studs, which are sometimes called **soldiers**. The form can be constructed to the full storey-height of the column, with cut-outs at the top to receive the incoming beam forms. The thickness of the sides does not generally provide sufficient bearing for the beam boxes, so the cut-outs have a margin piece fixed around the opening to provide extra bearing (see Fig. 10.3.3). However, it is general practice to cast the columns up to the underside of the lowest beam soffit, and to complete the top of the column at the same time as the beam, using make-up pieces to complete the column and receive the beam intersections. The main advantage of casting full-height columns is the lateral restraint provided by the beam forms; the disadvantage is the complexity of the formwork involved.

Column forms are held together with collars of timber (called **yokes**) or metal (called **clamps**). Timber yokes are purpose-made, whereas steel column clamps are adjustable within the limits of the blades (see Fig. 10.3.3).

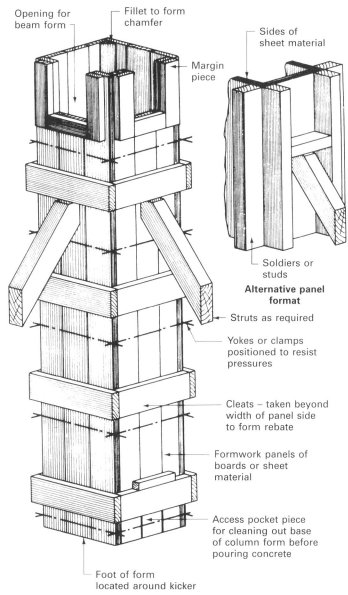

Opening for beam form

Fillet to form chamfer

Sides of sheet material

Margin piece

Soldiers or studs

Alternative panel format

Struts as required

Yokes or clamps positioned to resist pressures

Cleats – taken beyond width of panel side to form rebate

Formwork panels of boards or sheet material

Access pocket piece for cleaning out base of column form before pouring concrete

Foot of form located around kicker

Figure 10.3.3   Column formwork principles.

The spacing of the yokes and clamps should vary with the anticipated pressures, the greatest pressure occurring at the base of the column box. The actual pressure will vary according to:

- rate of placing;

- type of mix being used – generally, the richer the mix, the greater the pressure;

- method of placing – if vibrators are used, pressures can increase up to 50 per cent over placing and compacting by hand;

- air temperature – the lower the temperature, the slower the hydration process and consequently higher pressures are encountered.

Some preliminary raking strutting is required to plumb and align the column forms in all situations. Free-standing columns will need permanent strutting until the concrete has fully cured and strengthened, but with tied columns the need for permanent strutting must be considered for each individual case.

Shaped columns will need special yoke arrangements unless they are being formed using a patent system. Typical examples of shaped column forms are shown in Fig. 10.3.4.

There are also proprietary systems of formwork such as developed by the PERI company. Fig. 10.3.5 shows a typical column form known as 'trio', provided by PERI Ltd.

BEAM FORMWORK

A beam form consists of a three-sided box, which is supported by cross-members called **headtrees**, propped to the underside of the soffit board. In the case of framed buildings, support to the beam box is also provided by the column form. The soffit board should be thicker than the beam sides, because this member will carry the dead load until the beam has gained sufficient strength to be self-supporting. Soffit boards should be fixed inside the beam sides so that the latter can be removed at an early date: this will enable a flow of air to pass around the new concrete and speed up the hardening process, and will also release the formwork for reuse at the earliest possible time. Generally the beam form is also used to support the slab formwork, and the two structural members are then cast together. The main advantage of this method is that only one concrete operation is involved, although the complexity of the formwork is increased. If the beams and slabs are carried out as separate operations, there is the possibility of a shear plane developing between the beam and floor slab; it would be advisable to consult the engineer before adopting this method of construction. An example of a traditional timber beam form is shown in Fig. 10.3.6.

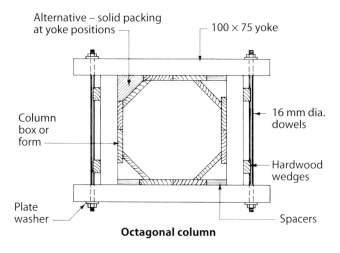

Alternative – solid packing at yoke positions

100 × 75 yoke

Column box or form

16 mm dia. dowels

Hardwood wedges

Plate washer

Spacers

Octagonal column

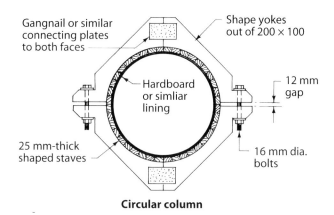

Gangnail or similar connecting plates to both faces

Shape yokes out of 200 × 100

Hardboard or simliar lining

12 mm gap

25 mm-thick shaped staves

16 mm dia. bolts

Circular column

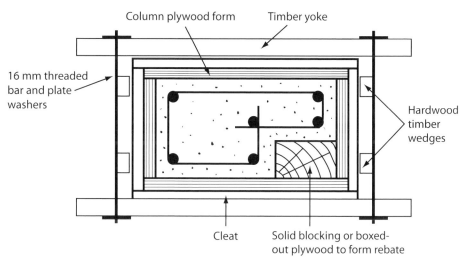

Column plywood form

Timber yoke

16 mm threaded bar and plate washers

Hardwood timber wedges

Cleat

Solid blocking or boxed-out plywood to form rebate

Figure 10.3.4   Shaped column forms and yokes.

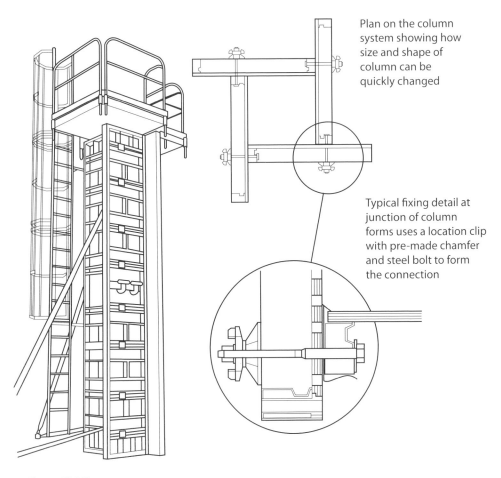

Plan on the column system showing how size and shape of column can be quickly changed

Typical fixing detail at junction of column forms uses a location clip with pre-made chamfer and steel bolt to form the connection

Figure 10.3.5   Proprietary 'PERI' system column formwork.

SLAB FORMWORK

Floor or roof slab formwork consists of panels of a size that can be easily handled. In small-scale developments, such as a two-storey shop unit or accommodation block, it is usually more cost-effective to use traditional timber formwork. In traditional timber formwork, plywood panels can be framed or joisted and are supported on timber bearers on intermediate steel props. The support of the horizontal panels is usually provided by the falsework to the supporting permanent floor beams (see Fig. 10.3.7). Adjustment for levelling purposes can be carried out using small timber folding wedges between the joists or framing and the beam box.

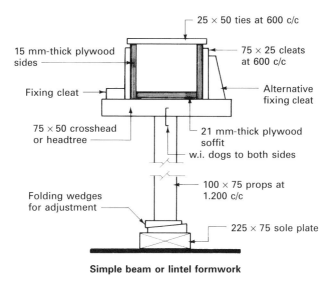

25 × 50 ties at 600 c/c

15 mm-thick plywood sides

75 × 25 cleats at 600 c/c

Fixing cleat

Alternative fixing cleat

75 × 50 crosshead or headtree

21 mm-thick plywood soffit

w.i. dogs to both sides

Folding wedges for adjustment

100 × 75 props at 1.200 c/c

225 × 75 sole plate

Simple beam or lintel formwork

21 mm-thick plywood soffit

15 mm-thick plywood beam sides

75 × 32 strut

75 × 50 cleat

Brace

75 × 32 runner or stringer

150 × 50 soffit support joists at 600 c/c

150 × 75 props at 1.200 c/c on folding wedges and sole plate

100 × 75 crosshead or headtree

75 × 32 brace

Edge beam and slab formwork

Figure 10.3.6 Typical timber beam formwork.

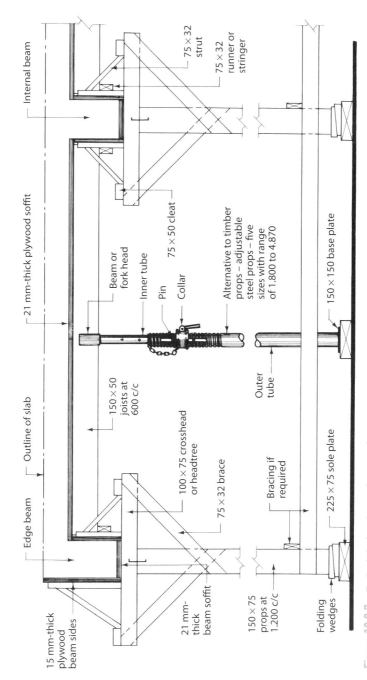

Internal beam

75 × 32 strut

75 × 32 runner or stringer

21 mm-thick plywood soffit

Beam or fork head

Inner tube

Pin

Collar

75 × 50 cleat

Alternative to timber props – adjustable steel props – five sizes with range of 1.800 to 4.870

150 × 150 base plate

Outline of slab

Edge beam

150 × 50 joists at 600 c/c

100 × 75 crosshead or headtree

75 × 32 brace

Outer tube

225 × 75 sole plate

15 mm-thick plywood beam sides

21 mm-thick beam soffit

150 × 75 props at 1.200 c/c

Bracing if required

Folding wedges

Figure 10.3.7 Typical timber beam formwork.

In multi-storey framed buildings where there are large areas of reinforced concrete floor slab to construct, the formwork will typically be a proprietary system comprising steel props and composite aluminium and steel formwork decking (see Fig. 10.3.8). As the floors progress upwards, it is important to be able to strike the formwork as early as possible to use on the floor above; however, floors will need to be 'back-propped', with the props left in place to support the construction of the upper floors. To achieve this, there are systems which incorporate a 'drophead' detail at the top of the steel props. This allows the slab formwork plates to be lowered and disengaged from the prop heads while still providing support of the slab at the propping location. The advantage of this is that, once the slab has gained initial strength, the slab formwork plates can be taken down and reused while the props remain in position until full strength of the slab has been achieved. As with most proprietary formwork systems, safe methods of working are usually incorporated via the use of cantilevered working platforms and edge protection barriers, to ensure the safety of operatives during the concreting operations.

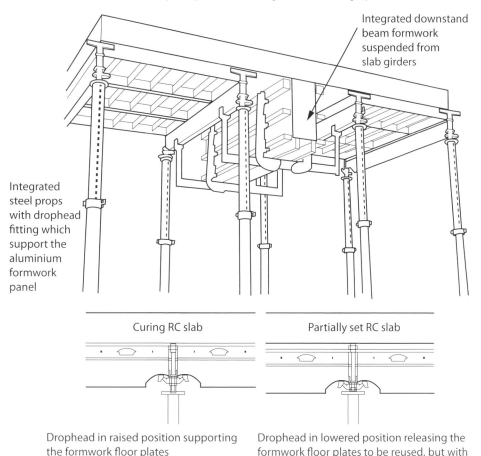

Integrated downstand beam formwork suspended from slab girders

Integrated steel props with drophead fitting which support the aluminium formwork panel

Curing RC slab

Partially set RC slab

Drophead in raised position supporting the formwork floor plates

Drophead in lowered position releasing the formwork floor plates to be reused, but with the vertical prop still in position

Figure 10.3.8 Proprietary 'PERI skydeck' system slab formwork.

When the formwork has been fabricated and assembled, the interior of the forms should be cleared of all rubbish and dirt. This is usually done by blowing the debris with an air hose clear of the form. Operatives must take care when carrying out this work and should wear appropriate goggles and mask. Forms should also be inspected for signs of grease, as this can affect the finish of the concrete. All joints and holes should be checked to ensure that they are grout-tight, and any gaps should be sealed with tape. The setting-out engineers also need to check that any cast-in bolts are correctly positioned and that there are polystyrene or cardboard bolt cones in place to allow for fixing tolerances.

Great care must be taken when applying some mould oils or emulsions, because over-oiling may cause some retardation of the setting of the cement. Emulsions should not be used in conjunction with steel forms as they encourage rusting because of the water content in the emulsion. Note that, generally, mould oils and emulsions also act as release agents, and therefore it is essential that the oil or emulsion is applied only to the formwork and not to the reinforcement, as this may cause a reduction of bond.

The distance from the mixer to the formwork should be kept as short as possible to maintain the workability of the mix and to avoid double-handling as far as practicable. Care must be taken when placing and compacting the concrete to ensure that the reinforcement is not displaced. The depth of concrete that can be placed in one lift will depend upon the mix and section size. If vibrators are used as the means of compaction, this should be continuous during the placing of each batch of concrete until the air expulsion has ceased. Care must be taken, as over-vibrating concrete can cause segregation of the mix.

The method of curing the concrete will depend on climatic conditions, type of cement used, and the average temperature during the curing period. The objective is to allow the concrete to cure and obtain its strength without undue distortion or cracking. It may be necessary to insulate the concrete by covering it with polythene sheeting or an absorbent material, kept damp to control the surface temperature and prevent the evaporation of water from the surface. Under normal conditions, using ordinary Portland cement and with an average air temperature of over 10 °C, this period would be two days, rising to four days during hot weather and days with prolonged drying winds.

The striking or removal of formwork should only take place on instruction from the engineer or agent. The appropriate time at which it is safe to remove formwork can be assessed by tests on cubes taken from a similar batch mixed at the time the concrete was poured and cured under similar conditions. The characteristic cube strength should be 10 N/mm^2, or twice the stress to which the structure will then be submitted, whichever is the greater, before striking the formwork. If test cubes are not available, Table 10.3.1 from the superseded BS 8110 can be considered as indicative where ordinary Portland cement is used.

Table 10.3.1 Approximate times for striking formwork.

Location	Surface or air temperature	
	16 °C	7 °C
Vertical formwork	12 hours	18 hours
Slab soffits (props left under)	4 days	6 days
Removal of props to slab	10 days	15 days
Beam soffits (props left under)	10 days	15 days
Removal of props to beam	14 days	21 days

In very cold weather the above minimum periods should be doubled, and when using rapid-hardening Portland cement the above minimum periods can generally be halved.

Formwork must be removed slowly, as the sudden removal of the wedges is equivalent to a shock load being placed on the partly hardened concrete. Materials and/or plant should not be placed on the partly hardened concrete without the engineer's permission. In timber formwork that has been struck, projecting nails can cause serious injuries; great care must be taken when handling these forms, and all nails should be removed or hammered down to avoid puncture injuries. When the formwork has been removed it should be carefully cleaned to remove any concrete adhering to the face before being reused. If the forms are not required for immediate reuse they should be carefully stored and stacked horizontally to avoid twisting. The stacked formwork should be covered with tarpaulins or plastic sheet in such a way as to provide both protection and ventilation.

Precast concrete frames 10.4

The overall concept of a precast concrete frame is the same as for any other framing material. Single- or multi-storey frames can be produced on the skeleton or box frame principle. Single- and two-storey buildings can also be produced as portal frames. Most precast concrete frames are produced as part of a 'system' building, and therefore it is only possible to generalise in an overall study of this method of framing.

The typical components of a precast concrete frame building are shown in Fig. 10.4.1. Note that, although the elements are made of precast concrete,

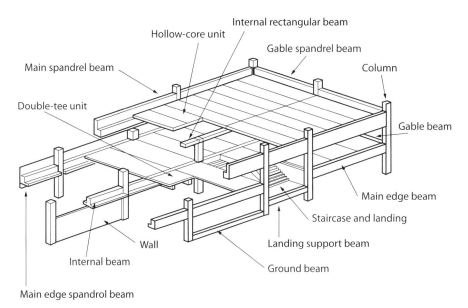

Figure 10.4.1   Precast concrete frame terminology.

the connections between the elements are generally made using in-situ reinforced concrete grouted connections. This ensures that the frame has rigid and braced lateral stability. The foundations for precast concrete frames are generally cast in-situ for the same reason.

Advantages

- Mixing, placing and curing of the concrete is carried out under factory-controlled conditions, which results in uniform and accurate units. The casting, being an 'off-site' activity, will release site space that would have been needed for the storage of cement and aggregates, mixing position, timber store and fabrication area for formwork and the storage, bending and fabrication of the reinforcement.

- Repetitive standard units reduce costs. It must be appreciated that the moulds used in precast concrete factories are precision-made, resulting in high capital costs. These costs must be apportioned over the number of units to be cast.

- Better quality control in the factory is possible in the mixing, placing, compacting and curing of the concrete, with less waste generated.

- Precast concrete can be erected quickly in a timeframe that is comparable to the use of steel-frame buildings. However, time must be allowed for the in-situ connections to be fully grouted at the joints.

- Frames can be assembled on site in cold weather, which helps with the planning, programming and progressing of the building operations. This is important to the contractor, because delays can result in the monetary penalty clauses, for late completion of the contract, being invoked.

- In general, the frames can be assembled using semi-skilled labour. With the high turnover rate of labour within the building industry, operatives can be recruited and quickly trained to carry out these activities.

Disadvantages

- System building is less flexible in its design concept than purpose-made structures. Note that there is a wide variety of systems available to the designer, so most design briefs can be fulfilled without too much modification to the original concept.

- Mechanical lifting plant will be needed to position the units. This can add to the overall contracting costs, as generally larger plant is required for precast concrete structures than for in-situ concrete structures.

- Programming may be restricted by controls on delivery and unloading times laid down by the police and local authority planning conditions. Restrictions on deliveries must be established at the tender period so that the tender programme can be formulated with a degree of accuracy, and the cost of any special requirements can be included in the unit rates for pricing.

- Structural connections between the precast concrete units can present both design and contractual problems. The major points to be considered are protection against weather, fire and corrosion, appearance and the method of construction. The latter should be issued as an instruction to site, setting out in detail the sequence, temporary supports required and full details of the joint.

METHODS OF CONNECTION

FOUNDATION CONNECTIONS

Precast columns are connected to their foundations by one of two methods, depending mainly on the magnitude of the load. For light and medium loads, the foot of the column can be placed in a pocket left in the foundation. The column can be plumbed and positioned by fixing a collar around its perimeter and temporarily supporting the column from this collar using raking adjustable props. Wedges can be used to give added rigidity while the column is being grouted into the pocket (see Fig. 10.4.2). The alternative method is to cast or weld on a base plate to the foot of the column and use holding-down bolts to secure the column to its foundation in the same manner as described in detail for structural steelwork (see Fig. 10.4.2).

COLUMN CONNECTIONS

The main principle involved in making column connections is to ensure continuity, and this can be achieved by a variety of methods. In simple connections a direct bearing and grouted dowel joint can be used, the dowel being positioned in the upper or lower column. Where continuity of reinforcement is required, the reinforcement from both upper and lower columns is left exposed and either lapped or welded together before completing the connection with in-situ concrete. A more complex method is to use a stud and plate connection, in which one set of threaded bars is connected through a steel plate welded to a set of bars projecting from the lower column; again the connection is completed with in-situ concrete. Typical column connections are shown in Fig. 10.4.3. Column connections should be made at floor levels but above the beam connections, a common dimension being 600 mm above structural floor level. The columns can be of single- or multi-storey height, the latter having provisions for beam connections at the intermediate floor levels.

BEAM CONNECTIONS

As with columns, the main emphasis is on continuity within the joint. Three basic methods are used:

- a projecting concrete haunch is cast on to the column with a locating dowel or stud bolt to fix the beam;
- a projecting metal corbel is fixed to the column, and the beam is bolted to the corbel;

- column and beam reinforcement, generally in the form of hooks, are left exposed. The two members are hooked together and covered with in-situ concrete to complete the joint.

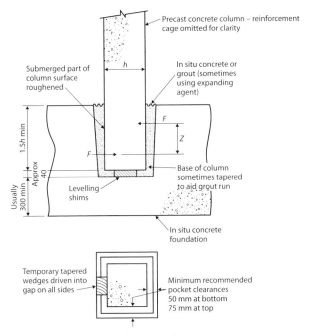

Precast concrete column – reinforcement cage omitted for clarity

In situ concrete or grout (sometimes using expanding agent)

Submerged part of column surface roughened

Base of column sometimes tapered to aid grout run

Levelling shims

In situ concrete foundation

1.5h min

Approx 40

Usually 300 min

Temporary tapered wedges driven into gap on all sides

Minimum recommended pocket clearances 50 mm at bottom 75 mm at top

Typical pocket base connection

Precast concrete column – reinforcement cage omitted for clarity

Base plate equal in size, or less than column size

Four corner pockets with anchor bars welded to base plate

Alternative detail

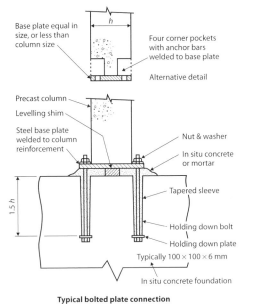

Precast column

Levelling shim

Steel base plate welded to column reinforcement

Nut & washer

In situ concrete or mortar

Tapered sleeve

1.5 h

Holding down bolt

Holding down plate Typically 100 × 100 × 6 mm

In situ concrete foundation

Typical bolted plate connection

Figure 10.4.2   **Precast concrete column to foundation connections.**

In all of these connections it is important that grout holes are cast in the precast member to enable a good grouted connection between the steel locating dowels or studs and the surrounding precast concrete.

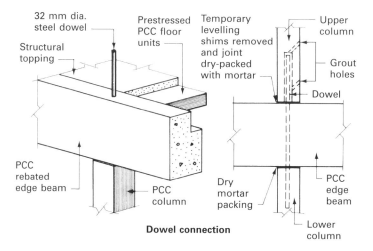

Dowel connection

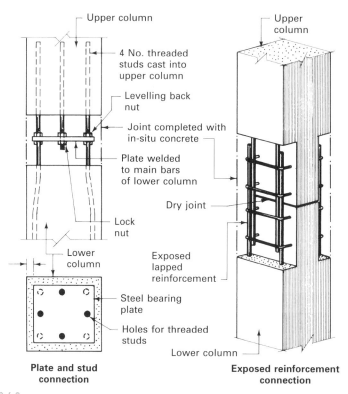

Plate and stud connection

Exposed reinforcement connection

Figure 10.4.3 Precast concrete column connections.

With most beam-to-column connections lateral restraint is assisted by leaving projecting reinforcement from the beam sides to bond with the floor slab or precast concrete floor units (see Fig. 10.4.4).

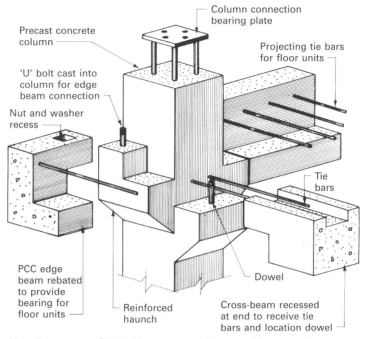

Note: Beam recess filled with cement grout to complete connection

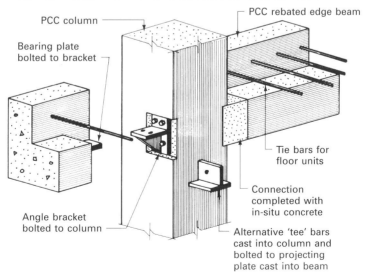

Figure 10.4.4   Typical precast concrete beam connections.

Structural steelwork frames 10.5

The design, fabrication, supply and erection of a structural steel frame are normally placed in the hands of a specialist subcontractor. The steelwork is usually referenced by a grid system that uniquely identifies each steel column and beam section. In this way the steel can be set out accurately, in the correct positions and to the correct levels.

The tolerances for steelwork can vary according to the type of external cladding system used, but normally +/- 10 mm over a distance of 10 m is acceptable for most low- to medium-rise structures. The foundation pads or bases for the steel columns are usually constructed by the specialist groundworks contractor or, on smaller projects, the main contractor. They must carefully position within the foundation excavation the necessary holding-down fixing bolts so that they can be cast solidly into the concrete when it is placed and compacted.

STRUCTURAL STEELWORK SECTIONS

The structural designer will analyse the structural elements according to the principles of limit state design. This is where the applied loads are multiplied by statistical safety factors, and capacities and resistances are determined using the design strength of the material, again with other statistical safety factors applied. The individual beam or column selected is then checked to see if it is safe or performs acceptably under a number of limiting states, beyond which the structure becomes unfit for its intended use. The main limiting state is of course checking the element will not collapse, but other limiting states include deflection, durability and fire resistance. This structural design is complex and now has to be done according to the relevant Eurocode – in this case Eurocode 3: *Design of steel structures BS EN 1993*. The standard hot-rolled sections selected are given in BS 4 and BS EN 10067 and in the *Handbook of Structural Steelwork* published jointly by the British Constructional Steelwork Association and the Steel Construction Institute. The basic types of standard hot-rolled sections are shown in Fig. 10.5.1 and are described below.

- **Universal beams** (UB) These are a range of sections supplied with tapered or parallel flanges, and are designated by their serial size × mass in kilograms per metre run. This means that within one 'serial size' there may be three or four different beams according to their different depths (D), breadth (B), flange thickness (T) and their subsequent mass per metre length – for example:

  '254 × 102' × 28 kg   where D = 260.4 mm, B = 102.1 mm, T = 10 mm;

  '254 × 102' × 25 kg   where D = 257.0 mm, B = 101.9 mm, T = 8.4 mm;

  '254 × 102' × 22 kg   where D = 254.0 mm, B = 101.6 mm, T = 6.8 mm.

  To facilitate the rolling operation of universal beam sections the inner profile is a constant dimension for any given serial size. The serial size as can be seen in inverted commas is therefore only an approximate width and breadth, and is given in millimetres. Generally, the heavier the mass per metre, the stronger the steel section.

- **Rolled Steel Joists** (RSJ) These are a range of small-size beams that have tapered flanges and are useful for lintels and small frames around openings. For joists, the serial size is the overall nominal dimension. They are mostly available in one mass per metre length, unlike UBs.

- **Universal columns** (UC) These members are rolled with parallel flanges, and are designated in the same manner as UBs. It is possible to design a column section to act as a beam (for example, in cases where headroom is an issue and shallow steel section is required); and, conversely, a beam section to act as a column, such as in portal frame columns that require a greater lateral strength to resist the horizontal thrusts from the portal rafters (see Part 10.6 on pages 536–538).

- **Channels** These are rolled with tapered flanges and designated by their nominal overall dimension × mass per metre run. They can be used for trimming and bracing members, or as a substitute for joist sections. Two channels bolted back to back can perform the same structural function as a UB, but are much lighter to install, which reduces the manual handling problems associated with lifting heavier beams. This type of substitution is important to comply with the design requirements of CDM regulations.

- **Angles** These are light framing and bracing sections with parallel flanges. The flange or leg lengths can be equal or unequal, and the sections are designated by the nominal overall leg lengths × nominal thickness of the flange.

- **T bars** These are used for the same purposes as angles, and are available as rolled sections with a short or long stalk; alternatively, they can be cut from a standard universal beam or column section. Designation is given by the nominal overall breadth and depth × mass per metre run.

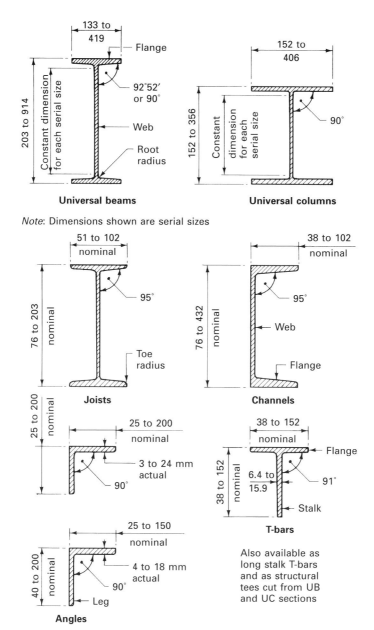

Universal beams

133 to 419

203 to 914 — Constant dimension for each serial size

Flange

92°52′ or 90°

Web

Root radius

Universal columns

152 to 406

152 to 356 — Constant dimension for each serial size

90°

Note: Dimensions shown are serial sizes

Joists

51 to 102 nominal

76 to 203 nominal

95°

Toe radius

Channels

38 to 102 nominal

76 to 432 nominal

95°

Web

Flange

Angles

25 to 200 nominal

25 to 200 nominal

3 to 24 mm actual

90°

T-bars

38 to 152 nominal

38 to 152 nominal

6.4 to 15.9

Flange

91°

Stalk

Also available as long stalk T-bars and as structural tees cut from UB and UC sections

25 to 150 nominal

40 to 200 nominal

4 to 18 mm actual

90°

Leg

Figure 10.5.1 Typical hot-rolled steel sections.

Cold-rolled steel sections to BS EN 10162 have become a popular alternative. They make more effective use of steel to permit selection of profiles more appropriate to the design loading. They are particularly useful in lightweight situations where standard hot-rolled sections would be excessive. These include prefabricated construction of modular units for 'slotting' into the main structural frame and for other factory-made units, such as prefabricated bathroom pods or larger whole-building modular construction.

Zed and Sigma sections have proved to be very successful alternatives to the traditional use of steel angles for truss purlins. They have many advantages, including:

- they are more cost-effective per unit of weight;
- deck fixings are shorter and simpler to locate around the upper flange;
- a shear centre is contained within the section to provide greater resistance to twisting.

Several examples of standard cold-rolled sections are shown in Fig. 10.5.2, with Zed and Sigma profiles in Fig. 10.5.3.

CASTELLATED UNIVERSAL SECTIONS

These are formed by flame-cutting a standard universal beam or column section along a castellated line; the two halves so produced are welded together to form an open-web beam. The resultant section is one and a half times the depth of the section from which it was cut (see Fig. 10.5.4). This increase in depth gives greater resistance to deflection without adding extra weight, but will reduce the clear headroom under the beams unless the overall height of the building is increased. Castellated sections are economical when used to support lightly loaded floor or roof slabs, and the voids in the web are convenient for housing service ducts and pipes within the structural floor space, as they can be fed through the pre-cut holes. With this form of beam, the shear stresses at the supports can be greater than the resistance provided by the web; in these cases, one or two voids are filled in by welding metal blanks into the voids.

CONNECTIONS

Connections in structural steelwork are classified as either shop connections or site connections, and can be made by using bolts or by welding. Riveted connections can be seen on many older structures, such as rail bridges, but are seldom specified for modern connections due to the cost, health and safety, and availability of more advanced forms of bolted or welded connection.

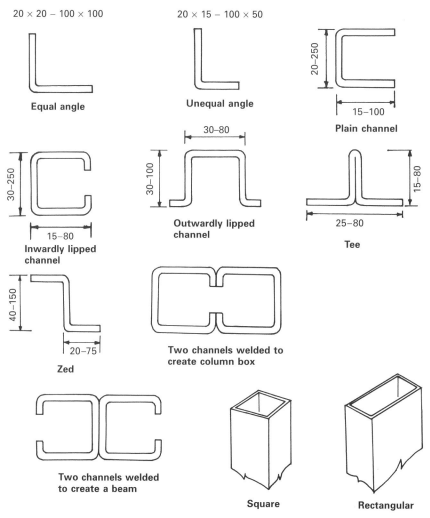

20 × 20 – 100 × 100

Equal angle

20 × 15 – 100 × 50

Unequal angle

20–250

15–100

Plain channel

30–250

15–80

Inwardly lipped channel

30–80

30–100

Outwardly lipped channel

25–80

15–80

Tee

40–150

20–75

Zed

Two channels welded to create column box

Two channels welded to create a beam

Square

Rectangular

Figure 10.5.2   Cold-rolled steel sections.

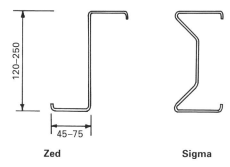

120–250

45–75

Zed

Sigma

Figure 10.5.3   Cold-rolled steel purlins.

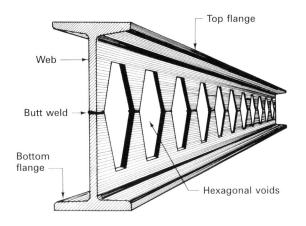

Top flange

Web

Butt weld

Bottom flange

Hexagonal voids

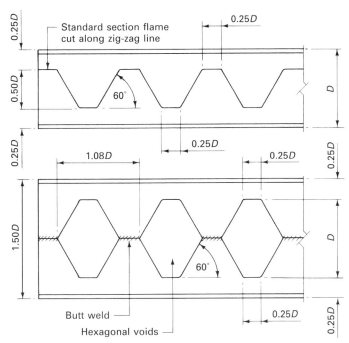

0.25D

0.50D

Standard section flame cut along zig-zag line

0.25D

60°

D

0.25D

0.25D

1.08D

0.25D

0.25D

1.50D

60°

D

Butt weld

Hexagonal voids

0.25D

0.25D

Note: Castellated joists, universal columns and zed sections also available

Figure 10.5.4   **Castellated beams.**

BOLTS

Ordinary or black bolts

These are the cheapest form of bolt available and can be either hot or cold forged, the thread being machined on to the shank. There are three grades of ordinary bolts – 4.6, 8.8 and 10.9 – all to BS EN ISO 4034. The two-figure reference given corresponds to the following classification: the first is one-hundredth of the minimum tensile stress in N/mm^2, and the second is the ratio between the minimum yield stress and the minimum tensile strength expressed as a percentage. For example, grade 8.8 bolt indicates a maximum tensile strength of 800 N/mm^2 and a yield stress of 640 N/mm^2 (0.8 x 800). The allowable shear stresses for grade 4.6 bolts are low, and therefore they should be used only for end connections of secondary beams or in conjunction with a seating cleat that has been designed and fixed to resist all the shear forces involved. The clearance in the hole for this form of bolt is usually specified as 2 mm over the diameter of the bolt. These bolts are manufactured to comparatively wide tolerances, with diameters ranging from 5 to 68 mm inclusive.

Precision or bright bolts

Sometimes called turned and fitted bolts, these have a machined shank and are therefore of greater dimensional accuracy, fitting into a hole with a small clearance allowance. However, these are not as common as the more widely available ordinary bolts.

Both precision bolts and ordinary bolts are generally used in non-preloaded situations: they are only hand-tightened with a spanner or long-handled podger.

High-strength friction-grip (HSFG) bolts

These are manufactured from high-tensile steels and are used in conjunction with high-tensile steel nuts and tempered washers. These bolts are very popular for both shop and site connections as fewer bolts are needed and hence the connection size is reduced. The object of this form of bolt is to tighten it to a predetermined shank tension in order that the clamping force thus provided will transfer the loads in the connecting members by friction between the parts and not by shear in, or bearing on, the bolts. Generally a torque-controlled spanner or pneumatic impact wrench is used for tightening or 'preloading' the bolt; other variations to ensure the correct torque are visual indicators, such as a series of pips under the head or washer, which are flattened when the correct amount of shank tension has been reached. HSFG bolts are produced as two grades: 'general grade', which is similar to the 8.8 grade, and the 'higher grade', which is comparable to the 10.9 grade ordinary bolt. Nominal standard diameters available are from 12 to 36 mm, with lengths ranging from 40 to 500 mm, as recommended in BS 4395-1 and 2.

General

The holes to receive bolts should always be drilled in a position known as the **back mark** of the section. The back mark is the position on the flange where the removal of material to form a bolt hole will have the least effect upon the section properties. The size of the hole is made 2 mm bigger for bolts up to 24 mm diameter, and 3 mm bigger for bolts larger than 24 mm in diameter. Actual dimensions and recommended bolt diameters are given in the *Handbook of Structural Steelwork*.

WELDING

Welding is considered primarily as a shop connection, as the cost, together with the need for inspection (which can be difficult on site), generally makes this method uneconomic for site connections. Offsite welded jointing provides a number of advantages, including:

- the opportunity for more innovative and aesthetic structural solutions;
- less weight than in bolted connections, because less end plates or seating cleats are required;
- protection against fire and corrosion, easily built into the design with welded connections;
- easy to introduce stiffening elements.

The basic methods of welding are oxyacetylene and electric arc. A blowpipe is used for oxyacetylene, which allows the heat from the burning gas mixture to raise the temperature of the surfaces to be joined. A metal filler rod is held in the flame, and the molten metal from the filler rod fuses the surfaces together. For most structural steel applications, the oxyacetylene method is limited to cutting.

In the alternative method, an electric arc is struck between a metal rod connected to a suitable low-voltage electrical supply and the surface to be joined, which must be earthed or resting on an earthed surface. The heat of the arc causes the electrode or metal rod to melt, and the molten metal can be deposited in layers to fuse the pieces to be joined together. With electric arc welding the temperature rise is confined to the local area being welded, whereas the oxyacetylene method causes a rise in metal temperature over a general area.

Welds are classified as either **fillet** or **butt** welds. Fillet welds are used on the edges and ends of members, and form a triangular fillet of welding material. Butt welds are used on chamfered end-to-end connections (see Fig. 10.5.5).

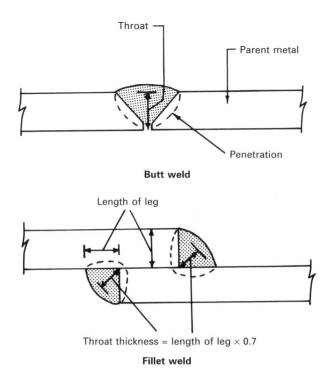

Throat

Parent metal

Penetration

Butt weld

Length of leg

Throat thickness = length of leg × 0.7

Fillet weld

Figure 10.5.5   Types of weld.

BASE CONNECTIONS

These take one of two forms: the base plate and the gusset base. In both methods a steel base plate is required to spread the load of the column on to the foundation. The end of the column and the upper surface of the base plate should be machined to give a good interface contact when using a base plate. The base plate and column can be connected together using cleats, or by fillet welding (see Fig. 10.5.6).

The gusset base is composed of a number of members that reduce the thickness of the base plate and can be used to transmit a high bending moment to the foundations. A machined interface between the column and base plate will enable all the components to work in conjunction with one another but, if this method is not adopted, the connections must transmit all the load to the base plate (see Fig. 10.5.6). The base is joined to the foundation by holding-down bolts, which must be designed to resist the uplift and tendency of the column to overturn. The bolt diameter, bolt length and size of the plate washer are therefore important. To allow for fixing tolerances the bolts are initially housed in a void or pocket, which is filled with grout at the same time as the base is grouted on to the foundation. To level and plumb the columns, steel wedges are inserted between the underside of the base plate and the top of the foundation (see Fig. 10.5.6).

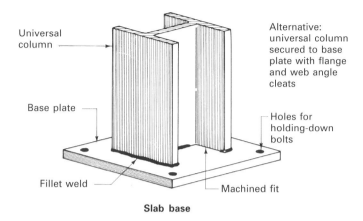

Universal column

Base plate

Fillet weld

Alternative: universal column secured to base plate with flange and web angle cleats

Holes for holding-down bolts

Machined fit

Slab base

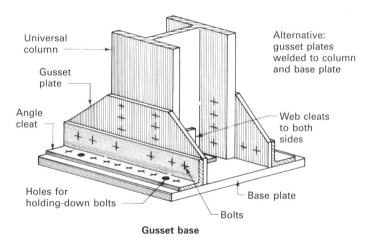

Universal column

Gusset plate

Angle cleat

Holes for holding-down bolts

Alternative: gusset plates welded to column and base plate

Web cleats to both sides

Base plate

Bolts

Gusset base

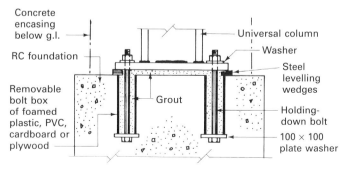

Concrete encasing below g.l.

RC foundation

Removable bolt box of foamed plastic, PVC, cardboard or plywood

Grout

Universal column

Washer

Steel levelling wedges

Holding-down bolt

100 × 100 plate washer

Figure 10.5.6 Structural steel column bases.

BEAM-TO-COLUMN CONNECTIONS

These can be designed as simple connections, where the whole of the load is transmitted to the column through a seating cleat. This is an expensive method, requiring heavy sections to overcome deflection problems. The usual method employed is the semi-rigid connection, where the load is transmitted from the beam to the column by means of top cleats and/or web cleats; for ease of assembly an erection cleat on the underside is also included in the connection detail (see Fig. 10.5.7). A fully rigid connection detail, which gives the greatest economy on section sizes, is made by welding the beam to the column (see Fig. 10.5.7). The uppermost beam connection to the column can be made by the methods described above; alternatively, a bearing connection can be used, which consists of a cap plate fixed to the top of the column to which the beams can be fixed, either continuously over the cap plate or with a butt joint (see Fig. 10.5.7).

COLUMN SPLICES

Column splices are generally needed where lengths in excess of 27.4 m are required, as any column over this length will have difficulty being transported independently on the public highway without a police escort. The use of splices enables sections of the same or different cross-sections to be bolted together once on site. Generally column splices are made at floor levels but above the beam connections at a height of 1 m, to allow for temporary edge protection to be fitted for safe access. The method used will depend upon the relative column sections (see Fig. 10.5.8).

BEAM-TO-BEAM CONNECTIONS

The method used will depend upon the relative depths of the beams concerned. Deep beams receiving small secondary beams can have a shelf angle connection, whereas other depths will need to be connected by web cleats (see Fig. 10.5.9) or end plates through which site bolts are fixed.

FRAME ERECTION

The primary reference document for erecting steel frames is the *Code of Practice for Erection of Multi-story Buildings* produced by the British Constructional Steelwork Association and endorsed by the Health and Safety Executive (HSE).

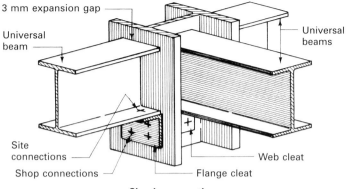

3 mm expansion gap

Universal beam

Universal beams

Site connections

Web cleat

Shop connections

Flange cleat

Simple connection

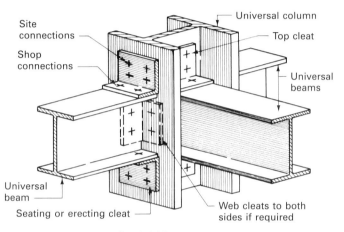

Universal column

Site connections

Top cleat

Shop connections

Universal beams

Universal beam

Web cleats to both sides if required

Seating or erecting cleat

Semi-rigid connection

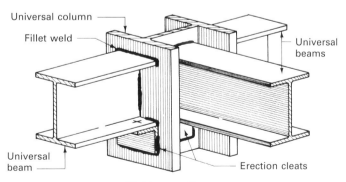

Universal column

Fillet weld

Universal beams

Universal beam

Erection cleats

Rigid connection

Figure 10.5.7 Structural steel beam to column connections.

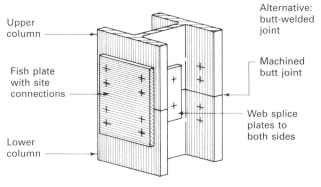

Upper column

Alternative: butt-welded joint

Fish plate with site connections

Machined butt joint

Web splice plates to both sides

Lower column

Columns with equal sections

Note: For columns of same serial size but of different sections splice is made using 4 No. fish plates fixed on the inside of flanges

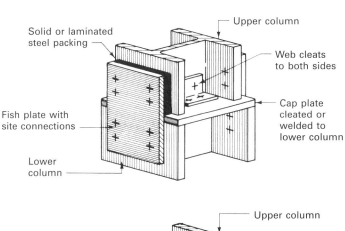

Solid or laminated steel packing

Upper column

Web cleats to both sides

Fish plate with site connections

Cap plate cleated or welded to lower column

Lower column

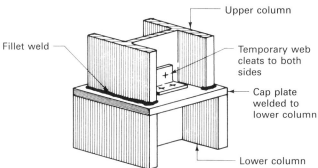

Upper column

Fillet weld

Temporary web cleats to both sides

Cap plate welded to lower column

Lower column

Alternative methods for columns of unequal sections

Note: Splices made at floor level but above beam connections

Figure 10.5.8   Structural steel column splices.

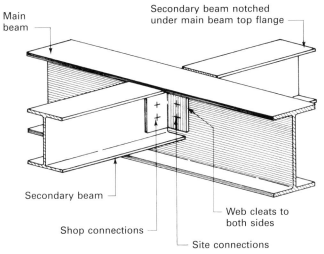

Main beam

Secondary beam notched under main beam top flange

Secondary beam

Shop connections

Web cleats to both sides

Site connections

Beam-to-beam connections

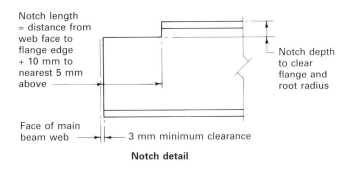

Notch length = distance from web face to flange edge + 10 mm to nearest 5 mm above

Notch depth to clear flange and root radius

Face of main beam web

3 mm minimum clearance

Notch detail

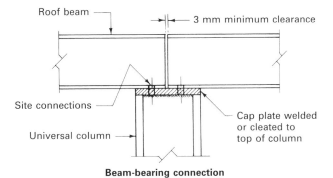

Roof beam

3 mm minimum clearance

Site connections

Universal column

Cap plate welded or cleated to top of column

Beam-bearing connection

Figure 10.5.9   Structural steel beam-to-beam connections.

Before any work is undertaken, the Management of Health and Safety Regulations 1999 require that a detailed risk assessment method statement is produced by a specialist steel frame contractor, and that it is checked by the main contractor. The risk assessment must clearly identify the hazards present and specify suitable control measures to reduce those risks to an acceptable level. This would include employing operatives with the correct training and qualifications, and that they are supervised by competent managers. The entire operation, from the delivery of the steel sections to the grouting of the base plates and de-propping, need to be meticulously planned and co-ordinated, with a clear method statement that is understood by the operatives who are carrying out the erection of the frame. Particular consideration needs to be given to the weather conditions: if windy, the sections will need carefully guying with tag lines. Although the safe workable wind speed varies from crane to crane, as a guide, the maximum wind speed at which lifting operations can take place is 10 metres per second (22 mph).

In certain cases, steel erection may need to be postponed until the weather improves.

Care must be taken to ensure that plant for lifting, handling or access is located on properly prepared ground. All lifting operations need to be carried out in accordance with the Lifting Operations and Lifting Equipment Regulations (LOLER) 1998 and BS 7121 *Code of practice for the safe use of cranes*. There should be sufficient hard-standing areas for the safe manoeuvring of either mobile cranes or mobile elevating working platforms (MEWPs). The bearing capacity beneath the crane outriggers needs to be sufficient to withstand the loads of the lifted steel sections. This includes positioning the outriggers safely away from buried services, pipes and inspection chambers. Furthermore, all steel sections must be correctly slung by a qualified slinger – an operative trained in the correct use of chains, hooks, webs and shackles – to suspend the load correctly in a stable position.

This erection of the frame will not normally be started until all the foundation bases have been cast and checked, because the structural steelwork contractor will need a clear site for manoeuvring the steel members into position. The steel beam or column sections will arrive on site each with a unique reference, based on the agreed structural grid, that clearly identifies the location and orientation of the section within the frame. A crane is used to lift the sections, which are then bolted into position by the steel erectors, who are on a safe working platform such as a mobile elevating working platform. A typical example (a Genie S-40) is shown in Fig. 10.5.10, together with its reach data sheet. Some of the larger MEWPs can reach up to 40 m high.

Edge protection, which provides a temporary barrier to prevent falling, is often prefixed to the high-level steel beams. This provides a safe and efficient method of installing edge protection, which benefits all the following trades as well as the steel erectors. In areas where these cannot be used, beam

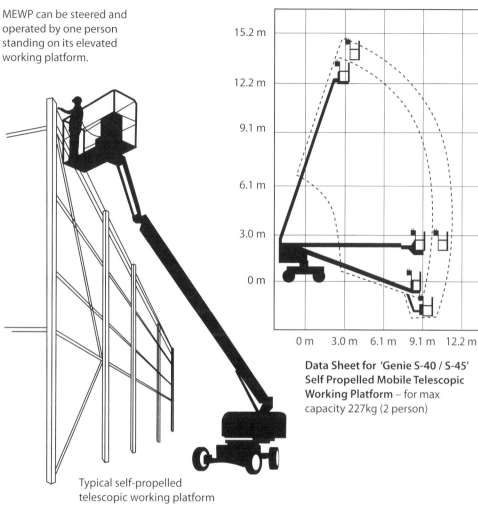

MEWP can be steered and operated by one person standing on its elevated working platform.

15.2 m

12.2 m

9.1 m

6.1 m

3.0 m

0 m

0 m   3.0 m   6.1 m   9.1 m   12.2 m

Data Sheet for 'Genie S-40 / S-45' Self Propelled Mobile Telescopic Working Platform – for max capacity 227kg (2 person)

Typical self-propelled telescopic working platform

Figure 10.5.10 **Operation and use of a typical MEWP.** Source: www.genielift.co.uk

straddling is an option. With beam straddling, the beam itself provides a working platform, with the steel erector in a harness and lanyard clipped into pre-drilled holes in the beam or temporary eyelets affixed to the top flange of the steel work. This is used in conjunction with nylon safety nets rigged to the bottom flange spanning between the steel beams; erectors should take additional care when straddling steelwork to ensure that the net attachment devices do not become dislodged. Under the Work at Height Regulations 2005, communal prevention systems like scaffolds, mobile towers and even safety nets are preferred over individual systems like personal harnesses. The procedure is to erect two storeys of steelwork before final plumbing and levelling take place. The grouting of the base plates and holding-down bolts is usually left until the whole structure has been finally levelled and plumbed. The grout can be a neat cement, a cement/sand mixture or a two-part chemical grout, depending on the gap to be filled. Typically for cement grouts that are mixed on site these mixes are:

- 12–25 mm gap – stiff mix of neat cement;
- 25–50 mm gap – fluid mix of 1:2 cement/sand and tamped;
- Over 50 mm gap – stiff mix of 1:2 cement/sand and rammed.

With large base plates a grouting hole is sometimes included, but with smaller plates three sides of the base plate are sealed with bricks or formwork and the grout is introduced through the open edge on the fourth side. To protect the base from corrosion, it should be encased with concrete up to the underside of the floor level, giving a minimum concrete cover of 75 mm to all the steel components.

FLOORS IN STEEL-FRAMED BUILDINGS

The typical method of forming a floor in a steel-framed building or as a mezzanine floor within a portal structure is to use either of the following:

- precast concrete planks (see Part 5 on page 239 ff);
- composite steel deck permanent formwork and in-situ concrete screed.

The term **permanent formwork** applies to formwork or shuttering designed to be left in place. It usually applies to reinforced concrete upper floor construction but can apply to walls and beams, where an attached facing material combines as outer cladding and formwork.

Application to floor slabs is with profiled galvanised steel sheeting. These metal profiles are a self-supporting shuttering system, functioning initially as troughed floor slab decking to receive an in-situ concrete topping. There are two basic types depending on the depth of profiles corrugations: shallow and deep.

- Shallow corrugated steel decking is fixed to a flange of supporting steel beam with an in-situ concrete topping reinforced with structural mesh. This method is used predominantly in medium- to high-rise steel frame structures where short shear bolts are welded or shot-fired to the top flange of the steel beam to provide an additional shear key into the concrete topping, so that the steel deck and concrete slab act compositely (see Fig. 10.5.11). Side and edge trims are produced to contain the wet concrete around the slab periphery and prevent grout leakage.
- Deep corrugated steel decking is fixed to the lower flange of supporting steel beam in Corus' Slimflor System, with an in-situ concrete topping reinforced with structural mesh (see Fig. 10.5.11). This method is for heavily loaded sections or where a flat floor soffit is required to reduce the height of the building.

These steel deck systems in most cases do not require temporary propping, or only require minimal temporary propping during the curing process of the hardening concrete. They have a number of other advantages over precast and in-situ concrete flooring systems:

- speed of construction – in excess of 400 m² of decking can be installed in one day by one team of operatives;

- safe method of construction – the decking acts as a safe working platform and as a canopy for workers working below;

- less weight than precast or in-situ concrete alternatives – requiring less materials in the main structure and foundations, and smaller cranage requirements;

- shallower construction – which can result in more room to accommodate services or, in high-rise, more storeys for the overall height of building (this is especially true for the Slimflor System, where the structural beams are hidden with the depth of the floor);

- ease of attachment of services to soffit – ducts and cable trays can be suspended from hangers on preformed dovetail hangers that slot into crenulations in the steel soffit.

The additional function is for the profiled and indented steel surface to bond with the wet concrete and combine as a composite floor slab when the concrete has matured.

FIRE PROTECTION OF STEELWORK

The Building Regulations, Approved Document B gives minimum fire resistance periods for steel structures and components according to a building's purpose group and its top-floor height above ground. The traditional method of fire protection is to encase the steel section with concrete. This requires formwork, and will greatly add to the structural dead load. Fifty mm cover will provide up to four hours' fire protection, whereas the more attractive hollow protection from brickwork will require at least 100 mm thickness for equivalent protection and will occupy more space. A number of lighter-weight and less bulky sheet materials based on plasterboard can be used to 'box' steel columns and beams, but because of their porous nature they are limited to interior use. They are also less practical where more than two hours' fire protection is required. A popular solution where aesthetics or appearance is not important is the covering of the section with a vermiculite concrete spray, usually to a thickness of 20 mm. This can provide a fire-resistance of up to four hours and is relatively cheap to apply. Some basic examples of solid and hollow protection to steelwork are shown in Fig. 10.5.12. For additional details and specific treatment of the subject of fire protection, consult the complementary volume, *Advanced Construction Technology*.

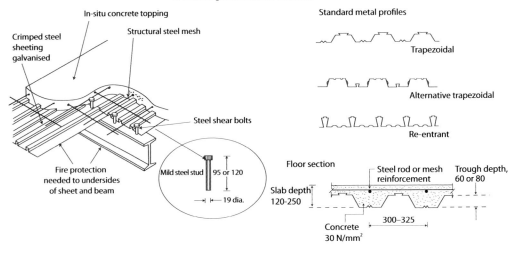

a) Shallow galvanised steel deck floor

In-situ concrete topping

Structural steel mesh

Crimped steel
sheeting
galvanised

Standard metal profiles

Trapezoidal

Alternative trapezoidal

Re-entrant

Steel shear bolts

Fire protection
needed to undersides
of sheet and beam

Mild steel stud 95 or 120

19 dia.

Floor section

Slab depth
120-250

Steel rod or mesh
reinforcement

Trough depth,
60 or 80

300–325

Concrete
30 N/mm²

b) Deep galvanised steel deck floor – Corus' Slimflor System

Asymmetrical steel
beam to support steel
corrugated troughs

Steel stop end to
prevent grout loss

In-situ concrete
topping reinforced
with structural mesh

Trough reinforced
with main steel
supported on
steel beam

Figure 10.5.11   Composite steel deck systems.

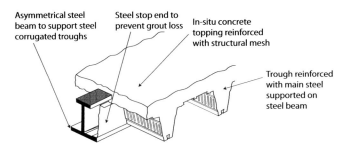

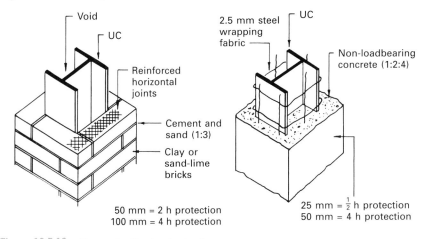

Void

UC

Reinforced
horizontal
joints

2.5 mm steel
wrapping
fabric

UC

Non-loadbearing
concrete (1:2:4)

Cement and
sand (1:3)

Clay or
sand-lime
bricks

50 mm = 2 h protection
100 mm = 4 h protection

25 mm = $\frac{1}{2}$ h protection
50 mm = 4 h protection

Figure 10.5.12   Fire protection to steel columns.

Low-rise steel buildings for industrial and commercial uses 10.6

This section will look at two popular forms of construction used to create modern industrial buildings for retail, storage or manufacturing. Here we will introduce the design and construction issues with regard to the use of steel, as this has tended to be the most popular material in low-rise short-to-medium-span industrial and commercial buildings. In *Advanced Construction Technology*, we look at other types of material used for similar purposes, including precast concrete and timber.

There are a number of methods for providing buildings with a large, unobstructed floor area and high floor-to-ceiling heights that are often required for industrial and commercial purposes. The two most commonly used forms of construction for low-rise industrial or commercial building are:

- steel truss and lattice girder arrangements;
- portal frame structures.

The structure of a steel building – especially of an industrial or commercial building – is quickly erected and clad, providing a weatherproof envelope that enables the floor and installation of services and internal finishes to proceed at an early stage. Since the construction schedule is always tied to the earliest handover date fixed by production planning, time saved in construction is usually very valuable.

FORMS OF CONSTRUCTION

The function of any building is to provide a protective covering to the envelope of the structure, and the roof span is critical to the sizing of the building. The roof structure must have sufficient strength to support its own weight and the load of the coverings, together with any imposed loadings such as snow and wind pressures, without collapse or excessive deflection. Typical roof spans and hence sizes of industrial buildings are given in Table 10.6.1.

Table 10.6.1   Typical span/depth ratios.

Form of construction	Span of roof (L)	Spanning beam or member depth ratio
Parallel chord roof trusses	10–100 m	L/12–20
Pitched roof trusses	8–20 m	L/5–10
Light roof beams	6–60 m	L/18–30
Traditional lattice roof girders	5–20 m	L/12–15
Portal frame	9–60 m	L/35–40

Taken from Cobb F (2010), *Structural Engineer's Pocket Book*, 2nd Edition, Elsevier

SHORT-SPAN ROOFS

These are roofs ranging up to 10 m in span. They would generally be of traditional timber-truss construction, with a flat or pitched profile covered with small units such as tiles or slates, as in standard domestic construction. Alternatively, lightweight cold-rolled lattice beams could be used, creating a flat or mono-pitch structure clad with flexible sheet materials, such as fibre-cement, galvanised steel or aluminium.

MEDIUM-SPAN ROOFS

In span ranges from 10 m to 40 m, the usual roof structure is a cold-rolled lightweight roof truss, lattice girders or hot-rolled portal frame of standard steel sections.

The envelope of the structure would comprise steel, zinc or aluminium sheet cladding material, which may or may not be insulated (thermally or acoustically), depending on the function of the building.

Conventional steel roofs trusses are still in existence in many old industrial buildings spanning up to 24 m, with their fragile, corrugated asbestos cement sheeting. Such buildings comprised a plane frame consisting of a series of rigid triangles composed of compression and tension members. The compression members are known as **rafters** and **struts**, and the tension members are termed **ties**. Standard hot-rolled mild-steel angles were used as the structural members, and these were connected together, where the centre lines converge, with flat-shaped plates called **gussets**. They tended to be riveted, bolted or welded together to form a rigid, triangulated truss, as shown in Fig. 10.6.1. However, these are not as popular as they used to be due to their costly gusseted connection details, unusable roof space compared to the simpler portal frame structure and the tendency to accumulate dust on the flanges of angles and plates. Also the roof cladding was traditional asbestos-cement sheeting, which is now known to be toxic causing diseases such as asbestosis and lung cancer when its fibres are inhaled.

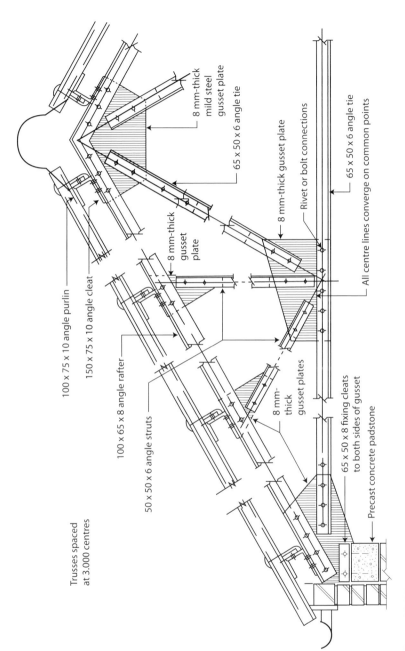

Trusses spaced
at 3.000 centres

100 x 75 x 10 angle purlin

150 x 75 x 10 angle cleat

100 x 65 x 8 angle rafter

50 x 50 x 6 angle struts

8 mm-thick
mild steel
gusset plate

65 x 50 x 6 angle tie

8 mm-thick
gusset
plate

8 mm-thick gusset plate

Rivet or bolt connections

65 x 50 x 6 angle tie

All centre lines converge on common points

8 mm-
thick
gusset plates

65 x 50 x 8 fixing cleats
to both sides of gusset

Precast concrete padstone

Figure 10.6.1   Traditional steel roof truss with fibre-cement roof.

LONG-SPAN ROOFS

These are roofs with a span over 40 m; they are generally designed by a specialist using long-span portal frame, lattice girder, space deck, grid shell or vaulting techniques. These roofs are covered in *Advanced Construction Technology*.

LIGHTWEIGHT STEEL TRUSS AND LATTICE GIRDER SYSTEMS

Trusses are generally considered to have pitched profiles while lattice girders have parallel top and bottom booms, which are suitable for flat or mono-pitch roofs. Typical profiles of these members are shown in Fig. 10.6.2. Truss and lattice girder arrangements can be fabricated from welded steel tubes, which are lighter in weight with a good strength-to-weight ratio, easier to transport

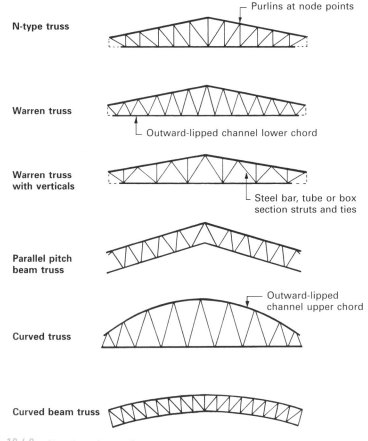

Figure 10.6.2   Chord section roof truss.

and erect, cleaner in appearance, and have less surface area on which to collect dust and therefore less surface area to protect with paint.

Another development from the traditional steel angle assembly is the use of cold-rolled outward-lipped channels or chord sections of high-yield steel with interlacing of steel bars, welded tubes or hollow square sections. This provides considerable design opportunities for different forms of lightweight lattice truss and for the simple fixing of various roof coverings. Fig. 10.6.3. shows standard chord sections with the possible inclusion of a timber insert to provide a simple means of attaching fixtures. Typical fixing and connection details for these lattice girders are shown in Fig. 10.6.4. These lattice girders can be supported on hot-rolled columns bolted to seating cleats or supported directly off loadbearing masonry walls, where the fixing is normally a standard holding-down bolt detail into a concrete padstone, cast or built into the wall.

A further advantage of this system is that, due to their slim profile, the beams can be incorporated at the head of partitioning and dry-lining systems without any requirement for separate boxing in or encasement, maximising the use of space and optimising the aesthetics.

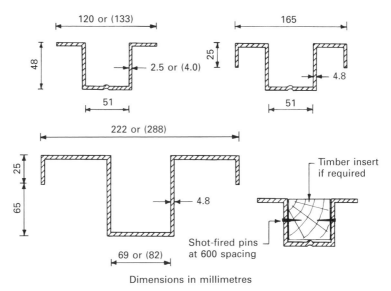

Dimensions in millimetres

Figure 10.6.3   Standard outward-lipped channels or chord sections.

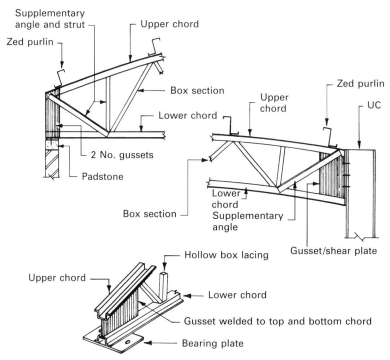

Figure 10.6.4   **Further assembly details for chord section trusses.**

STEEL PORTAL FRAMES

Steel portal frames can be fabricated from standard universal beam, column or rolled hollow sections. The basic form is a two-dimensional frame comprising, in its simplest form, a rigidly connected set of two steel columns connected to a pair of inclined steel rafters. The joints at the apex and the knee have to be designed to transfer bending moments in the rafters down into the column members. The base plate at the bottom of the column is fitted to the square-cut machined base of the column to ensure an even and full contact between the column and base plate. Due to the transfer of the bending moment, there is often a horizontal force that has to be restrained by the foundations as well as the vertical load of the weight of the structure.

Most systems employ welding techniques for the fabrication of the individual components undertaken within the factory. These components are then joined together on site using bolted connections for ease of assembly. The frames are designed to carry lightweight roof and wall claddings, and a range of systems can be used to meet thermal, acoustic and fire-resisting specifications.

If the building's use, irrespective of the framing material, is for an industrial process, the structure will have to comply with Part B of the Building Regulations: *Fire Safety*. The roof may also require purpose-made vents to relieve smoke logging and to reduce temperatures to ease firefighting access. The typical elements of a steel portal frame are shown in Fig. 10.6.5.

Steel has major advantages compared to other materials, particularly precast reinforced concrete. Its high strength-to-weight ratio and its high tensile and compressive strength enable steel buildings to be of relatively light construction. Steel is therefore the most suitable material for long-span roofs, where self-weight is of prime importance. Steel buildings can also be modified for extension or change of use due to the ease with which steel sections can be connected to existing work. Typical fixing details between the portal column and portal rafter are shown in Fig. 10.6.6. Also illustrated is the variety of different fixings for the various cladding systems that are available. These range from hot-rolled angle cleats welded to the portal frame with bolted angle purlins to cold-rolled Sigma and Zed purlins, which are shot-fired on site to the supporting portal frame.

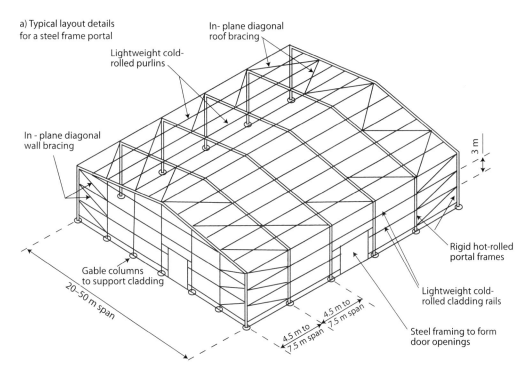

a) Typical layout details for a steel frame portal

In-plane diagonal roof bracing

Lightweight cold-rolled purlins

In-plane diagonal wall bracing

3 m

Rigid hot-rolled portal frames

Gable columns to support cladding

20–50 m span

4.5 m to 7.5 m span

4.5 m to 7.5 m span

Lightweight cold-rolled cladding rails

Steel framing to form door openings

b) Typical section for a steel frame portal

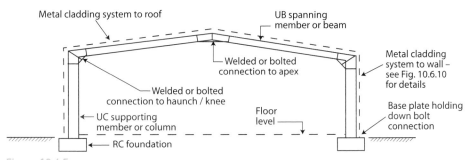

Metal cladding system to roof

UB spanning member or beam

Welded or bolted connection to apex

Metal cladding system to wall – see Fig. 10.6.10 for details

Welded or bolted connection to haunch / knee

Floor level

UC supporting member or column

Base plate holding down bolt connection

RC foundation

Figure 10.6.5   Typical portal frame elements.

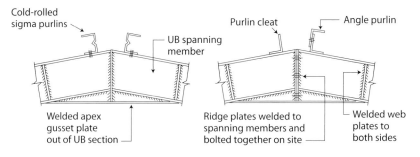

Typical apex details

Roof cladding omitted for clarity

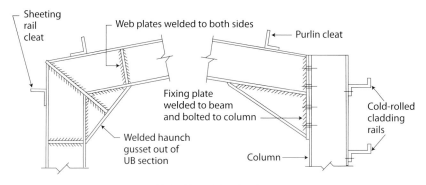

Typical knee joint details

Roof cladding omitted for clarity

Figure 10.6.6   **Portal frame fixing details.**

FOUNDATIONS AND FIXINGS

The foundation is usually a reinforced-concrete isolated base or pad foundation designed to suit loading and ground-bearing conditions. The connection of the frame for portals spanning up to 35 m is usually using a base plate connection. This is a traditional structural steelwork column-to-foundation connection using a slab or a gusset base, fixed to a reinforced concrete foundation with cast-in holding-down bolts (see Fig. 10.5.6). Where the portal frame has been designed to resist large bending moments at the foundations, a pocket connection may be specified. This is where the foot of the supporting member is inserted and grouted into a pocket formed in the concrete foundation, in a similar way to the precast concrete column foundation connection shown in Fig. 10.4.2. There are other forms of connection for large-span portal frames with no movement at the foundation connection, and these are known as pin or hinge connections. Details of these and the reasons why they are specified are covered in *Advanced Construction Technology*.

The basic requirements for covering materials to steel portal frames, trusses or lattice girders are:

- sufficient strength to support imposed wind- and snow-loadings;
- resistance to the penetration of rain, wind and snow;
- low self-weight, so that supporting members of an economic size can be used;
- acceptable standard of thermal insulation if habitable or occupational accommodation requiring space heating;
- acceptable fire resistance and resistance to spread of flame;
- durability to reduce the maintenance required during the anticipated life of the roof.

Not only is steel a versatile material for the structure of a building, but also a wide variety of cladding has been devised using the strength developed by folding thin sheets into profiled form. Insulated cladding systems with special coatings are now widely used for roofing and sidewall cladding. They have good appearance and durability, and are capable of being speedily fixed into position.

In a dry, closed environment, steel does not rust, and protection against corrosion is needed only for the erection period; for other environments, protection systems are available, which, depending on cost and suitable maintenance, prevent corrosion adequately.

Part L2 of the Building Regulations must be observed for industrial or commercial buildings if thermal insulation requirements apply. However, these are not applicable if the building can be demonstrated to have a low energy demand (that is, where the building is not heated or cooled), or is a temporary building with a life of less than two years. Where the building is not a low-energy-demand building, energy efficiency measures similar to those covered in Chapter 7 must be applied when the building's CO_2 emission rate (BER) is less than a calculated target emission rate (TER) for the building.

A further consideration for designers of industrial and commercial buildings is achieving accreditation by the Loss Prevention Certification Board (LPCB). The LPCB is part of BRE Global and is an accreditation body that approves construction products and services that protect exposure to 'unexpected' business risks, such as fire or attack on security and security systems, and has been developed in consultation with the Association of British Insurers. Often industrial and commercial buildings need a higher level of guaranteed protection to mitigate against commercial loss. Therefore building products such as roof or wall cladding materials have to have better fire resistance than specified in Part B of the Building Regulations. The LPS 1181 series of standards cover sandwich panels and built-up cladding systems including jointing and fixing methods, sealants, gaskets and flashings. The installation and fixing systems used for the claddings must also be such that it deters security breaches and theft. More information on the standards, products and services that are approved by the LPCB is available from www.redbooklive.com

There now follows a consideration of both uninstalled cladding systems used for agricultural and some storage situations, and insulated systems for use in retail and manufacturing purposes.

UNINSULATED INDUSTRIAL CLADDING

Corrugated roofing materials, if correctly applied, will provide a covering that will exclude the rain and snow, but will allow a small amount of wind penetration (unless the end laps are sealed with two ribbons of silicon or butyl mastic sealant). These coverings are designed to support normal snow loads, but are not usually strong enough to support the weight of operatives, and therefore a crawling ladder or board should be used. Owing to the poor thermal insulation value of these roofing materials, there is a risk of condensation occurring on the internal surface of the sheets. This risk can be reduced by using a suitable lining at rafter level or by a ceiling at the lower tie level. Unless a vapour control layer is included on the warm side of the lining, water vapour may pass through the lining and condense on the underside of the covering material, which could give rise to corrosion of the steel members. An alternative method is to form a 25 mm-wide ventilation cavity between the lining and the covering.

Fibre-cement profiled sheets

Fibre-cement sheets are made by combining natural and synthetic non-toxic fibres and fillers with Portland cement. Rolled to form the required profile, these sheets are pressed over templates. Typical profiles of these sheets are shown in Fig. 10.6.7. The finished product has a natural grey colour, but sheets with factory-applied surface coatings are available. These are unlike the now-obsolete asbestos-cement sheets, a hazardous building product that is no longer manufactured or specified due to the proven health risks. Great care must be taken if this is discovered in an old building; specialist waste contractors will be required to survey and dispose of the material without releasing its dangerous fibres into the atmosphere.

The sheets and fittings are fixed through the crown of the corrugation using either shaped bolts to steel purlins or drive screws to timber purlins. At least six fixings should be used for each sheet and, to ensure that a weathertight seal is achieved at the fixing positions, a suitable felt or lead pad with a diamond-shaped curved washer can be used. Alternatively, a plastic sealing washer can be employed (see Fig. 10.6.8). The sheets can be easily drilled for fixings, which should be 2 mm larger in diameter than the fixing and sited at least 40 mm from the edge of the sheet. Side laps should be positioned away from the prevailing wind, and end laps on low pitches should be sealed with a mastic or suitable preformed compressible strip.

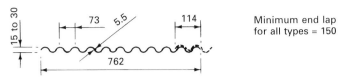

Minimum end lap
for all types = 150

Class 1: min pitch 10°; max purlin spacing 900 (symmetrical)

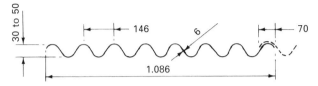

Class 2: min pitch 10°; max purlin spacing 1.400 (symmetrical)

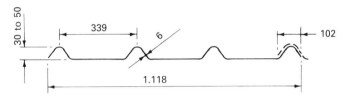

Class 2: min pitch 10°; max purlin spacing 1.400 (asymmetrical)

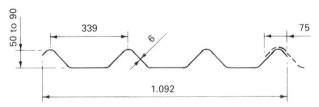

Class 3: min pitch 4°; max purlin spacing 1.680 (asymmetrical)

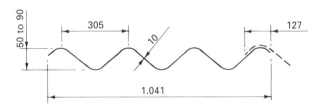

Class 4: min pitch 4°; max purlin spacing 1.980 (symmetrical)

Lengths	
Class 1	1.225 to 3.050
Class 2	1.525 to 3.050
Class 3	1.675 to 3.050
Class 4	1.825 to 3.050

Figure 10.6.7   Typical corrugated sheet profiles.

The U-value of fibre-cement sheets is high – generally about 6.0 W/m² K – so, if a higher degree of thermal resistance is required, it will be necessary to use a system of underlining sheets with an insulating material sandwiched between the profile and underlining sheet (see later).

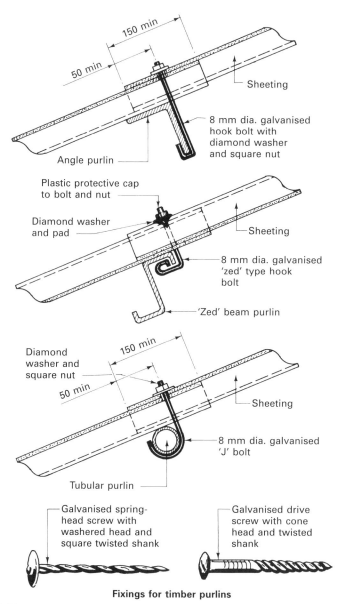

Fixings for timber purlins

Figure 10.6.8 Typical roof sheeting fixings.

Profiled aluminium sheeting

The sheets are normally made from an aluminium–manganese alloy resulting in a non-corrosive, non-combustible lightweight sheet (2.4–5.0 kg/m^2) to BS EN 485 or coated steel to BS EN 10326. Aluminium sheeting oxidises on the surface to form a protective film upon exposure to the atmosphere, and therefore protective treatments are not normally necessary. Fixings of copper or brass should not be used, because the electrolytic action between the dissimilar materials could cause harmful corrosion, and where the sheets are in contact with steelwork the steel members should be painted with at least two coats of zinc chromate or bituminous paint.

These are produced in a trapezoidal profile in varying depth and pitch dimensions, depending on the source of manufacture. Surface treatment may be galvanised for steel, painted or more commonly plastic-coated. Various types of plastic coating are possible, including acrylic, polyester and resins. All can be colour-pigmented to provide a wide range of choice to suit all types of building, including schools, hospitals, workshops, warehousing and offices. Fixings can be through the valley for direct connection to the purlin with self-drilling and tapping screws as shown in Fig. 10.6.9. Crown fixing, also shown, has self-tapping screws in pre-drilled holes with a butyl mastic (or equivalent) sealant to prevent ingress of rainwater.

INSULATED INDUSTRIAL CLADDING

There is now a range of insulated cladding materials for industrial and commercial buildings. The development of these products has led to their use in a variety of different situations where cost and speed of construction are important, such as schools, hospital and retail parks. The main materials used are aluminium or steel panels treated by galvanisation or plastic coatings. The insulation can come in a variety of forms, such as rigid-board closed-cell insulation in a composite sandwich or flexible glass-fibre quilts built up in layers within containing voids or trays.

Three basic types of wall cladding system are commonly used.

- **Built-up system with liner sheets** These systems consist of two separate cladding sheets: an external sheet, which is coloured and highly profiled, and a more lightly profiled inner liner sheet. The sheets are separated by spacer rails and insulation. The liner sheet and spacers are fixed to cladding rails that span between columns. The external sheet is fixed to the spacer. The normal method of attachment is by self-drilling, self-tapping screws.

- **Built-up system with liner trays** Deep U-shaped liner trays span between the structural columns fixed by bolted connections. The liner tray replaces the liner sheet, cladding rails and spacer rails of the above system. Quilt or rigid-board insulation is set within the liner trays, and the external cladding sheet is fixed directly on to the outer flanges of the tray.

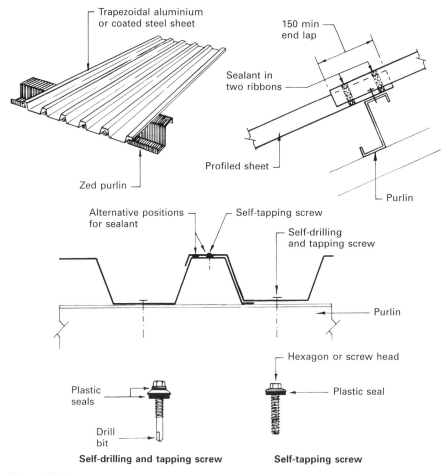

Figure 10.6.9   Profiled sheet metal roof covering.

■ **Composite panel system** Composite panels have a sandwich
   construction, comprising two steel or aluminium sheets bonded either side
   of an insulating core of rigid foam, mineral fibre or similar material. The
   bonded panel produces good stiffness, and both profiled and smooth
   surfaces are available.

Panels may be fixed using a variety of techniques including self-drilling,
self-tapping screws, and secret fixing brackets located within the interlocking
panel joints.

Typical details of these types of cladding system are shown in Fig. 10.6.10.

For details of typical industrial roof cladding systems, see Part 6 on page 284.

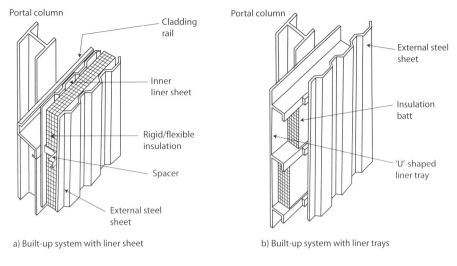

a) Built-up system with liner sheet

Portal column

Cladding rail

Inner liner sheet

Rigid/flexible insulation

Spacer

External steel sheet

b) Built-up system with liner trays

Portal column

External steel sheet

Insulation batt

'U'-shaped liner tray

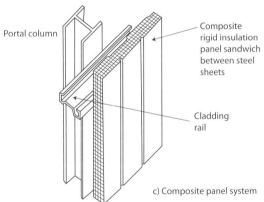

Portal column

Composite rigid insulation panel sandwich between steel sheets

Cladding rail

c) Composite panel system

Figure 10.6.10   Typical light steel cladding systems.

11

Services

Domestic water supply

11.1

An adequate supply of water is a basic requirement for most buildings, for reasons of personal drinking and hygiene or for activities such as cooking and manufacturing processes. In most areas a piped supply of water is available from the local utility company's water main.

The source of water is usually produced by condensation from within clouds, which falls to the ground as rain, snow or hail; it then either becomes surface water in the form of a river, stream or lake, or percolates through the subsoil until it reaches an impervious stratum, or is held in a water-bearing stratum.

The water authority extracts water from surface storage reservoirs or rivers, or through boreholes into a water-bearing aquifer. After this it is filtered, chlorinated to kill any bacteria, aerated, and often treated with fluoride for human consumption before distribution through a system of water mains under pressure.

The water company's mains are laid underground at a depth where they will be unaffected by frost or traffic movement, either of which can cause considerable damage to the water-main infrastructure. The layout of the system is generally a circuit with trunk mains feeding a grid of subsidiary mains for distribution to specific areas or districts. The materials used for main pipes are cast iron and uPVC. Cast iron is still in use but is slowly being replaced by the newer, plastic water-main pipes. The water main is owned by the water service authority up to the stopcock on a domestic property; past this point it is the owner's responsibility. Most water usage in the UK is being converted over to a metered supply.

Terminology

- **Main** A pipe for the general conveyance of water, as distinct from the conveyance to individual premises.

- **Service** A system of pipes and fittings for the supply and distribution of water in any individual premises.

- **Service pipe** A pipe in a service that is directly subject to pressure from a main, sometimes called the rising main, inside the building.

- **Communication pipe** That part of the service pipe that is owned by the water authority.

- **Distribution pipe** Any pipe in a service conveying water from a storage cistern.

- **Cistern** A container for water in which the stored water is under atmospheric pressure.

- **Storage cistern** Any cistern other than a flushing cistern.

- **Tank** A rectangular vessel completely closed and used to store water.

- **Cylinder** A closed cylindrical tank.

COLD WATER SUPPLY

From their ferrule connection on the water main, the water company will provide a communication pipe to a stop valve and surface box just outside the boundary. A long-sweep bend is included to relieve any stress likely to be exerted on the mains connection.

A service pipe is taken from this stop valve to an internal stop valve within the property, preferably located just above floor level and housed under the sink unit. The stop valve should have a drain-off valve incorporated in it, or just above it, so that the service pipe or rising main can be drained.

Care must be taken when laying a service pipe that it is not placed in a position where it can be adversely affected by frost, heavy traffic or building loads. The pedestrian footpath is an ideal location for the water mains supply pipework. A minimum depth of 750 mm is generally recommended for supplies to domestic properties as it is the depth that frost can not penetrate; where the pipe passes under a building it should be housed in a protective duct or pipe suitably insulated within 750 mm of the floor level (see Fig. 11.1.1).

Suitable materials for service pipes are high-performance polyethylene (HPPE), ductile iron, steel, and a range of composite metal/plastic piping. The pipe should be laid through a duct as it passes through the foundation external wall and up into the property. The duct is then sealed at floor level. HPPE pipe is favoured as it is resistant to both frost and corrosion, and it has largely superseded metal pipes in water distribution.

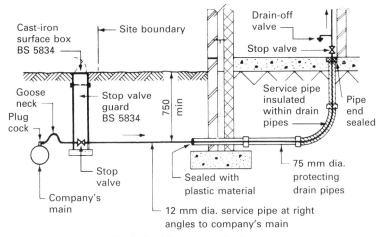

Cast-iron
surface box
BS 5834

Site boundary

Drain-off
valve

Stop valve

Goose
neck

Plug
cock

Stop valve
guard
BS 5834

750
min

Service pipe
insulated
within drain
pipes

Pipe
end
sealed

Stop
valve

Company's
main

Sealed with
plastic material

75 mm dia.
protecting
drain pipes

12 mm dia. service pipe at right
angles to company's main

Typical cold water service layout

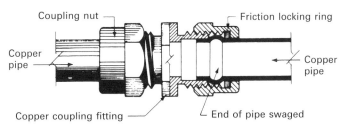

Coupling nut

Friction locking ring

Copper
pipe

Copper
pipe

Copper coupling fitting

End of pipe swaged

Typical manipulative compression joint

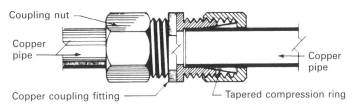

Coupling nut

Copper
pipe

Copper
pipe

Copper coupling fitting

Tapered compression ring

Typical non-manipulative compression fitting

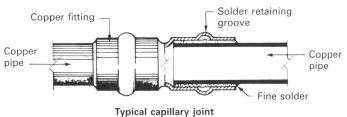

Copper fitting

Solder retaining
groove

Copper
pipe

Copper
pipe

Fine solder

Typical capillary joint

Figure 11.1.1   Supply pipe and copper pipe joint details.

WATER METER

This is a standard installation for all new buildings and conversions within buildings. It acts as a resource conservation measure, as the occupier is charged for every litre used. Existing premises may have meters installed at the water authority's discretion. Where practicable, the meter is located on the service pipe, in a small compartment below ground and just inside the property boundary. If this proves impossible owing to lack of space, the meter may be positioned at the base of the rising main with an external readout mounted on the wall outside.

DIRECT COLD WATER SUPPLY

With this system the whole of the cold water to the sanitary fittings and taps is supplied directly from the incoming service pipe. The direct system is used mainly in areas where large, high-level reservoirs provide a good mains supply and high pressure. With this system only a small cold water storage cistern to feed the hot water tank is required; this can usually be positioned below the roof ceiling level within an airing cupboard, giving a saving on pipe runs to the roof space and eliminating the need to insulate the pipes against frost (see Fig. 11.1.2).

Another advantage of the direct system is that fresh drinking water is available from several outlet points. The main disadvantage is the lack of reserve should the mains supply be cut off for repairs; also there can be a lowering of the supply pressure during peak demand periods.

When sanitary fittings are connected directly to a mains supply there is always a risk of contamination of the mains water by back-siphonage. This can occur if there is a negative pressure on the mains and any of the outlets are submerged below the water level, such as a hand spray connected to the taps.

INDIRECT COLD WATER SUPPLY

In this system all the sanitary fittings, except for a drinking water outlet at the sink, are supplied indirectly from a cold water storage cistern positioned at a high level, usually in the roof space. This system requires more pipework, but it gives a reserve supply in case of mains failure, and it also reduces the risk of contamination by back-siphonage (see Fig. 11.1.2). Note that the local water authority determines the system to be used in the area. When used in a block of flats, a sufficient-sized storage tank and its loadings will need to be accommodated. Indirect storage provides a means of storing water without any effect to occupants when pressure fluctuations arise on the direct main.

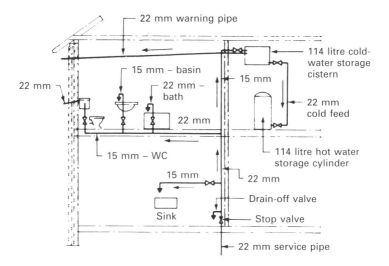

Direct system of cold water supply

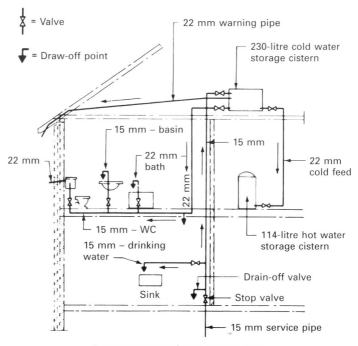

Indirect system of cold water supply

Note: All pipe sizes shown are standard copper outside diameters

Figure 11.1.2 **Cold water supply systems.**

PIPEWORK

There are many ways to transport and distribute water around a property. The most common is copper and plastic pipework, which can be installed without the use of any hot work.

COPPER PIPES

Copper pipes are the most common method of distribution around a dwelling. They have a smooth bore, giving low flow resistance; they are strong, can be easily jointed and bent, and will not corrode. Joints in copper pipes can be made by one of two methods:

- **compression joint** – this joint relies on the grip between a copper olive ring and the surrounding brass compression fitting, which is tightened against the pipe and olive forming a seal;
- **capillary joint** – oversized copper fitting is used that fits over the pipe ends to be joined. The application of heat makes the soft solder contained in a groove in the fitting flow around the end of the pipe, which has been cleaned and coated with a suitable flux to form a neat and rigid watertight joint.

Typical examples of copper pipe joints were shown in Fig. 11.1.1.

STEEL PIPES

Steel pipes for water supply can be obtained as black tube, galvanised or coated and wrapped for underground services. They tend to be used for commercial applications. The joint is usually made with a tapered thread and socket fitting and, to ensure a sound joint, a non-contaminating polytetrafluoroethylene (PTFE) seal tape is used (see Fig. 11.1.3).

POLYTHENE PIPE

Polythene pipe is very light in weight, easy to joint and non-toxic, and is available in long lengths, giving a saving on the number of joints required. Jointing of polythene pipes is generally of the compression type using a metal or plastic liner to the end of the tube (see Fig. 11.1.3). To prevent undue sagging, polythene pipes should be adequately fixed to the wall with saddle clips: recommended spacings are 14 times the outside diameter for horizontal runs and 24 times outside diameter for vertical runs. Coilless pipe is now available that will remain straight when it is installed from a coil and is considerably easier to pull into a void or floor space. The introduction of push-fit fittings has revolutionised the method of jointing. These fittings contain a sealing ring and a steel grab ring that secures onto the pipe as it is inserted into the fittings. The pressure of the water forces the sealing ring tight against the surrounding fitting, forming a seal. They have the advantage of being able to be reused, and are a fast, lead-free and reliable method of jointing.

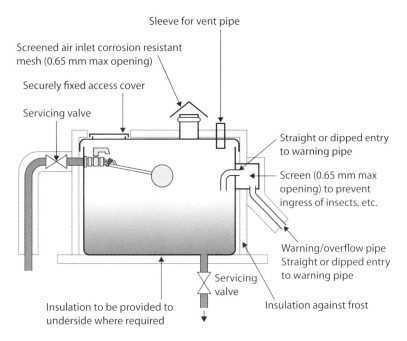

Sleeve for vent pipe

Screened air inlet corrosion resistant mesh (0.65 mm max opening)

Securely fixed access cover

Servicing valve

Straight or dipped entry to warning pipe

Screen (0.65 mm max opening) to prevent ingress of insects, etc.

Warning/overflow pipe
Straight or dipped entry to warning pipe

Servicing valve

Insulation against frost

Insulation to be provided to underside where required

Typical cold water storage cistern

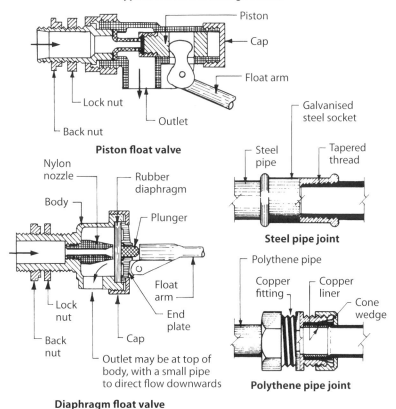

Piston

Cap

Float arm

Lock nut

Outlet

Back nut

Piston float valve

Nylon nozzle

Rubber diaphragm

Body

Plunger

Lock nut

Float arm

Back nut

End plate

Cap

Outlet may be at top of body, with a small pipe to direct flow downwards

Diaphragm float valve

Galvanised steel socket

Steel pipe

Tapered thread

Steel pipe joint

Polythene pipe

Copper fitting

Copper liner

Cone wedge

Polythene pipe joint

Figure 11.1.3 Cisterns, float valves and joints.

UNPLASTICISED PVC (uPVC)

This is plastic pipe for cold water services that is supplied in straight lengths up to 9,000 mm long and in standard colours of grey, blue or black. Blue is the water authorities' preferred colour, for easy identification when buried. Jointing can be by a screw thread, but the most common method is by solvent welding. This involves cleaning and chamfering the end of the pipe, which is coated with the correct type of adhesive and pushed into a straight coupling that has been similarly coated. The solvent will set within a few minutes, but the joint does not achieve its working strength for 24 hours. Heat fusion is also used, but mainly for the larger-diameter water authority mains. Here the pipe spigot and the coupling are heated with circular dies to melt the uPVC and effect a push-fit weld, which is completed when the pipe cools.

COLD WATER STORAGE CISTERNS

The size of cold water storage cisterns for dwelling houses will depend upon the reserve required and whether the cistern is intended to feed a hot water system. Minimum actual capacities recommended in model water byelaws are 100 litres for cold water storage only and 230 litres for cold and hot water services.

Cisterns should be adequately supported on a purpose-built tank stand and installed in accordance with the current Water Regulations. The cistern must be installed so that its outlets are above the highest discharge point on the sanitary fittings, as the flow to these is by gravity. If the cistern is housed in the roof, the pipes and cistern should be insulated against possible freezing of the water; preformed casings of suitable materials are available to suit most standard cistern sizes and shapes. These should have an access cover for maintenance of the cistern valve. The inlet and outlet connections to the cistern should be on opposite sides to prevent stagnation of water. A securely fitting cover must be provided to prevent ingress of dust, dirt and insects. To prevent a vacuum occurring as water is drawn, the cover is fitted with a screened vent. It also has a moulded sleeve/boss for adapting the hot water expansion pipe. The overflow or warning pipe is also fitted with a filter. Cisterns are available off the shelf that fully meet the stringent Water Regulations, required for water storage due to the possibility of contamination. A typical cistern installation is shown in Fig. 11.1.3.

Plastic cisterns have many advantages over the traditional galvanised mild-steel cisterns in that they are non-corrosive, rot-proof and frost-resistant, and have good resistance to mechanical damage. Materials used are polythene, polypropylene and glass fibre. These cisterns are made with a wall thickness to withstand the water pressure, and have an indefinite life. Some forms of polythene cistern can be distorted to enable them to be passed through an opening of 600 mm × 600 mm, which is a great advantage when planning access to a roof space. However, it is always better to deposit the cistern within the roof structure during construction, rather than have access difficulties later.

FLOAT VALVES

Every pipe supplying a cold water storage cistern must be fitted with a float valve to prevent an overflow of water, because of the subsequent damage it can cause. The float valve must be fitted at a higher level than the overflow to prevent it becoming submerged and creating the conditions where back-siphonage is possible. A float valve is designed to automatically regulate the supply of water by a floating ball closing the valve when the water reaches a predetermined level.

Two valves are in common use for domestic work: the piston valve and the diaphragm valve. The piston valve has a horizontal piston or plunger that closes over the orifice of a diameter to suit the pressure: high-, medium- and low-pressure valves are available (see Fig. 11.1.3). They tend to be manufactured out of brass. The diaphragm valve closes over an interchangeable nylon nozzle orifice, and the whole arrangement is normally manufactured from plastic. This type of valve is quieter in operation, easily adjustable, and less susceptible to the corrosion trouble caused by a sticking piston – this is one of the problems that can be encountered with the piston valve (see Fig. 11.1.3). Diaphragm valves now have the outlet at the top to increase the air gap between outlet and water level, thus reducing the possibility of back-siphonage if the cistern water level were to rise excessively.

HOT WATER SUPPLY

The supply of hot water to domestic sanitary fittings is usually taken from a hot water tank or cylinder, or from a combination boiler. The source of heat is usually in the form of a gas-fired, oil-fired or solid-fuel boiler; alternatives are a back boiler to an open fire or an electric immersion heater fixed into the hot water storage tank. When a quantity of hot water is drawn from the storage tank, it is immediately replaced by cold water from the cold water storage cistern. Two main systems are used to heat the water in the tank: the **direct** and **indirect** systems. Modern condensing combination boilers now make the storage of hot water redundant as they operate when the tap is turned on. This creates a pressure drop that fires up the boiler and heats the water directly as it is required, reducing the need for any storage cylinder.

DIRECT HOT WATER SYSTEM

This is the simplest and cheapest system. The cold water flows through the water jacket in the boiler, where its temperature is raised by combustion of the fuel and convection currents are induced, which causes the water to rise and circulate around the system. The hot water leaving the boiler is replaced by colder water descending from the hot water cylinder by gravity, thus setting up the circulation. The hot water cylinder must be insulated to avoid excessive heat loss and lower hot water temperatures. The hot water supply is drawn off from the top of the cylinder by a horizontal pipe at least 450 mm long to prevent 'one pipe' circulation being set up in the vent or expansion pipe. This pipe is run vertically from the hot water distribution pipe to a discharge position over the cold water storage cistern (see Fig. 11.1.4).

The main disadvantage of the direct system is its unsuitability for supplying a central heating circuit, as in hard water areas the pipes and cylinders will become furred with lime deposits. This precipitation of lime occurs when hard water is heated to temperatures of between 50 and 70 °C, which is the ideal temperature range for domestic hot water supply. As the system relies on the continual replenishment of water from the mains, it slowly builds up lime and chalk deposits that eventually block the pipework – a process called **furring**.

INDIRECT HOT WATER SYSTEM

This system is designed to overcome the problems of furring, which occurs with the direct hot water system. The basic difference is in the cylinder design, which now becomes a heat exchanger by using a closed-loop system. The cylinder contains a coil or annulus that is connected to the flow and return pipes from the boiler; at no point does the system replenish itself within this closed loop, only on the circulation system side. A transfer of heat takes place within the cylinder; after the initial precipitation of lime within the primary circuit and boiler, there is no further furring as fresh cold water is not being constantly introduced into the boiler circuit.

The supply circuit from the cylinder works in the same way as the direct hot water system, but a separate feed and expansion system is required for the boiler and primary circuit for initial supply, and for any necessary topping up due to evaporation. The feed cistern is similar to a cold water storage cistern but of a much smaller capacity, as it does not have to operate once the primary fill has taken place. The water levels in the two cisterns should be equal so that equal pressures act on the indirect cylinder, so they must be at the same level.

A gravity heating circuit can be taken from the boiler, its distribution being governed by the boiler output (see Fig. 11.1.4). Alternatively a small-bore forced system of central heating may be installed.

HOT WATER STORAGE CYLINDERS

Copper cylinders are produced to the recommendations of BS 1566. This standard recommends sizes, capacities and positions for screwed holes for pipe connections.

To overcome the disadvantage of the extra pipework involved when using an indirect cylinder, a single-feed indirect or 'primatic' cylinder can be used. This form of cylinder is entirely self-contained, and is installed in the same manner as a direct cylinder, but functions as an indirect cylinder.

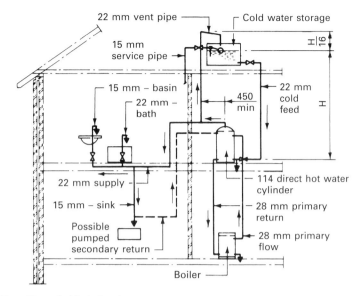

Note: Not suitable in hard water areas

Direct hot water system

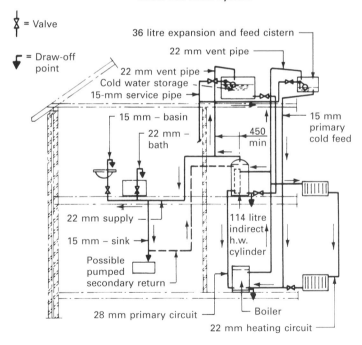

Indirect hot water system

Note: Pipe sizes shown are for standard copper outside diameters

Figure 11.1.4 Hot water system.

It works on the principle of the buoyancy of air, which is used to form seals between the primary and secondary water systems. When the system is first filled with water, the cylinder commences to fill and fully charges the primary circuit to the boiler with water. When the cylinder water capacity has been reached, two air seals will have formed, the first in the upper chamber of the primatic unit and the second in the air vent pipe. These volumes of air are used to separate the primary and secondary water. When the water is heated in the primary system, expansion displaces some of the air in the upper chamber to the lower chamber. This is a reciprocating action: the seals transfer from chamber to chamber as the temperature rises and falls.

Any excess air in the primary system is vented into the secondary system, which will also automatically replenish the primary system should this be necessary. As with indirect systems, careful control over the heat output of the boiler is advisable to prevent boiling and consequent furring of the pipework. Typical examples of cylinders are shown in Fig. 11.1.5.

FAULTS IN HOT WATER SYSTEMS

Unless a hot water system is correctly designed and installed a number of faults may occur, such as airlocks and noises. Airlocks are pockets of trapped air in the system, which will stop or slow down the circulation. Air suspended in the water will be released when the water is heated, and rise to the highest point. Good design will provide access points where the air can be bled from the system or an auto vent can be installed on the highest radiator to remove any trapped air. The most common positions for airlocks are sharp bends and the upper rail of a towel rail; the only cure for the latter position is for the towel rail to be vented.

Noises from the hot water system usually indicate a blocked pipe caused by excessive furring or corrosion. The noise is caused by the imprisoned expanded water, and the faulty pipe must be descaled or removed, or an explosion may occur.

Inadequate bracketing of pipes is another common cause of plumbing noise, particularly in the rising main, where pressures are higher than elsewhere. This is known as **water hammer**.

With the modern condensing boilers the condensates drain can freeze in winter as they discharge externally, and can stop the boiler operating. A condensate drain catches any condensation formed by the operation of the boiler and discharges it externally through a 22 mm drain pipe. If the water in the pipe freezes, this has to be removed and thawed out to restart the boiler. The higher pressure on the central heating side may cause leaks within older radiators where it is an upgrade to an old heating system.

MAINS HOT WATER SUPPLY

This has become a popular alternative to the traditional indirect installation by the use of highly efficient gas-fuelled condensing combi boilers.

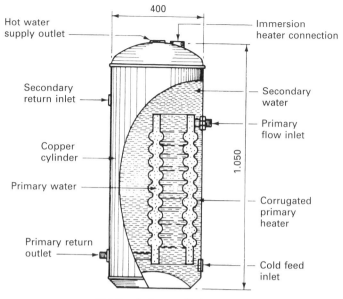

114 litre indirect cylinder

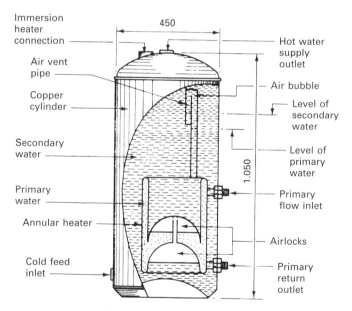

135 litre single-feed or 'Primatic' cylinder

Figure 11.1.5 Typical hot water cylinders.

The word 'combi' indicates the combining of the hot water and central heating systems into one unit. It is very economic in pipework, space requirements and installation time as conventional expansion facilities are not required. The system is mains-fed and sealed, with expansion accommodated in a specially designed vessel containing an air 'cushion'. The central heating system has to be filled using mains pressure to over one bar. Safety facilities are essential and include both pressure and thermal relief valves. This type of boiler is available with a common efficiency rating of over 90 per cent. Fig. 11.1.6 indicates the principles of operation.

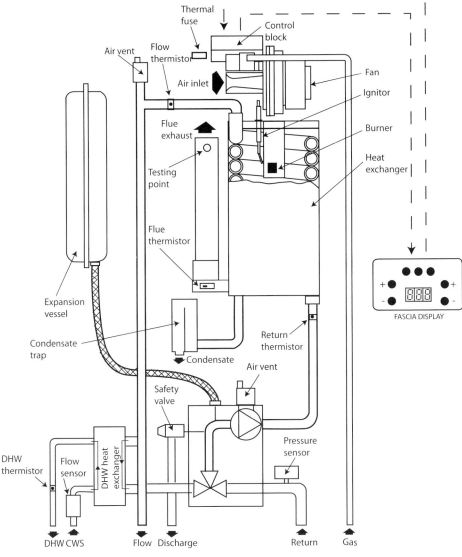

Figure 11.1.6 Schematic of mains-fed hot water supply.

Sanitary fittings and pipework 11.2

Sanitary fittings are essential for homes to function efficiently. They enable waste water to be removed from the home and can be considered under two main headings:

- **soil fitments** – those that are used to remove soil water and waste, such as water closets (WCs) and urinals;
- **waste water fitments** – those that are used to remove the waste water from washing and the preparation of food, including appliances such as washbasins, baths, showers and sinks.

All sanitary appliances should be made from materials that do not absorb water, are quiet in operation, are easy to clean, provide an aesthetical appearance and are fixed at a suitable height. A number of materials are available for most domestic sanitary fittings.

VITREOUS CHINA

This is a white clay body that is vitrified and permanently fused with a vitreous glazed surface when fired at a very high temperature, generally to the recommendations of BS 3402. Appliances made from this material are non-corrosive, hygienic and easily cleaned with a mild detergent or soap solution.

GLAZED FIRECLAY

This consists of a porous ceramic body glazed in a similar manner to vitreous china; it is exceptionally strong and resistant to impact damage, but will allow water penetration of the body if the protective glazing is damaged. Like vitreous china, these appliances are non-corrosive, hygienic and easily cleaned.

VITREOUS ENAMEL

This is a form of glass that can be melted and used to give a glazed protective coating over a steel or cast iron base. Used mainly for baths, sinks and draining boards, it produces a fitment that is lighter than those produced from a ceramic material, is hygienic and easy to clean, and has a long life. The finish, however, can be chipped and is subject to staining, especially from copper compounds from hot water systems.

PLASTIC MATERIALS

Acrylic plastics, glass-reinforced polyester resins and polypropylene sanitary fittings made from these plastics require no protective coatings and are very strong, light in weight and chip-resistant, but generally cost more than ceramic or vitreous enamel products. Care must be taken with fitments made of acrylic plastics as they become soft when heated: therefore they should be used for cold water fitments or have thermostatically controlled mixing taps. Plastic appliances can be easily cleaned using warm soapy water, and any dullness can be restored by polishing with ordinary domestic polishes.

STAINLESS STEEL

This is made from steel containing approximately 18 per cent chromium and 8 per cent nickel, which gives the metal a natural resistance to corrosion. Stainless steel appliances are very durable and relatively light in weight; stainless steel is mainly used for domestic and commercial kitchen sinks and drainers. Two finishes are available: polished or 'mirror' finish and 'satin' finish; the latter has a greater resistance to scratching.

GENERAL CONSIDERATIONS

The factors to be considered when selecting or specifying sanitary fitments are:

- cost – initial outlay, fixing and maintenance;
- hygiene – inherent hygiene and ease of cleaning;
- appearance – size, colour and shape;
- function – suitability, speed of operation and reliability;
- weight – additional support required from wall and/or floor;
- design – ease with which it can be included into the general services installation;
- water saving – how far the fitting reduces the amount of water required.

Building Regulation Approved Document G deals with sanitary conveniences and washing facilities – that is, water closets and washbasins. The quantity of

WCs and sanitary appliances in various locations is detailed in BS 6465: *Sanitary installations*. This Standard is in two parts: Part 1 is the code of practice for provision and installation, and Part 2 relates to spatial layout and design for sanitary accommodation. The provision of disabled sanitary accommodation must also be carefully considered within a domestic application, if required.

The subject of water conservation with respect to the use of sanitary fitments is very important for sustainability, with manufacturers endeavouring to adapt their products and develop new ideas in line with conservation and environmental issues. Dual-action flush devices on cisterns now save water, as do specially designed low-flush WCs. The Code for Sustainable Homes now sets standards for the use of water within new developments and provides further reading on this topic.

WATER CLOSETS

This simplest of devices enables high standards in public health to be maintained within the domestic home. From its simple concept it is now available in a variety of designs, colours and patterns, from wall-hung WCs to concealed cisterns. Most water closets are made from a ceramic base to the requirements of BS EN 33 or 37 with a horizontal outlet. The complete water closet arrangement consists of the pan, seat, flush pipe and flushing cistern. The cistern can be fixed as high level, low level (BS EN 37) or close coupled (BS EN 33); the latter arrangement dispenses with the need for a flush pipe, as the cistern sits directly on the WC. A typical arrangement is shown in Fig. 11.2.1. The British Standard water closet is termed a washdown type and relies on the flush of water to wash the contents of the bowl round the trap and into the soil pipe. The trap is an effective barrier to foul air from the soil pipe connection. An alternative form is the siphonic water closet, which is more efficient and quieter in operation but has a greater risk of blockage if misused. Two types are produced: the single-trap and the double-trap.

The single-trap siphonic water closet has a restricted outlet that serves to retard the flow when flushed, so that the bore of the outlet connected to the bowl becomes full and sets up a siphonic flushing action, completely emptying the contents of the bowl (see Fig. 11.2.1). With the double-trap siphonic pan, the air is drawn from the pocket between the two traps; when the flushing operation is started this causes atmospheric pressure to expel the complete contents of the bowl through both traps into the soil pipe (see Fig. 11.2.1).

The pan should be fixed to the floor with brass screws and covers and the connection to the soil pipe is normally made with a soil-pipe connector featuring a series of rubber seals that fits over the WC outlet and connects to the uPVC soil pipe. The flush pipe is invariably connected to the pan with a special plastic or rubber one-piece connector.

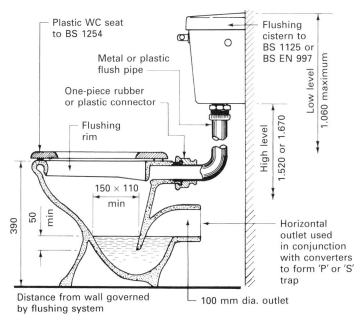

BS EN 37 ceramic washdown WC pan

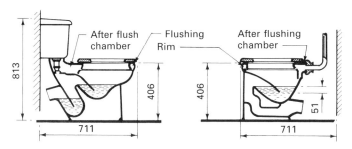

Typical siphonic WC pans

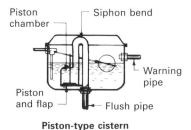

Piston-type cistern

Figure 11.2.1   WC pans and cisterns.

Flushing cisterns together with the flush pipes are usually constructed to the recommendations of BS EN 14055 and BS EN 997, and are usually made from ceramic ware or plastic materials. The current modern cistern uses the piston action to lift a body of water and start a syphonic action from the water reservoir in the cistern. The piston-type cistern is activated by a lever or dual-flush button. When activated, the disc- or flap-valve piston is raised and with it the water level, which commences the siphonage (see Fig. 11.2.1). The level of the water in the cistern is controlled by a float valve, and an overflow or warning pipe of a larger diameter than the inlet is fitted to discharge so that it gives a visual warning, usually in an external position. Current cistern technology has enabled an integral warning overflow to be built into the valve assembly that allows the overflow to pass through the WC. The capacity of the cistern will be determined by local water board requirements, the most common being 5, 6, 7 and 9 litres.

FLUSHING VALVES

Flushing valves have been a popular marine plumbing installation for many years. In compact lavatory compartments they have proved effective in restricted space, and they are economic with the use of limited water resources. They are becoming a credible alternative for use in buildings, particularly where usage is frequent, as they can flush at any time without delay. They comprise a large equilibrium valve, which functions on the principle of water displacement and pressure equalisation limiting the flow. The Water Supply (Water Fittings) Regulations give details of the pressure flushing valves and their use.

WASHBASINS

Washbasins for domestic work are usually made from a ceramic material, such as vitreous china. A wide variety of designs, sizes, types and colours is available, the choice usually being one of personal preference. BS 1188 and BS 5506 provide recommendations for ceramic washbasins and pedestals, and specify two basic sizes: 635 mm × 457 mm and 559 mm × 406 mm. Small wall-hung hand-rinse basins are also available for use in limited space, such as downstairs toilets. These are produced to BS EN 111. All these basins are a one-piece fitment, having an integral overflow, separate waste outlet and generally pillar taps (see Fig. 11.2.2).

Wash basins can be supported on wall-mounted cantilever brackets, leg support frames or pedestals. The pedestals are made from identical material to the washbasin, and are recessed at the back to receive the supply pipes to the taps and the waste pipe from the bowl. Although pedestals are designed to fully support the washbasin, most manufacturers recommend that small wall-mounted safety brackets are also used for stability.

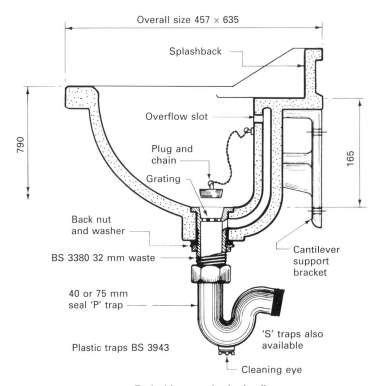

Overall size 457 × 635

Splashback

Overflow slot

Plug and
chain

Grating

790

165

Back nut
and washer

BS 3380 32 mm waste

Cantilever
support
bracket

40 or 75 mm
seal 'P' trap

'S' traps also
available

Plastic traps BS 3943

Cleaning eye

Typical lavatory basin details

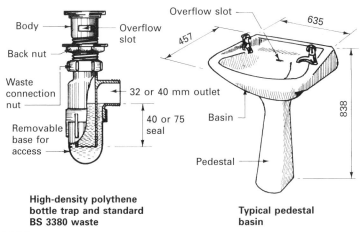

Body

Overflow
slot

Overflow slot

635

457

Back nut

Waste
connection
nut

32 or 40 mm outlet

40 or 75
seal

Basin

Removable
base for
access

838

Pedestal

**High-density polythene
bottle trap and standard
BS 3380 waste**

**Typical pedestal
basin**

Figure 11.2.2   Basins, traps and wastes.

BATHS AND SHOWERS

Baths are available in a wide variety of designs and colours, made from porcelain-enamelled cast iron, vitreous-enamelled sheet steel, 8 mm cast acrylic sheet, or 3 mm cast acrylic sheet reinforced with a polyester resin/glass fibre laminate. The latter is the most common material used by manufacturers as it is lightweight, which is an advantage when installation is undertaken. Most bath designs today are rectangular in plan for their outside dimensions in order to fit against internal walls. Internally they can be designed with different shapes and contouring. Special corner and free-standing baths are also available. They are made as flat bottomed as practicable, with just sufficient fall to allow for gravity emptying and resealing of the trap. Baths are supplied with one or two holes for pillar taps or mixer fittings, or none for wall-mounted taps, and holes for the waste outlets and overflow. Options include handgrips, built-in soap and sponge recesses, and overflow outlets. The overflow pipe captures the water level when the bath has been overfilled because the taps have not been turned off. It will not provide a visible warning as it is designed to connect with the waste trap beneath the bath (see Fig. 11.2.3). Support for baths is usually by adjustable feet for cast iron and steel, and by a strong cradle for the acrylic baths, coupled with a plywood base. The bath can be enclosed by including a subframe to which panels of enamelled hardboard or moulded high-impact polystyrene or glass fibre are fixed. These panels can be fixed by using stainless steel or aluminium angles, or they may be included as an integral part of the bath design.

Shower sprays can be used in conjunction with a bath by fitting a rigid plastic shower screen or flexible curtain to one end of the bath and using a shower mixer tap. However, a separate shower fitment is considered preferable. Essentially a shower requires a method of providing a high-level spray of mixed water and a tray to capture the run-off, with a sealed door top to prevent overspray. Showers require less space than the conventional bath, use less hot water and are considered to be more hygienic as the used water is being continuously discharged (see Fig. 11.2.3). Materials available are similar to those described for baths. The spray outlet is normally fixed to the wall, and is connected to a mixing valve so that the water temperature can be controlled. Spray outlets are available that directly connect to the hot and cold water supply or that are electric and just connect to the cold supply.

SINKS

Sinks are used mainly for the preparation of food and washing of dishes within a kitchen, and for the washing of clothes, and are usually positioned at the first drinking water supply outlet off the incoming water main. Their general design follows that described for basins, except that they are larger in area

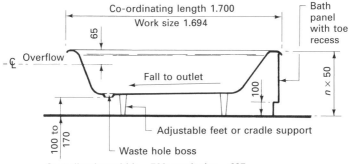

Co-ordinating width = 700; work size = 697
Co-ordinating sizes can be varied within range of
$n \times 100$ mm, where n = any natural number including unity

BS 1189 (cast iron), BS 4305 (acrylic) and BS 1390 (vitreous enamelled steel) baths. Also, BS EN 60335-2-60 whirlpool baths

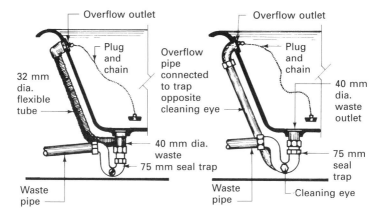

Alternative overflow connections

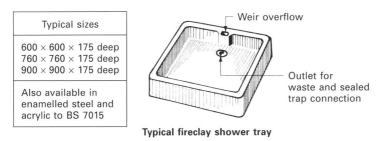

Typical sizes
$600 \times 600 \times 175$ deep $760 \times 760 \times 175$ deep $900 \times 900 \times 175$ deep
Also available in enamelled steel and acrylic to BS 7015

Typical fireclay shower tray

Figure 11.2.3 Baths and shower trays.

and deeper due to their intended function. Any material considered suitable for sanitary appliance construction can be used for the manufacture of a sink. Designs range from the simple Belfast sink with detachable draining boards of metal, plastic or timber to the combination units consisting of integral draining boards and twin bowls. Draining boards can now be integrated into worktop materials, such as granite, where the drainer is machined and polished into the worktop. Support can be wall-mounted cantilever brackets, framed legs or a purpose-made cupboard unit; typical details are shown in Fig. 11.2.4.

The layout of domestic sanitary appliances is governed by size of fitments, personal preference, pipework system being used and the space available. Building Regulations Approved Document G details requirements and minimum facilities to be made available in a dwelling, and the need for sanitary accommodation to be separated from kitchen areas.

DISCHARGE PIPEWORK

Building Regulations Approved Document H: *Sanitary pipework and drainage* sets out in detail the recommendations for soil pipes, waste pipes and ventilating pipes. These regulations govern such things as minimum diameters of soil pipes, material requirements, provision of adequate water seals by means of an integral trap or non-integral trap, the positioning of soil pipes on the inside of a building, overflow pipework and ventilating pipes.

PIPEWORK SYSTEMS

The most common domestic system in use today is the single-stack system.

SINGLE-STACK SYSTEM

This system was developed by the Building Research Establishment, and is fully described in their Digest No. 249. It is also fully detailed in BS EN 12056-2, and is now adapted into the Building Regulations Approved Document H. It is a simplification of the one-pipe system using deep seal traps, relying on venting by the discharge pipe and placing certain restrictions on basin waste pipes, which have a higher risk of self-siphonage than other appliances. A diagrammatic layout is shown in Fig. 11.2.5.

Materials that can be used for domestic stack pipework include galvanised steel prefabricated stack units (BS 3868), cast iron (BS 416) and uPVC (BS 4514). The latter is standard contemporary practice, being light and easy to install with simple push-fit or solvent-welded joints. The former would be used for commercial applications such as hospital wards. Branch waste pipes can be of a variety of plastics, including uPVC, polyethylene and polypropylene, or of copper.

Most manufacturers of soil pipes, ventilating pipes and fittings produce a variety of special components for various plumbing arrangements and

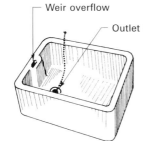

Weir overflow

Outlet for 40 mm waste and trap connection

BS 1206 fireclay sinks
Belfast pattern – wide range of sizes
from 457 × 380 × 200 deep up to
1.219 × 610 × 305 deep. Sinks
supported on cantilever brackets,
legs and bearers or on brick
dwarf walls. Fixing heights 850
to 920 to top of sink

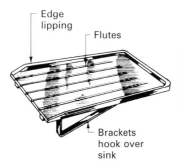

Edge
lipping

Flutes

Clip-on draining boards
Reversible and available in
hardwood, stainless steel,
aluminium alloy and vitreous
enamel in range of sizes
from 560 × 406 to 762 × 457

Brackets
hook over
sink

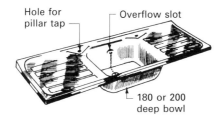

Hole for
pillar tap

Overflow slot

Stainless steel sinks
Generally made to the
requirements of BS EN
13310 available with single
or double drainers in a
range of sizes from 1.070 ×
460 to 1.600 × 530. Sink
tops supported on
cantilever brackets

180 or 200
deep bowl

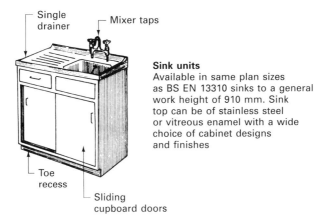

Single
drainer

Mixer taps

Sink units
Available in same plan sizes
as BS EN 13310 sinks to a general
work height of 910 mm. Sink
top can be of stainless steel
or vitreous enamel with a wide
choice of cabinet designs
and finishes

Toe
recess

Sliding
cupboard doors

Note: Waste fittings to BS EN 274-1

Figure 11.2.4   Sinks and draining boards.

appliance layouts. These fittings have the water closet socket connections, bosses for branch waste connections and access plates for cleaning and maintenance arranged as one prefabricated assembly, to ease site work and ensure reliable and efficient connections to the discharge or soil pipe.

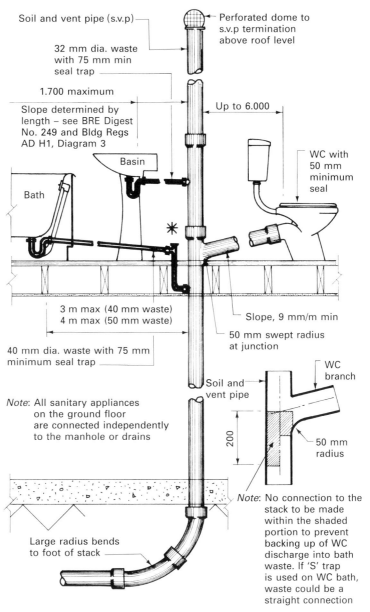

Figure 11.2.5 Single-stack system.

SMALL-DIAMETER PUMPED DISCHARGE SYSTEMS

A macerator and pump unit may be connected to the WC pan outlet to collect, process and discharge soil and waste from the WC and other sanitary appliances. The discharge is conveyed in a relatively small-diameter pipe to the nearest convenient soil stack. Fig 11.2.6 shows the possible installations from the box attachment at the back of a WC pan, through vertical and near-horizontal pipes. Structural disruption is minimal, and the small discharge pipes can be used for considerable distances between WC and stack in accordance with the manufacturer's instructions.

Macerator installation is permitted in a dwelling only as a supplement to an existing gravity-discharging WC. See Building Regulation G: Hygiene, Approved Document G, *Sanitary conveniences and washing facilities*.

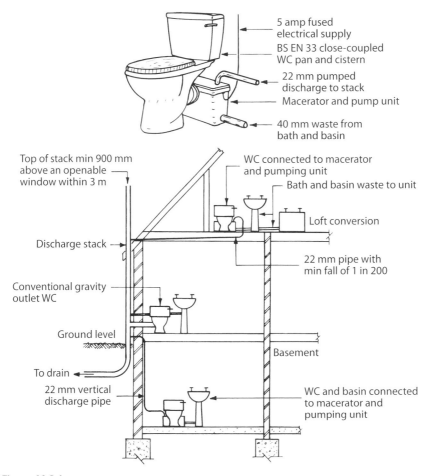

Figure 11.2.6 **Macerator/pump installation.**

This requirement is there to make sure that, in the event of an electrical power failure, the conventional gravity-discharge WC will still function.

BS EN 12050 contains further guidance on macerator construction and testing.

Most applications are for loft conversions, extensions and basements, or for other situations where it is impractical to install a normal discharge branch pipe from the new sanitary fitting. The Care Standards Act has also produced a need for additional facilities. Care home proprietors must provide their residents with access to suitable and sufficient conveniences. This places greater emphasis for en-suite facilities, particularly in care homes for elderly people.

Installation guidelines for using a macerator pumped system

- By application to, and at the discretion of, the local authority building control department.
- Building Regulations approval will apply only where appliance specification satisfies strict test standards, such as that set by the British Board of Agrément (BBA).
- Discharge pipework:
  - 22 or 28 mm outside diameter copper or equivalent in stainless steel or polypropylene;
  - long radius 'pulled' bends at direction changes, not proprietary elbow fittings;
  - minimum fall or gradient, 1 in 200;
  - up to 50 m horizontally and 5 m vertically.
- Not suitable for use with siphonic WC pans.
- Short spigot, 100 mm diameter horizontal outlet pans to BS EN 33 or 37 required.
- Electrical requirements:
  - 230 volts single phase;
  - 5 amp fused non-switched neon light spur box;
  - 450–650 watts pump specification.

Drainage   11.3

Drainage is the system of pipework, installed below ground level, to convey the foul and surface water to a suitable disposal installation. The usual method of disposal is to connect the pipework to the public sewer, which will convey the discharges to a local authority sewage treatment plant for processing. Alternatives are a small self-contained treatment plant on site or a cesspool; the latter is a collection tank to hold the discharge until it can be collected in a special tanker lorry and taken to the local authority sewage treatment installation for disposal.

PRINCIPLES OF GOOD DRAINAGE

- Materials should have suitable strength and durability for their location.

- Diameter of drain to be as small as practicable: for soil drains, the minimum diameter allowed is 100 mm; for surface water, the minimum diameter is 75 mm; generally 100 mm branch and 150 mm main runs are used.

- Every part of a drain should be accessible for the purposes of inspection, maintenance and cleaning.

- Drains should be laid in straight runs as far as possible to avoid issues with bends.

- Drains must be laid to a gradient that will render them efficient. The fall or gradient should be calculated according to the rate of flow, the velocity required and the diameter of the drain. Individual domestic buildings have an irregular flow, and tables to accommodate this are provided in the Building Regulations Approved Document H. Alternatively, Maguire's rule can be used to give a gradient with reasonable velocity. This well-established rule of thumb calculates the gradient by dividing the pipe diameter (mm) by 2.5. Thus a 100 mm drain can have a fall of 1 in 40, a

150 mm drain 1 in 60, and so on. Velocity of flow should be at least 0.75 m/s to ensure self-cleansing, with a drain capacity not more than 0.75 proportional depth (i.e. not more than 0.75 times the depth of the internal pipe) to ensure clear flow conditions.

- Every drain inlet should be trapped to prevent the entry of foul air into the building; the minimum seal required is 50 mm. Reference to Table 1 of the Building Regulations Approved Document H gives further details on trap sizes. The trap seal is provided in many cases by the sanitary fitting itself: for example, a WC. Rainwater drains need not be trapped unless they connect with a soil drain or sewer.

- A rodding access point should be located at the head or start of each drain run.

- Inspection chambers, manholes, rodding eyes or access fittings should be placed at changes of direction and gradient if these changes would prevent the drain from being readily cleansed.

- Inspection chambers must also be placed at a junction, unless each run can be cleared from an access point. Table 11 of Approved Document H gives inspection chamber sizes.

- A change of drainpipe size will also require access.

- At junctions, an oblique angle must be used; this is pre-manufactured into the fitting.

- Avoid placing drains under foundations and building substructures as it places loading on their integrity. Drains in this position may be subject to a build-over agreement with the water supply authority. This will stipulate the measures that have to be undertaken when building over or near a drain run.

- Drains that are within 1 m of the foundations to the walls of buildings and below the foundation level must be backfilled with concrete up to the level of the underside of the foundations. Drains more than 1 m from the foundations are backfilled with concrete to a depth equal to the distance of the trench from the foundation less 150 mm.

The depth of cover to a drain must be carefully considered and will vary with the location, be it a field or garden, or a main trunk road. Tables 8 and 9 of Approved Document H give minimum depths of cover for various drainage applications.

DRAINAGE SCHEMES

The scheme or plan layout of drains will depend upon a number of factors, including:

- number of discharge points connecting to the drain;
- relative positions of discharge points;

- drainage system and location of the local authority sewers;
- internal layout of sanitary fittings;
- external positions of rainwater pipes;
- function of buildings connecting to drain system;
- topography of the area to be served.

Drainage systems must be designed within the limits of the terrain, so that the discharges can flow by gravity from the point of origin to the point of discharge without the need for pumping systems. The pipe sizes and gradients must be selected to provide sufficient capacity to accommodate maximum flows and any expansion to the system, while at the other extreme they must have adequate self-cleansing velocity at minimum flows to prevent debris from accumulating. Economic and construction factors control the depth to which drains can be laid. Therefore, in flat and opposingly inclined areas, it may be necessary to provide pumping stations to raise the drainage discharge to higher-level sewers.

There are three drainage systems used by local authorities in the UK, and the method employed by any particular authority will determine the basic scheme to be used for the drain runs from individual premises.

COMBINED SYSTEM

With this system, all the drains discharge into a common or combined sewer. It is a simple and economic method as there is no duplication of drains. This method has the advantage of easy maintenance, all drains are flushed when it rains and it is impossible to connect to the wrong sewer. The main disadvantage is that all the discharges must pass through the sewage treatment installation, which could be costly and prove to be difficult with periods of heavy rain when surface water run-off to the sewer is at its maximum.

TOTALLY SEPARATE SYSTEM

This is the most common method employed by local authorities. Two sewers are used in this method. One sewer receives the surface water discharge and conveys this direct to a suitable outfall, such as a river, where it is discharged without treatment, as it is basically rainwater. The second sewer receives all the soil or foul discharge from baths, basins, sinks, showers and toilets; this is then conveyed to the sewage treatment installation. More drains are required, and it is often necessary to cross drains one over the other, which may prove difficult. There is a risk of connection to the wrong sewer, and the soil drains are not flushed during heavy rain, but the savings on the treatment of a smaller volume of discharge leads to an overall economy for the local authority.

PARTIALLY SEPARATE SYSTEM

This is a compromise between the other two systems, and is favoured by some local authorities because of its flexibility. Two sewers are used, one to carry surface water only and the other to act as a combined sewer. The amount of surface water to be discharged into the combined sewer can be adjusted according to the capacity of the sewage treatment installation.

Soakaways, which are pits below ground level designed to receive surface water and allow it to percolate into the soil, are sometimes used to lessen the load on the surface water sewers. Typical examples of the three drainage systems are shown in Fig. 11.3.1.

PRIVATE SEWERS

A sewer can be defined as a means of conveying waste, soil or rainwater below the ground that has been collected from the drains, and conveying it to the final disposal point. If the sewer is owned and maintained by the local authority it is generally called a **public sewer**, whereas one owned by a single person or a group of people and maintained by them can be classed as a **private sewer**.

When planning the connection of houses to the main or public sewer, one method is to consider each dwelling in isolation, but important economies in design can be achieved by the use of a private sewer. A number of houses are connected to the single sewer, which in turn is connected to the public sewer. Depending upon the number of houses connected to the private sewer, and the distance from the public sewer, the following savings are possible:

- total length of drain required, as each house is on a single branch and not an individual connection to the main sewer;
- number of connections to the public sewer, as there is only one;
- amount of openings in the roads;
- number of inspection chambers.

However, as the sewer is in private ownership, the main disadvantage is that, should any work be required on it due to settlement or damage, the cost is equally divided between all parties that it serves. A comparative example is shown in Fig. 11.3.2.

DRAINAGE MATERIALS

Drainpipes are considered as either rigid or flexible according to the material used in their manufacture. Clay is a major material used for rigid drainpipes in domestic work; flexible drainpipes are produced from unplasticised PVC.

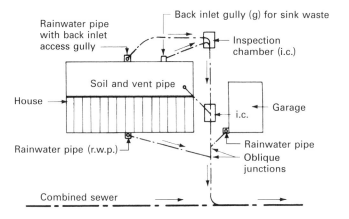

Combined system

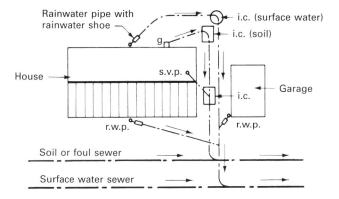

Separate system

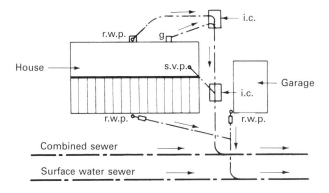

Partially separate system

Figure 11.3.1   Drainage systems.

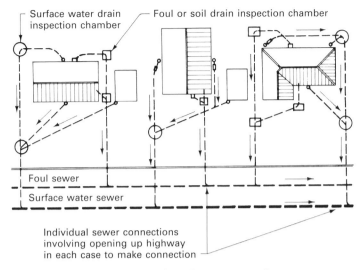

Individual sewer connections
involving opening up highway
in each case to make connection

Individual drain and sewer connections

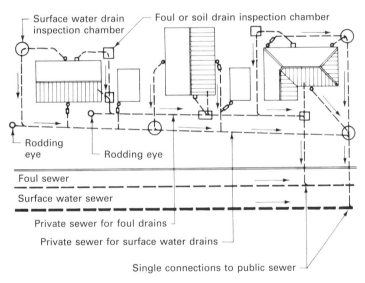

Private sewer for foul drains

Private sewer for surface water drains

Single connections to public sewer

Note: Generally 20 houses can be connected to a 100 mm dia.
private sewer at a gradient of 1:70 and 100 houses
can be connected to a 150 mm dia. private sewer
laid to a fall of 1:150

Figure 11.3.2   Example of a private sewer arrangement.

CLAY PIPES

Clay is the traditional material used for domestic drainage, with current manufacturing standards in accordance with BS 65 and BS EN 295. The high standards of manufacture combined with quality, dense materials produce a drain that is impervious to fluids. Several qualities of pipes are produced, ranging from standard pipes for general use to surface water pipes and pipes of extra strength to be used where heavy loadings are likely to be encountered. The type and quality of pipes are marked on the barrel so that they can be identified after firing.

Clay pipes are produced in a range of diameters from 100 to 300 mm nominal bore for general building work. Lengths vary between manufacturers up to 1,750 mm for plain-end pipes, with spigot and socketed pipes produced in shorter lengths down to 600 mm. They can be obtained with sockets and spigots prepared for rigid or flexible jointing. Rigid jointing is all but obsolete except for repair work on old drainage systems. Most pipes are supplied plain-ended for use with push-fit polypropylene sleeve couplings, which slide over the pipe with the use of a lubricant. A wide variety of fittings for use with clay pipes is manufactured to give flexibility when planning drainage layouts and means of access. Typical examples of clay pipes, joints and fittings are shown in Figs. 11.3.3 and 11.3.4.

Clay pipes are resistant to attack by a wide range of acids and alkalis; they are therefore suitable for all forms of domestic drainage.

UNPLASTICISED PVC PIPES

These pipes and fittings are made from polyvinyl chloride plus additives that are needed to facilitate the manufacture of the polymer and produce a sound, durable pipe. BS 3506 gives the requirements for pipes intended for industrial and commercial purposes, and BS 4660 covers the pipes and fittings for domestic applications. Standard outside diameters are 110 and 160 mm (100 and 150 mm nominal bore) with non-standard outside diameters of 82, 200, 250 and 315 mm being available from most manufacturers. Long lengths of 3 m and 6 m can be easily cut using a hacksaw. BS EN 1401-1 is the specification for uPVC sewer pipes in nominal diameters of 200–630 mm. The pipes are obtainable with socket joints for either a solvent-welded joint or a ring-seal joint. uPVC pipes have a smooth bore, are light, durable and easy to handle; long lengths reduce the number of joints required. They can be jointed and laid in all weathers.

DRAIN LAYING

Domestic drains are laid in trenches that are excavated and if necessary supported in a similar manner to that described for foundations trenches (see page 80); the main difference is that drain trenches are excavated to the required fall or gradient. It is good practice to programme the work to enable the activities of excavation, drain laying and backfilling to be carried out in quick succession so that the excavations remain open for the shortest possible time.

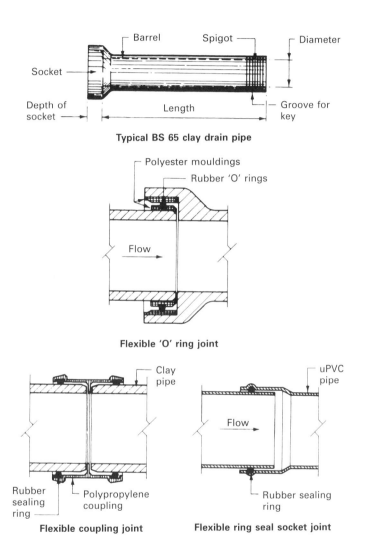

Typical BS 65 clay drain pipe

Flexible 'O' ring joint

Flexible coupling joint

Flexible ring seal socket joint

Figure 11.3.3   Pipes and joints.

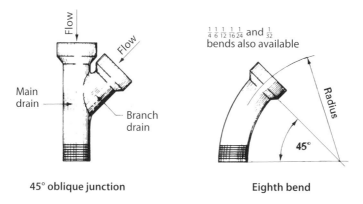

Flow

Main drain

Branch drain

45° oblique junction

$\frac{1}{4}$ $\frac{1}{6}$ $\frac{1}{12}$ $\frac{1}{16}$ $\frac{1}{24}$ and $\frac{1}{32}$ bends also available

Radius

45°

Eighth bend

$\frac{1}{2}$ and $\frac{3}{4}$ section channel junctions and bends similar

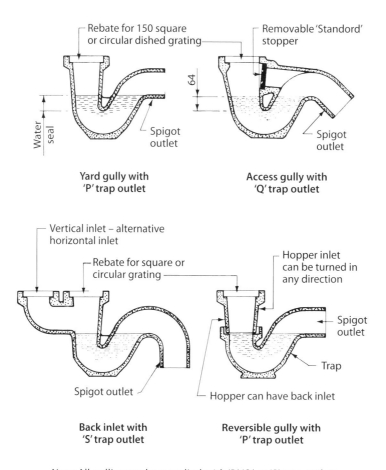

Rebate for 150 square or circular dished grating

64

Water seal

Spigot outlet

Yard gully with 'P' trap outlet

Removable 'Standord' stopper

Spigot outlet

Access gully with 'Q' trap outlet

Vertical inlet – alternative horizontal inlet

Rebate for square or circular grating

Spigot outlet

Back inlet with 'S' trap outlet

Hopper inlet can be turned in any direction

Spigot outlet

Trap

Hopper can have back inlet

Reversible gully with 'P' trap outlet

Note: All gullies can be supplied with 'P', 'Q', or 'S' trap outlets

Figure 11.3.4 **Clay drainpipe fittings and gullies.**

Approved Document H recommends drains to be of sufficient strength and durability and so jointed that the drain remains watertight under all working conditions, including any differential movement between the pipe and ground. The coupling that joins flexible pipes and clay pipes together has a limited amount of movement within its design while still remaining watertight.

There is no requirement to cover drainage with a protective layer of concrete unless the local situation requires this additional support. Both clay and uPVC drains can accommodate axial flexibility and extensibility by combining flexible jointing with a granular flexible bedding medium, such as pea gravel. Several examples of bedding specifications are provided in the Building Regulations Approved Document H and BS EN 752: *Drain and sewer systems outside buildings*. Two popular applications of drainage trenches and bedding mediums are shown in Fig. 11.3.5.

The selected material required for the granular bedding and surround and for tamping around pipes laid on a jointed concrete base must be of the correct quality. Pipes depend to a large extent upon the support bedding for their strength and must therefore be uniformly supported on all sides by a material that can be hard-compacted. Generally a non-cohesive granular material with a particle size of 5–20 mm is suitable; this can be spread along the trench and levelled to give the required fall to the pipework.

Pipes with socket joints are laid from the bottom of the drain run with the socket end laid against the flow, each pipe being aligned and laid to the correct fall. The collar of the socket is laid in a prepared 'hollow' in the bedding, and the bore is centralised. Mechanical or flexible joints are self-aligning. Most flexible joints require a special lubricant to ease the jointing process, and those that use a coupling can be laid in any direction.

MEANS OF ACCESS

Drains require access for testing, maintenance and clearance of blockages. Four possibilities exist:

- rodding eye;
- inspection chamber;
- shallow access fitting;
- manhole.

RODDING EYES

These are generally located at the head of a drain and are effectively a swept extension of the drain up to ground level, with a sealed access plate. With local authority approval they can replace a more expensive inspection chamber to provide means for clearance in one direction only.

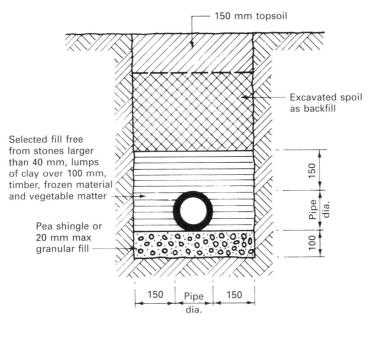

150 mm topsoil

Excavated spoil as backfill

Selected fill free from stones larger than 40 mm, lumps of clay over 100 mm, timber, frozen material and vegetable matter

Pea shingle or 20 mm max granular fill

150

Pipe dia.

100

150 Pipe dia. 150

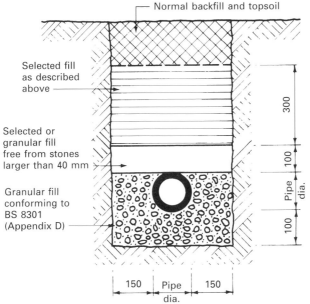

Normal backfill and topsoil

Selected fill as described above

Selected or granular fill free from stones larger than 40 mm

Granular fill conforming to BS 8301 (Appendix D)

300

100

Pipe dia.

100

150 Pipe dia. 150

Figure 11.3.5 Typical pipe bedding details.

ACCESS FITTINGS

These provide access to the drain run, for example, through a removable access and rodding cover within a gully pot.

INSPECTION CHAMBERS

These are used to invert depths of up to 1 m. They provide for limited access, and contain facilities for a few junctions and branch connections. They are smaller than a traditional brick manhole chamber and are constructed circular on plan. Materials include plastics, clay, and precast concrete. Fig. 11.3.6 shows examples of means of drain access, with the plastic inspection chamber containing optional adaptors, which can be cut to suit 100 or 150 mm pipes at various approach angles. Raising sections bring the inspection chamber base up to ground level; any adjustment can be accommodated by cutting one raising ring. Finally a secure lid made of concrete, plastic or cast aluminium frame and cover is fitted.

MANHOLES

These are inspection chambers over 1 m to invert. A manhole is a compartment containing half or three-quarter section round channels to enable the flow to be observed, and to provide a drain access point for cleansing and testing. Both inspection chambers and manholes are positioned to comply with the access recommendations of Approved Document H (see Fig. 11.3.1).

Simple domestic drainage is normally concerned only with shallow manholes up to an invert depth of 1,800 mm. The internal sizing is governed by the depth to invert, the number of branch drains, the diameter of branch drains and the space required for a person to work within the manhole. A general guide to the internal sizing of inspection chambers and manholes is given in Approved Document H, Tables 11 and 12.

Manholes can be constructed of brickwork or of rectangular or circular precast concrete units, a faster method of construction (see Fig. 11.3.7). The access covers used in domestic work are generally of cast iron or pressed steel, and are light-duty as defined in BS EN 124. Care should be taken with the siting of the manhole as a light-duty cover will not withstand the loading from vehicular traffic. Manholes have a single seal, which should be bedded in grease to form an airtight joint; double seal covers would be required if the access was situated inside the building. Concrete access covers are available for use with surface water manholes.

LOCATION OF DRAIN ACCESS POINTS

The location of access points within a drain is laid out within Approved Document Part H. This states the maximum distances between any access point on a drain run. Long straight drain runs must also be avoided, otherwise

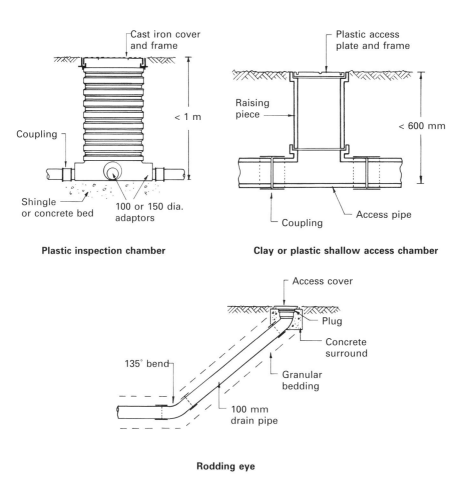

Plastic inspection chamber

Cast iron cover and frame

< 1 m

Coupling

Shingle or concrete bed

100 or 150 dia. adaptors

Clay or plastic shallow access chamber

Plastic access plate and frame

Raising piece

< 600 mm

Coupling

Access pipe

Rodding eye

Access cover

Plug

Concrete surround

135° bend

Granular bedding

100 mm drain pipe

Figure 11.3.6   Drains – means of access up to 1 m deep.

parts of the drain would be very difficult to access by cleaning rods in the event of a blockage. For domestic drainage (up to 300 mm diameter) the maximum spacing between different means of access provision are as listed in Table 11.3.1 (taken from Table 13 of Approved Document H).

VENTILATION OF DRAINS

To prevent foul air from the combined drains escaping and causing a nuisance, all drains should be vented by a flow of air. A ventilating pipe should be provided at or near the head of each main drain and any branch drain exceeding 10.000 in length. The ventilating pipe can be a separate pipe, or the soil discharge stack pipe can be carried upwards to act as a ventilating discharge stack or soil vent pipe. Ventilating pipes should be open to the outside air and carried up for at least 900 mm above the head of any window opening within a horizontal distance of 3.000 from the ventilating pipe, which should be finished with a cage or cover that does not restrict the flow of air.

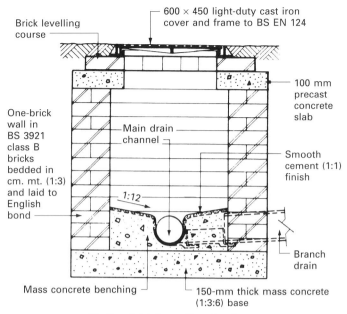

Brick levelling course

600 × 450 light-duty cast iron cover and frame to BS EN 124

100 mm precast concrete slab

One-brick wall in BS 3921 class B bricks bedded in cm. mt. (1:3) and laid to English bond

Main drain channel

Smooth cement (1:1) finish

1:12

Branch drain

Mass concrete benching

150-mm thick mass concrete (1:3:6) base

Shallow brick inspection chamber

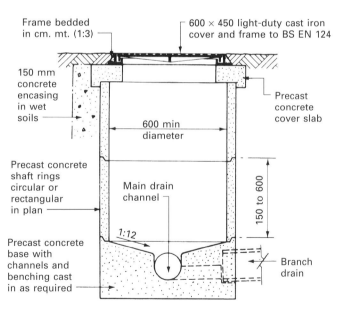

Frame bedded in cm. mt. (1:3)

600 × 450 light-duty cast iron cover and frame to BS EN 124

150 mm concrete encasing in wet soils

Precast concrete cover slab

600 min diameter

Precast concrete shaft rings circular or rectangular in plan

Main drain channel

150 to 600

Precast concrete base with channels and benching cast in as required

1:12

Branch drain

Precast concrete inspection chamber (BS 5911-4 and BS EN 1917)

Figure 11.3.7 Typical shallow inspection chambers.

Table 11.3.1 Maximum spacing of drainage access (m).

	To: Small access fitting	Large access fitting	Junction	Inspection chamber	Manhole
From:					
Start of drain	12	12	–	22	45
Rodding eye	22	22	22	45	45
Small access fitting	–	–	12	22	22
Large access fitting	–	–	22	45	45
Inspection chamber	22	45	22	45	45
Manhole	–	–	–	45	90 (note 2)

Notes

1. Small access fitting: 150 mm × 100 mm or 150 mm diameter. Large access fitting: 225 mm × 100 mm.

2. Up to 200 m is permitted if the manhole is of substantial proportions, i.e. sufficient to allow a person entry space to work.

Conventional venting through the stack is not always necessary. An air admittance valve can be fitted in up to four consecutive dwellings of no more than three storeys, if the fifth dwelling is conventionally vented. Systems with air admittance valves have the following advantages:

■ ventilating stack can terminate inside the building (above highest spillover level), typically in the roof space, as the valve only works one way and does not permit foul air into the building;

■ greater design flexibility;

■ adaptable to plastic or metal pipework;

■ visually unobtrusive, as there is no projecting external stack pipe on the outside of the building.

The principles of application and operation are shown in Fig. 11.3.8. A discharge of water in the stack creates a slight negative pressure, sufficient to open the valve and admit air. After discharge a return to atmospheric pressure allows the spring to re-seal the unit to prevent foul air escaping. To satisfy the Building Regulations, the device must have received design approval in accordance with the Board of Agrément and have a valid certificate.

RAINWATER DRAINAGE

A rainwater drainage installation is required to collect the discharge from roofs and paved areas, and convey it to a suitable drainage system or soakaway. Paved areas, such as garage forecourts or hardstands, are laid to a fall so as to direct the rainwater into a collecting yard gully, which is connected to the surface water drainage system. A rainwater installation for a roof consists of a collection

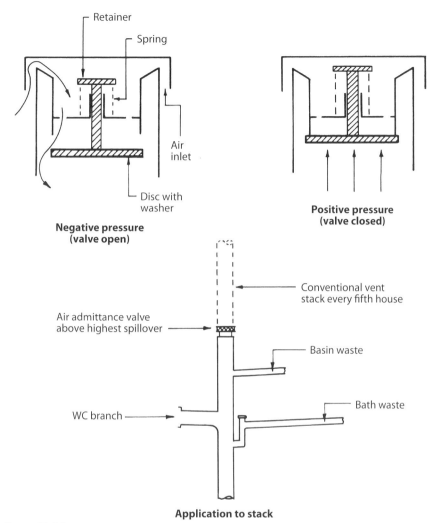

Retainer

Spring

Air
inlet

Disc with
washer

**Negative pressure
(valve open)**

**Positive pressure
(valve closed)**

Conventional vent
stack every fifth house

Air admittance valve
above highest spillover

Basin waste

WC branch

Bath waste

Application to stack

Figure 11.3.8   **Air admittance valve.**

channel called a **gutter**, which is connected to a vertical rainwater downpipe. The rainwater pipe is terminated at its lowest point by means of a rainwater shoe for discharge to a surface water drain, or a trapped gully if the discharge is to a combined drain (see Fig. 11.3.9). By using a shoe, any debris washed down the rainwater system will be caught by the grate of the gulley it enters and within the trap of the gully. If a separate system of drainage or soakaways is used, it may be possible to connect the rainwater pipe direct to the drains, provided there is an alternative means of access for cleansing should it become blocked.

The materials available for domestic rainwater installations are galvanised pressed steel, seamless aluminium, cast iron and uPVC. The usual material for domestic work is uPVC.

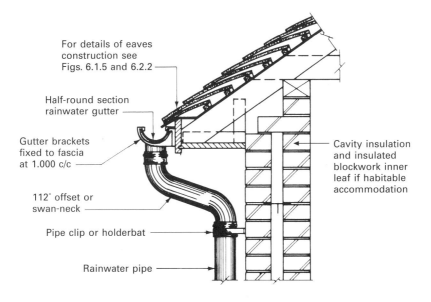

For details of eaves construction see Figs. 6.1.5 and 6.2.2

Half-round section rainwater gutter

Gutter brackets fixed to fascia at 1.000 c/c

112° offset or swan-neck

Pipe clip or holderbat

Rainwater pipe

Cavity insulation and insulated blockwork inner leaf if habitable accommodation

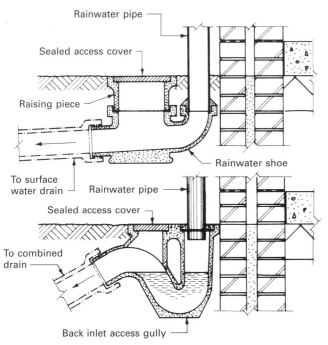

Rainwater pipe

Sealed access cover

Raising piece

To surface water drain

Rainwater shoe

Rainwater pipe

Sealed access cover

To combined drain

Back inlet access gully

Figure 11.3.9   Rainwater pipework and drainage.

CAST IRON RAINWATER GOODS

Cast iron rainwater pipes, gutters and fittings are generally made to the requirements of BS 460, which specifies a half-round section gutter with a socket joint in diameters from 75 to 150 mm and an effective length of 1,800 mm. This type of guttering is expensive when compared to uPVC alternatives and is mainly used in restoration and conservation work, where a like-for-like match is required. The gutter socket joint should be lapped in the direction of the flow and sealed with either putty or an approved sealing compound before being bolted together. The guttering should then be painted, or can be specified to have a powder-coated finish. The gutter is supported at 1,000–1,800 mm centres by means of mild steel gutter brackets screwed to the feet of rafters for open eaves, or to the fascia board with closed eaves.

Cast iron rainwater pipes are also produced to a standard effective length of 1,800 mm with a socket joint that is sealed with putty or run lead or, in many cases, dry jointed – the pipe diameters range from 50 to 150 mm. The downpipes are fixed to the wall by means of pipe nails and spacers when the pipes are supplied with ears, or with split-ring hinged holderbats when the pipes are supplied without ears cast on. A full range of fittings such as outlets, stopped ends and internal and external angles is available for cast iron half-round gutters, and, for the downpipes, fittings such as bends, offsets and rainwater heads are produced.

UNPLASTICISED PVC RAINWATER GOODS

The advantages of uPVC rainwater goods over cast iron are:

- easier jointing, gutter bolts are not required and the joint is self-sealing, generally by means of a butyl or similar strip;
- lighter to handle;
- more economical material costs;
- corrosion is eliminated;
- no decoration is required – several standard colours are available, including brown, black, white and grey;
- fewer breakages;
- better flow properties usually enable smaller sections and lower falls.

Half-round gutters are supplied in standard effective lengths up to 6 m, with diameter ranging from 75 to 150 mm. Downpipes are supplied in two standard lengths of 2 and 4 m with diameters of 50, 63, 75 and 89 mm. The gutters, pipes and fittings are generally produced to the requirements of BS 12200-1. Typical details of domestic rainwater gutters and pipework are shown in Fig. 11.3.9.

SIZING OF PIPES AND GUTTERS

The sizing of the gutters and downpipes to cater for the discharge from a roof effectively will depend upon:

- total area of roof to be drained;
- intensity of the rainfall;
- materials that the gutter and downpipe are manufactured from;
- fall within gutter, usually in the range 1:150 to 1:600;
- number, size and position of outlets.

The requirements for Building Regulations Approved Document H concerning rainwater drainage can be satisfied using the design guide tables contained in this document, which give guidance to sizing gutters and downpipes, and to the selection of suitable materials.

CONNECTIONS TO SEWERS

It is generally recommended that all connections to sewers shall be made so that the incoming drain or private sewer is joined to the main sewer obliquely in the direction of flow, and so that the connection will remain watertight and satisfactory under all working conditions. As the main sewer is generally situated within the main carriageway, the final connection is normally made by the local authority or under their direction and supervision by an approved contractor.

The method of connection will depend upon a number of factors:

- relative sizes of sewer and connecting drain or private sewer;
- relative invert levels of the main sewer and the connection;
- position of nearest inspection chamber on the sewer run;
- whether the sewer is existing or being laid concurrently with the drains or private sewers;
- whether stopped or joinder junctions have been built into the existing sewer (see Fig. 11.3.10);
- the shortest and most practicable route.

If the public sewer is of a small diameter (less than 225 mm), the practical method is to remove two or three pipes and replace with new pipes and an oblique junction to receive the incoming drain. If three pipes are removed it is usually possible to 'spring in' two new pipes and the oblique junction and joint in the usual manner, but if only two pipes are removed a collar connection will be necessary (see Fig. 11.3.10).

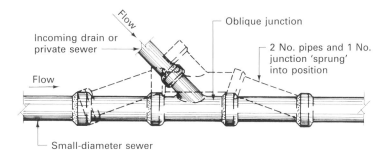

Removing 3 No. pipes and inserting oblique junction

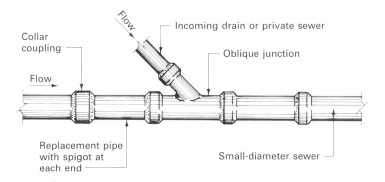

Removing 2 No. pipes and inserting oblique junction

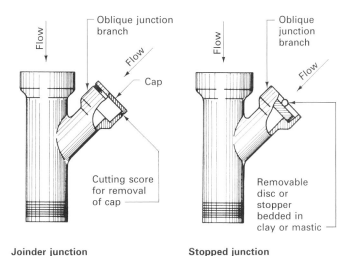

Joinder junction   Stopped junction

Figure 11.3.10   Connections to small-diameter sewers (traditional rigid-jointed spigot and socket clay pipes).

If new connections have been anticipated, stopped junctions or joinder junctions may have been included in the sewer design. A stopped junction has a disc temporarily secured in the socket of the branch arm, whereas the joinder has a cover cap as an integral part of the branch arm. Careful removal of the disc or cap is essential to ensure that a clean, undamaged socket is available to make the connection (see Fig. 11.3.10).

Connections to inspection chambers or manholes, whether new or existing, can take several forms, depending mainly upon the differences in invert levels. If the invert levels of the sewer and incoming drain are similar, the connection can be made in the conventional way using an oblique branch channel. Where there is a difference in invert levels the following can be considered:

■ a ramp formed in the benching within the inspection chamber;

■ a backdrop manhole or inspection chamber;

■ an increase in the gradient of the branch drain;

■ the provision of pumping if the sewer invert is above the drain invert.

The maximum difference between invert levels that can be successfully overcome by the use of a ramp is a matter of conjecture. The current standard BS EN 752 gives a maximum difference of 1.800 m. The generally accepted limit of invert level difference for the use of internal ramps is 700 mm. Typical constructional details are shown in Fig. 11.3.11.

Where the limit for ramps is exceeded, a backdrop manhole construction can be considered. This consists of a vertical 'drop' pipe with access for both horizontal and vertical rodding. If the pipework is of clay or concrete, the vertical pipe should be positioned as close to the outside face of the manhole as possible and encased in not less than 150 mm of mass concrete. Cast iron pipework is usually sited inside the chamber and fixed to the walls with rawbolts. Whichever material is used, the basic principles are constant (see Fig. 11.3.11).

Changing the gradient of the incoming drain to bring its invert level in line with that of the sewer requires careful consideration and design. Although simple in conception, the gradient must be such that a self-cleansing velocity is maintained and the requirements of Building Regulations Part H are not contravened.

Connections of small-diameter drains to large-diameter sewers can be made by any of the methods described above, or by using a saddle connection. A saddle is a short-socketed pipe with a flange or saddle curved to suit the outer profile of the sewer pipe. To make the connection, a hole must be cut in the upper part of the sewer to receive the saddle, ensuring that little or no debris is allowed to fall into the sewer. A small pilot hole is usually cut first, and this is enlarged to the required diameter by careful cutting and removing the debris outwards. The saddle connection is bedded onto the sewer pipe with a cement mortar, and the whole connection is surrounded with a minimum of 150 mm of mass concrete to form a secure joint (see Fig. 11.3.12).

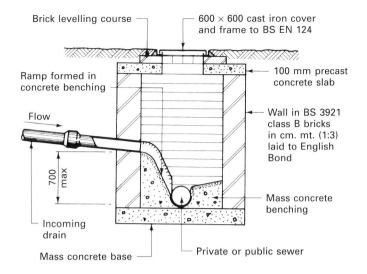

Brick levelling course

600 × 600 cast iron cover and frame to BS EN 124

Ramp formed in concrete benching

100 mm precast concrete slab

Flow

Wall in BS 3921 class B bricks in cm. mt. (1:3) laid to English Bond

700 max

Mass concrete benching

Incoming drain

Mass concrete base

Private or public sewer

Ramp connection

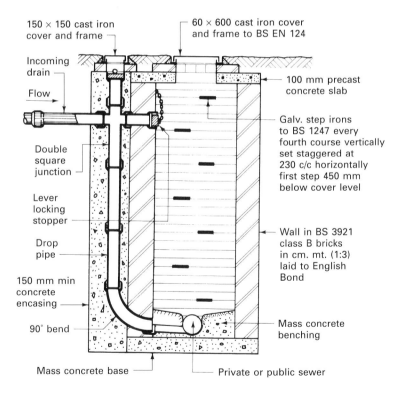

150 × 150 cast iron cover and frame

60 × 600 cast iron cover and frame to BS EN 124

Incoming drain

100 mm precast concrete slab

Flow

Galv. step irons to BS 1247 every fourth course vertically set staggered at 230 c/c horizontally first step 450 mm below cover level

Double square junction

Lever locking stopper

Drop pipe

Wall in BS 3921 class B bricks in cm. mt. (1:3) laid to English Bond

150 mm min concrete encasing

90° bend

Mass concrete benching

Mass concrete base

Private or public sewer

Backdrop connection

Figure 11.3.11   Manhole and inspection chamber sewer connections.

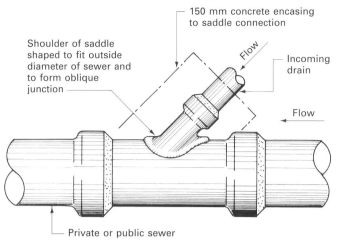

150 mm concrete encasing
to saddle connection

Shoulder of saddle
shaped to fit outside
diameter of sewer and
to form oblique
junction

Flow

Incoming
drain

Flow

Private or public sewer

Connection arrangement

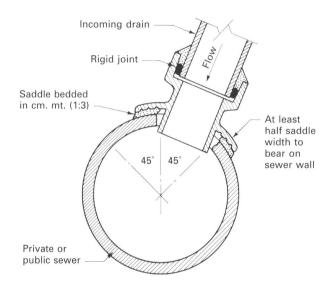

Incoming drain

Rigid joint

Flow

Saddle bedded
in cm. mt. (1:3)

45° 45°

At least
half saddle
width to
bear on
sewer wall

Private or
public sewer

Note: Saddle connection should be made in
the crown of the sewer pipe within 45°
on either side of the vertical axis

Typical section

Figure 11.3.12   Saddle connections to sewers.

To satisfy the Building Regulations requirements for watertightness of drains, reference to Approved Document H permits application of either a water or an air test. A smoke test could also be applied to determine the position of any apparent leakage revealed by the other tests. The Approved Document also makes reference to the recommendations contained in BS EN 752.

The local authority will carry out drain testing after the backfilling of the drain trench has taken place: therefore it is in the contractor's interest to test drains and private sewers before the backfilling is carried out, as the detection and repair of any failure discovered after backfilling can be time-consuming and costly.

TYPES OF TEST

There are two methods available for the testing of drains.

- **Water test** The usual method employed; carried out by filling the drain run being tested with water under pressure and observing whether there is any escape of water (see Fig. 11.3.13).
- **Air test** This is not a particularly conclusive test, but it is sometimes used in special circumstances, such as large-diameter pipes, where a large quantity of water would be required. If a failure is indicated by an air test, the drain should be retested using the more reliable water test (see Fig. 11.3.14).

The illustrations of drain testing have been prepared on the assumption that the test is being carried out by the contractor before backfilling has taken place.

In general, the testing of drains should be carried out between manholes as work proceeds. Manholes should be tested separately, and short branches of less than 6.000 m should be tested with the main drain to which they are connected; long branches would be tested in the same manner as a main drain.

A soakaway is a pit dug in permeable ground that receives the rainwater and surface water discharge from the roof and paved areas of a small building, and is so constructed that the water collected can percolate into the surrounding subsoil. This method saves resources on any treatment should the drain have been connected to a combined sewer. To function correctly and efficiently, a soakaway must be designed to take the following factors into account:

- permeability or rate of dispersion of the subsoil;
- volume to be dispersed;
- storage capacity required to accept sudden inflow of water, as encountered during a storm;
- local authority requirements as to method of construction and siting in relation to buildings;
- depth of water table.

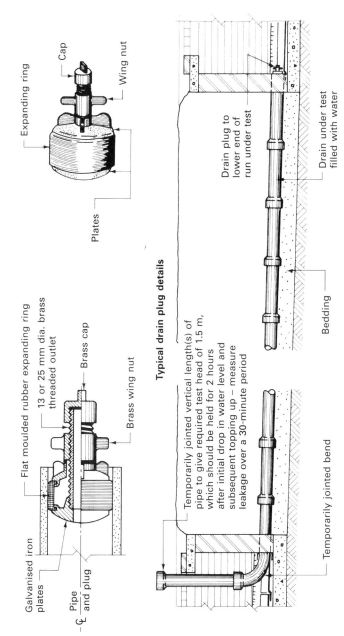

Expanding ring

Cap

Wing nut

Plates

Typical drain plug details

Flat moulded rubber expanding ring

13 or 25 mm dia. brass threaded outlet

Brass cap

Brass wing nut

Galvanised iron plates

Pipe and plug

C

Drain plug to lower end of run under test

Drain under test filled with water

Bedding

Temporarily jointed vertical length(s) of pipe to give required test head of 1.5 m, which should be held for 2 hours after initial drop in water level and subsequent topping up – measure leakage over a 30-minute period

Temporarily jointed bend

Figure 11.3.13   Water testing of drains.

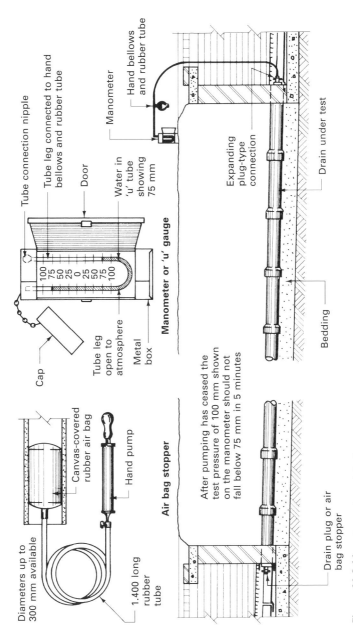

Tube connection nipple

Tube leg connected to hand
bellows and rubber tube

Door

Manometer

Hand bellows
and rubber tube

Water in
'u' tube
showing
75 mm

100
75
50
25
0
25
50
75
100

Expanding
plug-type
connection

Drain under test

Manometer or 'u' gauge

Cap

Tube leg
open to
atmosphere

Metal
box

Bedding

Diameters up to
300 mm available

Canvas-covered
rubber air bag

Hand pump

Air bag stopper

1.400 long
rubber
tube

After pumping has ceased the
test pressure of 100 mm shown
on the manometer should not
fall below 75 mm in 5 minutes

Drain plug or air
bag stopper

Figure 11.3.14   Air testing of drains.

Before any soakaway is designed or constructed, the local authority should be contacted to obtain permission and ascertain its specific requirements. Some authorities will permit the use of soakaways only as an outfall to a subsoil drainage scheme.

The rate at which water will percolate into the ground depends mainly on the permeability of the soil. Generally, clay soils are unacceptable for soakaway construction, whereas sands and gravels are usually satisfactory. An indication of the permeability of a soil can be ascertained by observing the rate of percolation. A borehole 150 mm in diameter should be drilled to a depth of 1.000 m. Water to a depth of 300 mm is poured into the hole, and the time taken for the water to disperse is noted. This gives an indication of the permeability of the soil that the proposed soakaway is to be constructed in. Several tests should be made to obtain an average figure, and the whole procedure should be repeated at 1.000 m stages until the proposed depth of the soakaway has been reached. A suitable diameter and effective depth for the soakaway can be obtained from design guidance in Building Research Establishment Digest 365.

TYPES OF SOAKAWAY

A soakaway is constructed by excavating a pit of the appropriate size and either filling the void with selected coarse granular material or lining the sides of the excavation with brickwork or precast concrete rings (see Figs. 11.3.15 and 11.3.16). A soakaway is a very green method of surface water disposal: it is carbon-efficient and just relies on gravity to disperse the water.

Filled soakaways are usually employed only for small capacities, because it is difficult to estimate the storage capacity, and the life of the soakaway may be limited by the silting-up of the voids between the granular filling material. Lined soakaways are generally more efficient and have a longer life. If access is provided, they can be inspected and maintained at regular intervals, with any build-up of silt removed.

Soakaways should be sited away from buildings so that foundations are unaffected by the percolation of water from the soakaway, which could lead to expansion of a clay soil placing pressure on a foundation. The minimum 'safe' distance is often quoted as 5.000 m, but local authority advice should always be sought. The number of soakaways required can be determined only by having the facts concerning total drain runs, areas to be drained and the rate of percolation for any particular site.

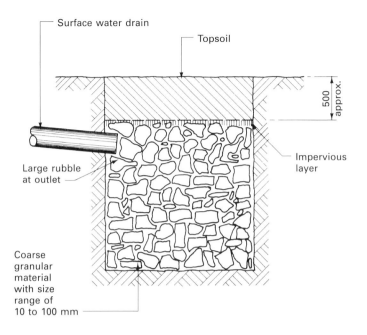

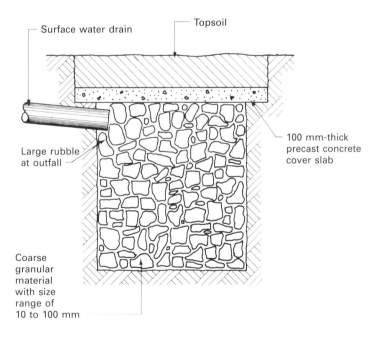

Suitable coarse granular materials include broken bricks, crushed sound rock and hard clinker

Figure 11.3.15   Filled soakaways.

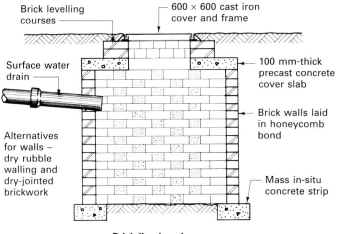

Brick levelling courses

600 × 600 cast iron cover and frame

Surface water drain

100 mm-thick precast concrete cover slab

Brick walls laid in honeycomb bond

Alternatives for walls –
dry rubble walling and dry-jointed brickwork

Mass in-situ concrete strip

Brick-lined soakaway

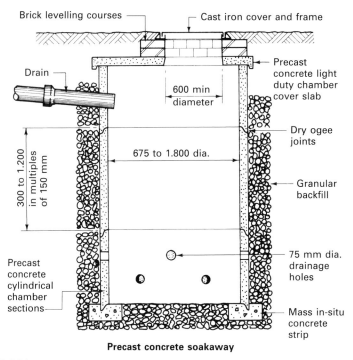

Brick levelling courses

Cast iron cover and frame

Drain

600 min diameter

Precast concrete light duty chamber cover slab

Dry ogee joints

675 to 1.800 dia.

300 to 1.200 in multiples of 150 mm

Granular backfill

75 mm dia. drainage holes

Precast concrete cylindrical chamber sections

Mass in-situ concrete strip

Precast concrete soakaway

Figure 11.3.16   Lined soakaways.

Domestic electrical installations 11.4

A simple definition of the term **electricity** is not possible, but it can be considered as a form of energy due to the free movement of tiny particles called electrons. If sufficient of these free or loose **electrons** move, an electric current is produced in the material in which they are moving. Materials that allow an electric current to flow readily (such as most metals and water) are called **conductors** and are said to have a low resistance. Materials that resist the flow of an electric current (such as rubber, glass and most plastics) are called **insulators**.

For an electric current to flow there must be a complete path or circuit from the source of energy through a conductor back to the source. Any interruption of the path will stop the flow of electricity. The following are common terms associated with electricity.

- The pressure that forces or pushes the current around the circuit is called the **voltage**, and is measured in volts.
- The **current** is the rate at which the electricity flows, and is measured in **amperes**.
- The **resistance** offered by the circuit to passage of electrons is measured in **ohms**.
- A **watt** is the unit of power, and is equal to the product of volts × amperes.

If the resistance within a circuit is altered by the use of wires and materials of different thicknesses, the energy that passes through the circuit will dissipate in the form of heat. The greater the resistance, the hotter it gets. Therefore a heating element can be constructed that is used in appliances such as cookers, irons and fires. The conductor in a filament bulb is a very thin wire of high resistance that becomes white hot, giving out light as well as heat.

Most domestic premises receive a single-phase supply of electricity from an area electricity authority via a distribution grid of pylons and high-voltage cables. This is then transformed down at a rating of 230 volts, and a frequency of 50 hertz, which can be used for domestic consumption. The area authority's cable, from which the domestic supply is taken, consists of four lines: three lines each carry a 230 volt (V) supply, and the fourth is the common return line or neutral, which is connected to earth at the transformer or substation as a safety precaution should a fault occur on the electrical appliance. Each line or phase is tapped in turn together with the neutral to provide the single-phase 230 V supply. Electricity is generated and supplied as an **alternating current**, which means that the current flows first one way then the other; the direction change is so rapid that it is hardly discernible in fittings such as lights. The cycle of this reversal of flow is termed **frequency**.

Fig. 11.4.1 shows a typical underground domestic electricity supply to an external meter box. This contains the electric meter, and an isolation fuse that is normally sealed to stop interference from occupiers. The consumer unit should be located as close as possible to the intake on the inside of the cavity wall as shown, or on an adjacent partition.

The conductors used in domestic installations are called **cables**, and consist of a conductor of low resistance, such as copper, surrounded by an insulator of high resistance, such as plastic. Cable sizes are known by the nominal cross-sectional area of the conductor, and those up to 2.5 mm^2 are usually of one strand. Larger cables consist of a number of strands to give them flexibility. All cables are assigned a rating in amperes, which is the maximum load the cable can carry without becoming overheated.

For domestic work, wiring drawings are not usually required; instead the positions of outlets, switches and lighting points are shown by standard convention and symbols on the plans. Specification of fittings, fixing heights and cables is given either in a schedule or in a written specification (see Fig. 11.4.2).

CIRCUITS

Domestic buildings are wired using a **ring main circuit**, as opposed to the older method of having a separate fused subcircuit to each socket outlet. Lighting circuits are carried out by using the **loop in** method.

The supply or intake cable will enter the building through below-ground ducts and be terminated in the area authority's fused sealing chamber, which should be sited in a dry, accessible position. From the sealing chamber the supply passes through the meter, which records the electricity consumed in units of kilowatt/hours, and then on to the consumer unit, which has an isolation switch controlling the supply to the circuit fuses or miniature circuit-breakers. These fuses or circuit breakers are a protection against excess current or overload of the circuit: should overloading occur, the fuse or circuit

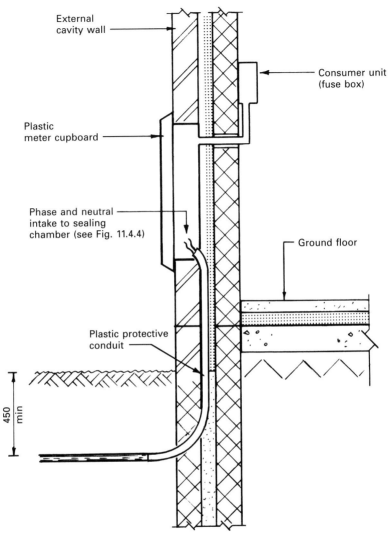

External
cavity wall

Consumer unit
(fuse box)

Plastic
meter cupboard

Phase and neutral
intake to sealing
chamber (see Fig. 11.4.4)

Ground floor

Plastic protective
conduit

450
min

Figure 11.4.1 **Underground domestic electrical supply.**

breaker will isolate the circuit from the supply. Modern consumer units are
fitted with a residual current device (RCD). This switches off the electricity
once a fault is detected and is potentially a lifesaver. This is now a
requirement under the provisions of BS 7671. Building Regulations Approved
Document Part P sets out some minimum standards for electrical wiring in
domestic homes which must be undertaken by a registered (with a
Government-approved scheme) and qualified Part P electrical contractor who
has passed the requirements of the Institute of Electrical Engineers' training
and examination.

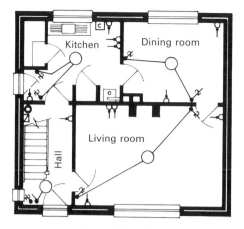

Ground floor plan

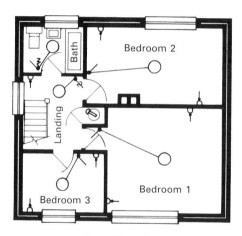

Upper floor plan

Symbols		
✓ One-way switch	⚡ Pendant switch	☑ Consumer unit
✗ Two-way switch	◯ Ceiling outlet	◉ Meter
▷— Switch socket outlet	⊂⊐ Immersion heater	⒞ Cooker control

Figure 11.4.2   Typical domestic electrical layout.

The number of fuseways or miniature circuit breakers contained in the consumer unit will depend upon the size of the building and the equipment to be installed, and therefore the number of circuits required. A separate ring circuit of 32 amp loading should be allowed for every 100 m^2 of floor area, and as far as practicable the number of outlets should be evenly distributed over the circuits. A typical domestic installation would have the following circuits from the consumer unit:

1. 6 amp: ground floor lighting up to 10 fittings or a total load of 6 amps;

2. 6 amp: upper floor lighting as above;

3. 16 amp: immersion heater;

4. 32 amp: ring circuit 1;

5. 32 amp: ring circuit 2;

6. 32 amp: ring circuit, kitchen;

7. 45 amp: cooker circuit.

Note: A further 40 or 45 amp circuit breaker or fuse may be installed for an electric shower unit.

The complete installation is earthed by connecting the consumer unit earth bar to the sheath of the supply cable, with an earthing terminal clip or by connection to a separate earth electrode. Fig. 11.4.3 shows a standard consumer unit with disposition of fuseways or miniature circuit breakers. A preferable alternative is the split-load consumer unit shown in Fig. 11.4.4. This contains specific protection to ground-floor sockets that could have extension leads attached for use with portable garden equipment, which need to be protected with an RCD.

For lighting circuits using insulated wiring a 1.5 mm^2 conductor is required, and therefore a twin with earth cable is used. The loop-in method of wiring is shown in Fig. 11.4.5. It is essential that lighting circuits are properly earthed, as most domestic fittings and switches contain metal parts or fixings which could become live should a fault occur. Lighting circuits using a conduit installation with single-core cables can be looped from switch to switch as shown in Fig. 11.4.5. Conduit installation consists of metal or plastic tubing together with various boxes for forming junctions and housing switches, which gives a protected, rewireable system. If plastic conduit is used, as it is non-conductive, the circuit must have an insulated earth conductor throughout.

A ring circuit for socket outlets consists of a twin 2.5 mm^2 earthed cable starting from and returning back to the consumer unit. The cables are looped into the outlet boxes, making sure that the correct cable is connected to the correct terminals (see Fig. 11.4.3). The number of outlets is unlimited if the requirement of one ring circuit per 100 m^2 of floor area has been adopted.

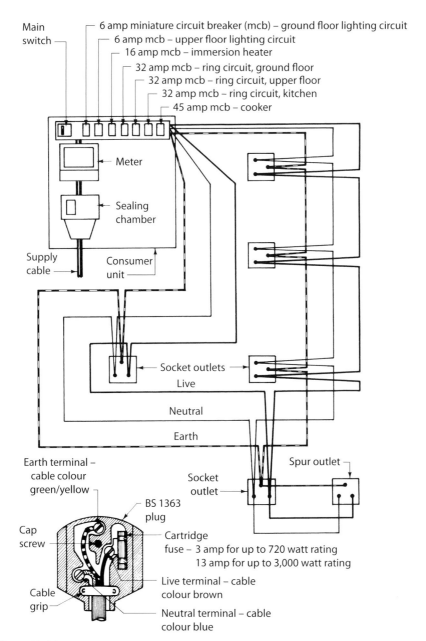

Main switch

6 amp miniature circuit breaker (mcb) – ground floor lighting circuit
6 amp mcb – upper floor lighting circuit
16 amp mcb – immersion heater
32 amp mcb – ring circuit, ground floor
32 amp mcb – ring circuit, upper floor
32 amp mcb – ring circuit, kitchen
45 amp mcb – cooker

Meter

Sealing chamber

Supply cable

Consumer unit

Socket outlets

Live

Neutral

Earth

Earth terminal – cable colour green/yellow

Spur outlet

Socket outlet

BS 1363 plug

Cap screw

Cartridge fuse – 3 amp for up to 720 watt rating
13 amp for up to 3,000 watt rating

Live terminal – cable colour brown

Cable grip

Neutral terminal – cable colour blue

Figure 11.4.3 Ring circuits and plug wiring.

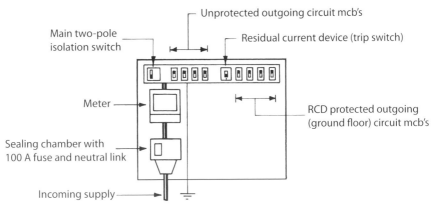

Unprotected outgoing circuit mcb's

Main two-pole
isolation switch

Residual current device (trip switch)

Meter

RCD protected outgoing
(ground floor) circuit mcb's

Sealing chamber with
100 A fuse and neutral link

Incoming supply

Figure 11.4.4   Split-load consumer unit.

Spur outlets leading off the main ring circuit are permissible provided that not more than two outlet sockets are on any one spur, and that not more than half the socket outlets on the circuit are on spurs. Socket outlets can be switch controlled and of single or double outlet; the double outlet is considered the best arrangement as it discourages the use of multiple adaptors, with associated risks of overloading. Fixed appliances, such as wall heaters, should be connected directly to a fused spur outlet to reduce the number of trailing leads. Movable appliances such as irons, radios and standard lamps should have a fused plug for connection to the switched outlet, conforming to the requirements of BS 1363. The rating of the cartridge fuse should be in accordance with the rating of the appliance. Appliances with a rating of not more than 720 watts should be protected by a 3 amp fuse, and appliances over this rating up to 3,000 watts should have a 13 amp fuse. As with the circuit, correct wiring of the plug is essential (see Fig. 11.4.3).

The number of outlets is not mandatory, but the minimum numbers recommended for various types of accommodation are:

- kitchens – 6 plus cooker control unit with one outlet socket;
- living rooms – 8;
- dining rooms – 4;
- bedrooms – 4;
- halls – 1;
- landings – 1;
- garages – 1;
- stores and workshops – 1.

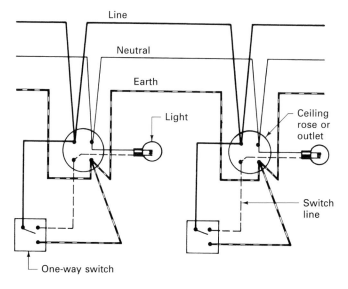

'Loop-in' method of wiring using sheathed cable

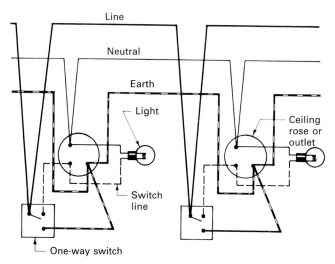

Single-core cable looped from switch to switch

Figure 11.4.5 Typical lighting circuits.

The outlets should be installed around the perimeter of the rooms in the most convenient and economic positions to give maximum coverage with minimum amount of trailing leads.

Cables sheathed with PVC can be run under suspended floors by drilling small holes on the neutral axis of the joists; where the cables are to be covered by wall finishes or floor screed they should be protected either by oval conduit or by means of small metal cover channels fixed to the wall or floor.

Systems using mineral-insulated covered cables that are used in fire alarm installations follow the same principles. This form of cable consists of single strands of copper or aluminium, all encased in a sheath of the same metal, which is densely packed with fine magnesium oxide insulation around the strands. This insulating material is unaffected by heat in a fire or by age and is therefore very durable, but it can absorb moisture. The sealing of the ends of this type of cable with special sealing 'pots' is therefore of paramount importance, so that moisture cannot enter the cable. Cables used in a conduit installation have adequate protection, but it is generally necessary to chase the walls of the building to accommodate the conduit, outlet socket boxes and switch boxes below the wall finish level. Surface-run conduit is normally secured to the backing using screwed, shaped clips called **saddles**.

Guidance for the installation of electrical circuits and electrical equipment is provided within the Building Regulations Approved Document P: *Electrical safety*. Further details are produced in the Institution of Electrical Engineers Regulations in accordance with BS 7671: *Requirements for Electrical Installations*.

With the exception of minor work such as the replacement of socket outlets and switches, all other proposed electrical work in dwellings requires notification to the local building control authority in order to meet the requirements of Part P.

Certification of a satisfactory installation is by:

- the inspection of the work by a qualified electrician appointed by the local building control office; or
- self-certification by a prescribed suitably registered and qualified competent person.

Domestic gas installations 11.5

Gas is a combustible fuel that burns with a luminous flame; it is used mainly in domestic installations as a source of heat energy, in appliances such as gas fires, cookers and condensing boilers. Historically, gas preceded electricity as a domestic energy utility, and was used to power lighting by the use of coal tar gas.

Natural gas is extracted from gas fields within the North Sea, Russia and other gas-producing continents and piped to process plants where it is prepared for domestic use. It can be stored in a liquid state in huge underground storage chambers, from where it is conveyed as a gas through a network of underground mains installed and maintained by the gas transporter, National Grid. Some gas is still stored in above-ground cylindrical holders, but these are gradually being phased out. Table 11.5.1 shows a comparison of the calorific values of natural and bottled gases, which are used when mains gas is unavailable. Most domestic appliances manufactured for use with mains gas can be converted for bottled gas by changing the burners.

Table 11.5.1 Calorific values of gases for domestic appliances.

Type of gas	Approx. calorific value (MJ/m³)
North Sea gas	38
Bottled gas (Propane)	96
Bottled gas (Butane)	122

Notes

1. Approximate values are given, as values will vary slightly depending on the source and quality.
2. Propane and butane are otherwise known as liquid petroleum gas (LPG). They are found under the North Sea bed and in other oil resources around the world. For transportation they are pressurised and liquefied to about 1/200 their gas volume.

The installation of gas supplies can be considered in three stages.

- **Main** This is the part of the gas distribution system that is connected to form a grid. The grid is an interconnection of main pipes, normally located beneath roads or footways. The arrangement of pipes enables maintenance and repairs to be undertaken in isolation, without a major disruption of supply to a large community, as areas can be closed off by the use of valves. For ease of identification, all new underground gas installation pipework is undertaken in a yellow plastic, not to be confused with water (blue), electricity (red) or drainage (orange).

- **Service pipe** This is the connection between the main and the consumer control valve positioned just before the meter, which is housed within an external enclosure cabinet. For domestic installations the service pipe diameter is between 25 and 50 mm according to the number and type of appliances being installed. The pipeline should be as short as possible, at right angles to the main, and laid at least 375 mm below the surface to avoid any damage. Fig. 11.5.1 shows a typical domestic installation with meter details and accessories.

- **Internal distribution** This commences at the consumer control valve and consists of a governor to stabilise gas pressure and volume, a meter to record the amount of gas consumed, and pipework to convey the gas supply to appliances. The meter and associated equipment are the property of the gas provider, but thereafter the installation is the responsibility of the building owner. A typical internal distribution to a house is shown in Fig. 11.5.2.

PIPEWORK

Pipes for internal distribution can be of mild steel (not galvanised) or solid drawn copper. Flexible tubing of metal-reinforced rubber may be used for the final connection to cookers and portable appliances, where they may need to be pulled out for maintenance. The size of pipes will depend on factors such as gas consumption of appliances, frictional losses due to pipe length, and number of bends. A typical domestic installation is in 22 mm copper from the meter, with a 15 mm branch to a boiler and 10 mm to individual low-rated appliances. Pipes must not be accommodated in the cavity of a wall, as a leak could go undetected, but may pass through a wall provided a sleeve of non-corrodible material protects the pipe against any differential movement or settlement, and this is sealed around the gas pipe both externally and internally using a flexible compound.

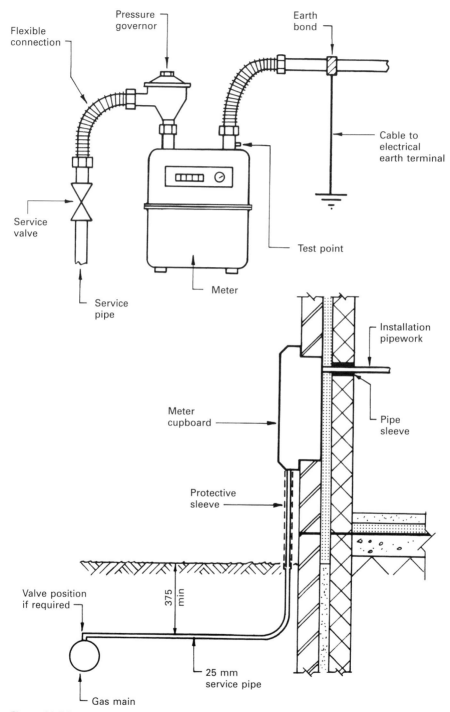

Flexible connection

Pressure governor

Earth bond

Cable to electrical earth terminal

Service valve

Test point

Service pipe

Meter

Installation pipework

Meter cupboard

Pipe sleeve

Protective sleeve

Valve position if required

375 min

Gas main

25 mm service pipe

Figure 11.5.1 Domestic gas supply.

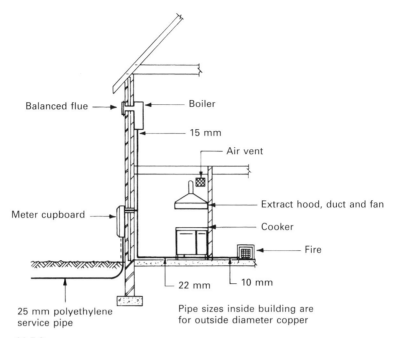

Figure 11.5.2 **Internal distribution to appliances.**

Gas appliances fall into two categories:

◼ gas supply only: cookers, radiant convector fires, decorative fuel effect fires;

◼ gas supply plus other services: central heating boilers, water heaters.

Any gas appliance must be fitted and installed by a competent person who is registered under the 'Gas Safe' scheme. The Gas Safety (Installation and Use) Regulations and the Building Regulations Approved Document J give extensive guidance on the regulations and rules concerning the fitting of gas appliances and the extraction of the combustion exhaust. A gas fire has a visible heat source, as the gas burns through the fragile fireclay radiant block within the gas fire. Gas fires can be set on a fireplace hearth with a closure backplate over the fire opening. A void in the plate allows for extraction of burnt gases directly up the chimney. This must be installed correctly in accordance with the Regulations. Precast concrete flue blocks can also be used.

Decorative fuel-effect (DFE) fires are a popular adaptation within traditional fireplace openings. Log or coal effects are designed to resemble a real fire by burning gas and discharging the combusted products indirectly into the open flue. For efficient combustion, most DFE fires will require a room air vent for the correct combustion of the gas. The manufacturer's installation details must be consulted along with the Regulations.

Gas cookers and instantaneous water heaters are an exception to the requirements for flues, but they should be installed and used only in well-ventilated rooms. The Building Regulations Approved Document J, Section 3 should be consulted for minimum areas of air vents relative to room volume and their position to the appliance.

Room-sealed or balanced flue heaters are appliances that have the heat exchanger sealed from the room by its connection to a flue. Air for combustion of gas is drawn in from an external duct at the back of the unit, and burnt gases discharge through an adjacent flue. Gas central heating consists of a wall-mounted or free-standing boiler connected to a flue. This is the heat source for either a ducted fan-assisted warm air circulation system or a pump-circulated hot water system.

The Health and Safety Executive, through the Gas Safety (Installation and Use) Regulations, requires the installation and maintenance of gas fittings, appliances and storage vessels to be undertaken only by a competent and qualified person. A facility for assessing and accrediting individuals to undertake the various aspects of gas fitting and maintenance has been established as the Gas Safe Register. Ref. Building Regulations Approved Document J, Section 3.2.

The installation of gas services, like that of electricity, is a specialist subject not normally developed within the framework of construction technology, as it is undertaken by specialist contractors using registered and qualified personnel.

Bibliography

Edwards, B. (2009) *Rough Guide to Sustainability – A Design Primer 3rd Edition*. London: RIBA Publishing

Emmitt, S. and Gorse, C.A. (2006) *Barry's Advanced Construction of Buildings*. Oxford: Blackwell Science

Emmitt, S. and Gorse, C.A. (2010) *Barry's Introduction to the Construction of Buildings*. Oxford: Blackwell Science

Evans, H (2010) *Guide to the Building Regulations*. London: RIBA Publications

Ferrett, E. and Hughes, P. *Introduction to Health and Safety at Work 4th Edition*. Oxford: Butterworth-Heinemann

Hurst, M. and Topliss, S. (2010) *BTEC National Construction and the Built Environment*. Oxford: Pearson

JTL (2005) *Plumbing level 3*. Oxford: Heinemann

JTL (2011) *Plumbing level 2 3rd edition*. Oxford: Pearson

Mackay, WB (2005) *Mackay's Building Construction*. Shaftsbury: Donhead Publishing Ltd

Stroud Foster, J. and Greeno, R. (2007) *Mitchell's Structure and Fabric Part 1 7th Edition*. Harlow: Pearson Prentice Hall

Stroud Foster, J., Harington, R. and J and Greeno, R. (2007) *Mitchell's Structure and Fabric Part 2 7th Edition*. Harlow: Pearson Prentice Hall

Taylor (2000) *Materials in Construction: An Introduction*. Harlow: Longman

Williams, A. (2008) *NBS Shortcuts: Book 1 – Structure and Fabric*. London: RIBA Enterprises Ltd

Williams, A. (2009) *NBS Shortcuts: Book 2 – Sustainability and Practice*. London: RIBA Enterprises Ltd

General

BRE Good Building Guides: IHS BRE Press

NHBC Technical Standards. Milton Keynes: National House Building Council

Websites

Health and Safety Executive: http://www.hse.gov.uk

Legislation: http://www.legislation.gov.uk/

Index